METHODS IN MOLECULAR BIOLOGY™

Series Editor
John M. Walker
School of Life Sciences
University of Hertfordshire
Hatfield, Hertfordshire, AL10 9AB, UK

For further volumes:
http://www.springer.com/series/7651

Sirtuins

Methods and Protocols

Edited by

Matthew D. Hirschey

Institute of Molecular Physiology, Duke University Medical Center, Durham, NC, USA

Humana Press

Editor
Matthew D. Hirschey
Institute of Molecular Physiology
Duke University Medical Center
Durham, NC, USA

ISSN 1064-3745 ISSN 1940-6029 (electronic)
ISBN 978-1-4939-6045-3 ISBN 978-1-62703-637-5 (eBook)
DOI 10.1007/978-1-62703-637-5
Springer New York Heidelberg Dordrecht London

Printed on acid-free paper

Humana Press is a brand of Springer
Springer is part of Springer Science+Business Media (www.springer.com)

Preface

The sirtuins are a family of genes encoding NAD(+)-dependent protein deacylases conserved in several organisms including yeast, worms, flies, and mammals. This family of genes has been shown to play crucial roles in the molecular pathways that regulate cell survival, metabolism, and the diseases of aging. Because of the diverse roles sirtuins play, research in the area of sirtuins has expanded into multiple disciplines at an accelerated pace, identifying new and unexpected roles for this family of genes. A diverse array of model organisms and scientific techniques are used to study the sirtuins, which are presented in a comprehensive manner in this volume of *Methods in Molecular Biology.*

After an introductory chapter, the first four chapters (Part I) focus on "Methods to generate sirtuin biology tools," one each for yeast, *C. elegans*, *Drosophila*, and mammalian model organisms. In addition to reviewing common tools, some techniques are presented along with these tools as examples for their use. The next five chapters (Part II) focus on "Methods to identify sirtuin substrates," covering several enrichment and proteomic strategies to measure changes in acylation of sirtuin substrates. The following five chapters (Part III) cover "Methods to measure sirtuin activity," which is one of the major challenges currently facing the sirtuin field. Both direct and indirect measurements of sirtuin activity are presented. Finally, the last five chapters (Part IV) on "Methods to study sirtuin biology" describe protocols to measure some of the major biological pathways controlled by the sirtuins, including metabolism, autophagy, genomic stability, circadian rhythms, and calorie restriction.

Together, the chapters in this volume present detailed protocols for sirtuin research that can be followed directly or modified to investigate new areas of sirtuin biology. We deeply thank the contributing authors to this volume and wish success to the scientists using these protocols for discoveries in this rapidly evolving and exciting field.

Durham, NC *Matthew D. Hirschey*

Contents

Part III Methods to Measure Sirtuin Activity

Part IV Methods to Study Sirtuin Biology

Contributors

ALISON H. AFFINATI • *Division of Endocrinology, Metabolism and Molecular Medicine, Feinberg School of Medicine, Northwestern University, Chicago, IL, USA; Department of Neurobiology, Northwestern University, Evanston, IL, USA*

JOSHUA L. ANDERSEN • *Department of Chemistry and Biochemistry, Brigham Young University, Provo, UT, USA*

KRISTIN A. ANDERSON • *Sarah W. Stedman Nutrition and Metabolism Center, Duke University Medical Center, Durham, NC, USA; Department of Pharmacology and Cancer Biology, Duke University Medical Center, Durham, NC, USA*

JOSEPH BASS • *Division of Endocrinology, Metabolism and Molecular Medicine, Feinberg School of Medicine, Northwestern University, Chicago, IL, USA; Department of Neurobiology, Northwestern University, Evanston, IL, USA*

ANTONIO BEDALOV • *Fred Hutchinson Cancer Research Center, Seattle, WA, USA*

RAFAEL DE CABO • *Translational Gerontology Branch, Intramural Research Program, National Institute on Aging, National Institutes of Health, Baltimore, MD, USA*

CARLES CANTÓ • *Nestlé Institute of Health Sciences, Lausanne, Switzerland*

YUE CHEN • *Ben May Department for Cancer Research, University of Chicago, Chicago, IL, USA*

GOZDE COLAK • *Ben May Department for Cancer Research, University of Chicago, Chicago, IL, USA*

JESSICA CURTIS • *Translational Gerontology Branch, Intramural Research Program, National Institute on Aging, National Institutes of Health, Baltimore, MD, USA*

WEIWEI DANG • *Huffington Center on Aging, Department of Molecular and Human Genetics, Baylor College of Medicine, Houston, TX, USA*

JOHN DOMINY • *Department of Cancer Biology, Dana-Farber Cancer Institute, Boston, MA, USA; Department of Cell Biology, Harvard Medical School, Boston, MA, USA*

TOREN FINKEL • *Center for Molecular Medicine, NHLBI, National Institutes of Health, Bethesda, MD, USA*

KRISTOFER S. FRITZ • *Department of Pharmaceutical Sciences, Skaggs School of Pharmacy and Division of Cardiology, School of Medicine, University of Colorado Denver Anschutz Medical Center, Aurora, CO, USA*

BRADFORD W. GIBSON • *Buck Institute for Research on Aging, Novato, CA, USA*

MICHELLE F. GREEN • *Sarah W. Stedman Nutrition and Metabolism Center, Duke University Medical Center, Durham, NC, USA*

LEONARD GUARENTE • *Glenn Lab for the Science of Aging and Department of Biology, Massachusetts Institute of Technology, Cambridge, MA, USA*

JASON M. HELD • *Buck Institute for Research on Aging, Novato, CA, USA*

STEPHEN L. HELFAND • *Department of Molecular Biology, Cell Biology, and Biochemistry, Brown University, Providence, RI, USA*

KATHLEEN A. HERSHBERGER • *Sarah W. Stedman Nutrition and Metabolism Center, Duke University Medical Center, Durham, NC, USA*

MATTHEW D. HIRSCHEY • *Sarah W. Stedman Nutrition and Metabolism Center, Duke University Medical Center, Durham, NC, USA; Department of Pharmacology and Cancer Biology, Duke University Medical Center, Durham, NC, USA; Department of Medicine, Duke University Medical Center, Durham, NC, USA*

HEE-KYUNG HONG • *Division of Endocrinology, Metabolism and Molecular Medicine, Feinberg School of Medicine, Northwestern University, Chicago, IL, USA; Department of Neurobiology, Northwestern University, Evanston, IL, USA*

BASIL P. HUBBARD • *Department of Genetics, Harvard Medical School, Boston, MA, USA*

SHIN-ICHIRO IMAI • *Department of Developmental Biology, Washington University School of Medicine, St. Louis, MO, USA*

BRIAN K. KENNEDY • *Buck Institute for Research on Aging, Novato, CA, USA*

ELIZABETH KWAN • *Department of Genome Sciences, University of Washington, Seattle, WA, USA; Fred Hutchinson Cancer Research Center, Seattle, WA, USA*

IN HYE LEE • *Center for Molecular Medicine, NHLBI, National Institutes of Health, Bethesda, MD, USA*

YU LI • *Cell Signaling Technology, Danvers, MA, USA*

HENING LIN • *Department of Chemistry and Chemical Biology, Cornell University, Ithaca, NY, USA*

DAVID B. LOMBARD • *Department of Pathology and Institute of Gerontology, University of Michigan, Ann Arbor, MI, USA*

BILIANA MARCHEVA • *Division of Endocrinology, Metabolism and Molecular Medicine, Feinberg School of Medicine, Northwestern University, Chicago, IL, USA; Department of Neurobiology, Northwestern University, Evanston, IL, USA*

BARBARA MARTINEZ-PASTOR • *The Massachusetts General Hospital Cancer Center, Harvard Medical School, Boston, MA, USA*

PALOMA MARTÍNEZ-REDONDO • *Chromatin Biology Laboratory, Cancer Epigenetics and Biology Program (PEBC), Bellvitge Biomedical Research Institute (IDIBELL), L'Hospitalet de Llobregat, Barcelona, Spain*

EOIN MCDONNELL • *Sarah W. Stedman Nutrition and Metabolism Center, Duke University Medical Center, Durham, NC, USA*

RAUL MOSTOSLAVSKY • *The Massachusetts General Hospital Cancer Center, Harvard Medical School, Boston, MA, USA*

JONATHAN MOTLEY • *Sarah W. Stedman Nutrition and Metabolism Center, Duke University Medical Center, Durham, NC, USA*

CLARA B. PEEK • *Division of Endocrinology, Metabolism and Molecular Medicine, Feinberg School of Medicine, Northwestern University, Chicago, IL, USA; Department of Neurobiology, Northwestern University, Evanston, IL, USA*

PERE PUIGSERVER • *Department of Cancer Biology, Dana-Farber Cancer Institute, Boston, MA, USA; Department of Cell Biology, Harvard Medical School, Boston, MA, USA*

KATHRYN MOYNIHAN RAMSEY • *Division of Endocrinology, Metabolism and Molecular Medicine, Feinberg School of Medicine, Northwestern University, Chicago, IL, USA; Department of NeurobiologyNorthwestern University, Evanston, IL, USA*

MATTHEW J. RARDIN • *Buck Institute for Research on Aging, Novato, CA, USA*

ALEX ROBESON • *Department of Pharmacology and Cancer Biology, Duke University Medical Center, Durham, NC, USA*

CARLOS SEBASTIAN • *The Massachusetts General Hospital Cancer Center, Harvard Medical School, Boston, MA, USA*

DAFNE M. SILBERMAN • *The Massachusetts General Hospital Cancer Center, Harvard Medical School, Boston, MA, USA*
JEFFREY C. SILVA • *Cell Signaling Technology, Danvers, MA, USA*
DAVID A. SINCLAIR • *Department of Genetics, Harvard Medical School, Boston, MA, USA*
MARY E. SKINNER • *Department of Pathology and Institute of Gerontology, University of Michigan, Ann Arbor, MI, USA*
XIAOYANG SU • *Department of Chemistry and Chemical Biology, Cornell University, Ithaca, NY, USA*
J. WILL THOMPSON • *Duke Proteomics Core Facility, Institute for Genome Sciences and Policy, Duke University Medical Center, Durham, NC, USA*
HEIDI A. TISSENBAUM • *Program in Gene Function and Expression, Program in Molecular Medicine, University of Massachusetts Medical School, Worcester, MA, USA*
SCOTT TSUCHIYAMA • *Buck Institute for Research on Aging, Novato, CA, USA*
ALEJANDRO VAQUERO • *Chromatin Biology Laboratory, Cancer Epigenetics and Biology Program (PEBC), Bellvitge Biomedical Research Institute (IDIBELL), L'Hospitalet de Llobregat, Barcelona, Spain*
MOHAN VISWANATHAN • *Department of Biology, Massachusetts Institute of Technology, Cambridge, MA, USA*
DONGNING WANG • *Sarah W. Stedman Nutrition and Metabolism Center, Duke University Medical Center, Durham, NC, USA*
RACHEL WHITAKER • *Department of Molecular Biology, Cell Biology, and Biochemistry, Brown University, Providence, RI, USA*
JASON G. WOOD • *Department of Molecular Biology, Cell Biology, and Biochemistry, Brown University, Providence, RI, USA*
JUN YOSHINO • *Center for Human Nutrition, Department of Medicine, Washington University School of Medicine, St. Louis, MO, USA*
JEANHO YUN • *Center for Molecular Medicine, NHLBI, National Institutes of Health, Bethesda, MD, USA; Department of Biochemistry and Mitochondria Hub Regulation Center, College of Medicine, Dong-A University, Busan, South Korea*
YINGMING ZHAO • *Ben May Department for Cancer Research, University of Chicago, Chicago, IL, USA*
LEI ZHONG • *The Massachusetts General Hospital Cancer Center, Harvard Medical School, Boston, MA, USA*
ANITA ZHU • *Department of Chemistry and Chemical Biology, Cornell University, Ithaca, NY, USA*

Part I

Methods to Generate Sirtuin Biology Tools

Chapter 1

Introduction: Sirtuins in Aging and Diseases

Leonard Guarente

Abstract

Over the past 15 years, the number of papers published on sirtuins has exploded. The initial link between sirtuins and aging comes from studies in yeast, in which it was shown that the life span of yeast mother cells (replicative aging) was proportional to the SIR2 gene dosage. Subsequent studies have shown that SIR2 homologs also slow aging in *C. elegans*, *Drosophila*, and mice. An important insight into the function of sirtuins came from the finding that yeast Sir2p and mammalian SIRT1 are NAD^+-dependent protein deacetylases. In mammals, there are seven sirtuins (SIRT1-7). Their functions do not appear to be redundant, in part because three are primarily nuclear (SIRT1, 6, and 7), three are mitochondrial (SIRT3, 4, and 5), and one is cytoplasmic (SIRT2). The past decade has provided an avalanche of data showing deacetylation of many key transcription factors. In this chapter, I will address the evidence that sirtuins mediate the effects of CR on physiology and will then turn to the evidence of a relationship between sirtuins and aging and life span. Finally, I will discuss the roles of sirtuins in diseases of aging and the prospects of translating these findings to novel therapeutic strategies to treat major diseases.

Key words Sirtuins, CR, Nicotinamide adenine dinucleotide, NAD, Aging, Life span

1 Sirtuins and CR

Sirtuins were first experimentally linked to CR by studies on replicative aging in yeast [1, 2], although strain specificity and experimental conditions appear to influence the relative importance of sirtuins for extension of yeast life span by CR [3]. Similar results have been reported in *C. elegans* and *Drosophila*, although again [4–6] results to the contrary have also been reported [7]. One surmises that in these lower organisms, the exact nature of the CR (lowering the glucose levels in yeast to varying degrees, or reducing the number of *E. coli* in the worm diet, or lowering the yeast extract in the fly food) is critical in determining the pathways engaged. Thus, it may be impossibly confusing to lump all of these different protocols under a single rubric and expect a simple conclusion about genetic requirements for life extension.

Matthew D. Hirschey (ed.), *Sirtuins: Methods and Protocols*, Methods in Molecular Biology, vol. 1077,
DOI 10.1007/978-1-62703-637-5_1, © Springer Science+Business Media, LLC 2013

In rodents, CR has been defined for 80 years, and consists of limiting the chow diet of the animals. In this system, there is a great deal of evidence for the importance of sirtuins in CR [8–10]. First, substrates of SIRT1, SIRT3, SIRT4, and SIRT5 are all involved in metabolic adaptations to energy limitation. One key aspect of this regulation is a decision to produce chemical energy (ATP) by oxidative metabolism of energy sources in mitochondria versus by glycolytic metabolism in the cytosol [11]. The above sirtuins facilitate the complete catabolism of carbohydrates, amino acids, and fat to CO_2 in mitochondria, accompanied by the maximum yield of ATP permitted by cellular energetics.

Second, the levels of sirtuin proteins SIRT1 and SIRT3 are induced by CR [12]. This increase would be accompanied by a rise in NAD^+ in both the nucleus and mitochondria driven by a slowing in glycolysis and acceleration in oxidative phosphorylation in the respective compartments. Thus, one expects a higher degree of deacetylation of sirtuin substrates during CR, and indeed this is what has been observed. Amazingly, SIRT5 appears not to be a protein deacetylase, but rather this sirtuin has high desuccinylation and demalonylation activities [13, 14].

Third, genetic data underscores the importance of sirtuins. Loss of function data shows that sirtuins are required for physiological effects of CR. For example, mice lacking SIRT1 do not live longer on CR, and mice lacking SIRT3 are not protected against hearing loss that occurs during aging [15, 16]. Gain of function data shows that transgenic mice over-expressing SIRT1 are protected against metabolic decline that occurs with normal aging or induced by a high fat diet [17–19].

Fourth, small molecule activators of SIRT1 have been described that are either natural products (resveratrol) or novel synthetic compounds [20, 21]. These compounds also protect mice against metabolic decline, and elicit transcriptional changes in cells that largely overlap with changes induced by CR [22]. Moreover, metabolic effects of these compounds appear to depend on SIRT1, most convincingly shown in mice in which SIRT1 can be deleted in adults [23].

All told, the above evidence linking sirtuins to CR is very strong. This case is further strengthened by the recently demonstrated effects of manipulating the cellular levels of the sirtuin cosubstrate NAD^+. Precursors of NAD^+, such as NMN or NR, have been shown to get into cells and raise the levels of intracellular NAD^+ [24, 25]. Importantly, NMN and NR also have protective effects when fed to mice, much like SIRT1 over-expression or the SIRT1 activator compounds. It will be interesting to observe whether yet stronger effects are observed when SIRT1 is over-expressed and animals are supplemented with NAD^+ precursors.

2 Sirtuins and Aging

Sir2p was first implicated in yeast mother cell aging when a mutation in the *SIR4* gene was found that extended the life span [26]. This *SIR4* mutation was shown to redirect Sir2p to the rDNA [27] and thereby extend the life span [28] by suppressing the production of extrachromosomal rDNA circles (ERCs) [29]. Indeed, a recent survey of yeast longevity determinants by QTL mapping indicated that *SIR2* is the most important QTL in the yeast genome [30]. Remarkably, the rDNA locus itself was found to be the second most important QTL in this recent study.

Studies in *C. elegans* and *Drosophila* consistently indicate that *SIR2* orthologs also could extend life span when over-expressed [5, 6, 31–34], although one study reports no extension [7]. The studies in *C. elegans* are of note, because the original work employed transgenic worms missing the bona fide *sir-2.1* promoter (which had not been mapped at the time). The degree of extension reported originally was reduced but not eliminated by extensive backcrossing of the original isolates [32]. Other transgenic worm strains made subsequently do contain the *sir-2.1* promoter and also extend the life span [33]. In Drosophila, it was also reported that extensive back crossing could eliminate the effects of *dSIR2* on life span [7]. However, at least one report shows life extension in a dox-inducible *dSIR2* system, which eliminates strain background effects [6].

Transgenic mice for the *SIR2* ortholog *SIRT1* display slower aging as measured by metabolic decline, for example, the acquisition of diabetes or bone loss, but do not display an extension in the life span [17, 19]. However, transgenic mice for another mammalian *SIR2* homolog *SIRT6* do display an extended life span in males [35]. Furthermore, small molecule SIR2 ortholog activators can extend life span in many organisms, with a robust 60 % extension in zebrafish. All told, the aging data makes a compelling case that sirtuins do slow aging across many different species.

3 Sirtuins and Disease

Mammalian sirtuins have broad physiological effects that make them attractive targets for several diseases. Among these are diabetes, neurodegenerative diseases, pro-inflammatory diseases, cardiovascular disease, and cancer. SIRT1 is the most studied of the seven mammalian sirtuins as a disease target. The links between SIRT1 and metabolism are many, as suggested in the CR section above. In brief, SIRT1 transgenic mice or mice dosed with SIRT1-activating compounds are protected against the deleterious metabolic effects of a high fat diet [36, 37]. In some cases, these mice are also protected against the metabolic decline induced by normal aging.

There are probably many mechanisms that contribute to the protective activity of SIRT1. These include activation of mitochondrial metabolism via deacetylation of PGC-1alpha [38] and PPARalpha [39], repression of inflammation via deacetylation of NF-kB [40], and repression of fat and cholesterol synthesis via deacetylation of SREBP1c [41, 42].

In addition, a pattern of neuro-protection has recently emerged, in which mice over-expressing SIRT1 in the brain are protected in models of Alzheimer's disease, Parkinson's disease, or Huntington's disease [43–47]. In each case, a different molecular mechanism for protection was suggested by the analysis. It will be of interest to systematically compare the effects of SIRT1 on these disease models to determine whether there might be more general neuro-protective mechanisms at work. In any case, this broad protection against neurodegenerative disorders is quite remarkable, and suggests that SIRT1-activating compounds that can penetrate the blood–brain barrier may be an important therapeutic approach.

Mice over-expressing SIRT1 are also protected from pathologies driven by inflammation. A good example is the inflammation-induced model of liver cancer, which is dramatically attenuated in SIRT1 transgenic mice [19]. At least two of the seven sirtuins SIRT1 and SIRT6 repress a critical driver of inflammation, NF-kB, either by deacetylating p65 (SIRT1) [40] or histones at NF-kB target genes (SIRT6) [48]. These and likely additional molecular mechanisms open new strategies to treat inflammatory disorders. Moreover, inflammation appears to contribute to many of the major diseases of aging, and the anti-inflammatory action of SIRT1 may explain at least a portion of its broad activity against diseases.

Recent studies have also revealed an important connection between the mitochondrial SIRT3 and cancer. Because this sirtuin deacetylates and activates enzymes for oxidative metabolism in mitochondria and at the same time promotes detoxification of reactive oxygen species (ROS), it may be at the center of aging and metabolism [12]. Most interestingly, SIRT3 may play a role in the Warburg effect, by which cancer cell growth is aided by an elevated level of glycolysis [49, 50]. Indeed, knocking out SIRT3 increases ROS and thus up-regulates HIF-1alpha, a positive regulator of glycolytic genes. Indeed between 20 and 40 % of human tumor cells displayed loss of SIRT3, indicating an important tumor suppressor function for this sirtuin. Remarkably, a similar tumor suppression has recently been described, in which SIRT6 also represses the Warburg effect by deacetylating histones at HIF-controlled genes [51].

Finally, numerous studies show a positive effect of SIRT1 activation on endothelial cell function [52–54]. This includes activation of nitric oxide synthase, to facilitate vasodilation, and regulation of notch signaling. Thus, SIRT1 activation has been demonstrated to protect against pathological cardiac hypertrophy.

Factoring in the protective effects on metabolism mentioned above, SIRT1 may be a valuable target for cardiovascular disease generally.

4 The Future?

SIRT1-activating compounds are being tested on patients for protection against metabolic or pro-inflammatory disorders. The pathway of drug development is always unpredictable. However, the deep and wide involvement of sirtuins in biological processes suggests that compounds that target them may exert profound effects on many aspects of human physiology. The fervent hope is these effects can be triggered with salutary consequences in treating diseases of aging, leading to an important role for sirtuin drugs moving forward.

References

1. Imai S, Armstrong CM, Kaeberlein M, Guarente L (2000) Transcriptional silencing and longevity protein Sir2 is an NAD-dependent histone deacetylase. Nature 403:795–800
2. Lin SJ, Defossez PA, Guarente L (2000) Requirement of NAD and SIR2 for life-span extension by calorie restriction in Saccharomyces cerevisiae. Science 289:2126–2128
3. Kaeberlein M, Kirkland KT, Fields S, Kennedy BK (2004) Sir2-independent life span extension by calorie restriction in yeast. PLoS Biol 2:E296
4. Wang Y, Tissenbaum HA (2006) Overlapping and distinct functions for a Caenorhabditis elegans SIR2 and DAF-16/FOXO. Mech Ageing Dev 127:48–56
5. Rogina B, Helfand SL (2004) Sir2 mediates longevity in the fly through a pathway related to calorie restriction. Proc Natl Acad Sci USA 101:15998–16003
6. Bauer J, Morris S, Chang C, Flatt T, Wood J, Helfand SL (2009) dSir2 and Dmp53 interact to mediate aspects of CR-dependent lifespan extension in D. melanogaster. Aging (Albany NY) 1:38–48
7. Burnett C, Valentini S, Cabreiro F, Goss M, Somogyvari M, Piper MD, Hoddinott M, Sutphin GL, Leko V, McElwee JJ, Vazquez-Manrique RP, Orfila AM, Ackerman D, Au C, Vinti G, Riesen M, Howard K, Neri C, Bedalov A, Kaeberlein M, Soti C, Partridge L, Gems D (2011) Absence of effects of Sir2 overexpression on lifespan in C. elegans and Drosophila. Nature 477:482–485
8. Haigis MC, Sinclair DA (2010) Mammalian sirtuins: biological insights and disease relevance. Annu Rev Pathol 5:253–295
9. Imai S, Guarente L (2010) Ten years of NAD-dependent SIR2 family deacetylases: implications for metabolic diseases. Trends Pharmacol Sci 31:212–220
10. Guarente L (2011) Franklin H. Epstein lecture: sirtuins, aging, and medicine. N Engl J Med 364:2235–2244
11. Hirschey MD, Shimazu T, Goetzman E, Jing E, Schwer B, Lombard DB, Grueter CA, Harris C, Biddinger S, Ilkayeva OR, Stevens RD, Li Y, Saha AK, Ruderman NB, Bain JR, Newgard CB, Farese RV Jr, Alt FW, Kahn CR, Verdin E (2010) SIRT3 regulates mitochondrial fatty-acid oxidation by reversible enzyme deacetylation. Nature 464:121–125
12. Bell EL, Guarente L (2011) The SirT3 divining rod points to oxidative stress. Mol Cell 42:561–568
13. Du J, Zhou Y, Su X, Yu JJ, Khan S, Jiang H, Kim J, Woo J, Kim JH, Choi BH, He B, Chen W, Zhang S, Cerione RA, Auwerx J, Hao Q, Lin H (2011) Sirt5 is a NAD- dependent protein lysine demalonylase and desuccinylase. Science 334:806–809
14. Peng C, Lu Z, Xie Z, Cheng Z, Chen Y, Tan M, Luo H, Zhang Y, He W, Yang K, Zwaans BM, Tishkoff D, Ho L, Lombard D, He TC, Dai J, Verdin E, Ye Y, Zhao Y (2011) The first identification of lysine malonylation substrates and its regulatory enzyme. Mol Cell Proteomics 10(M111):012658

15. Boily G, Seifert EL, Bevilacqua L, He XH, Sabourin G, Estey C, Moffat C, Crawford S, Saliba S, Jardine K, Xuan J, Evans M, Harper ME, McBurney MW (2008) SirT1 regulates energy metabolism and response to caloric restriction in mice. PLoS One 3:e1759
16. Someya S, Yu W, Hallows WC, Xu J, Vann JM, Leeuwenburgh C, Tanokura M, Denu JM, Prolla TA (2010) Sirt3 mediates reduction of oxidative damage and prevention of age-related hearing loss under caloric restriction. Cell 143:802–812
17. Banks AS, Kon N, Knight C, Matsumoto M, Gutierrez-Juarez R, Rossetti L, Gu W, Accili D (2008) SirT1 gain of function increases energy efficiency and prevents diabetes in mice. Cell Metab 8:333–341
18. Bordone L, Cohen D, Robinson A, Motta MC, van Veen E, Czopik A, Steele AD, Crowe H, Marmor S, Luo J, Gu W, Guarente L (2007) SIRT1 transgenic mice show phenotypes resembling calorie restriction. Aging Cell 6:759–767
19. Herranz D, Munoz-Martin M, Canamero M, Mulero F, Martinez-Pastor B, Fernandez-Capetillo O, Serrano M (2010) Sirt1 improves healthy ageing and protects from metabolic syndrome-associated cancer. Nat Commun 1:3
20. Howitz KT, Bitterman KJ, Cohen HY, Lamming DW, Lavu S, Wood JG, Zipkin RE, Chung P, Kisielewski A, Zhang LL, Scherer B, Sinclair DA (2003) Small molecule activators of sirtuins extend Saccharomyces cerevisiae lifespan. Nature 425:191–196
21. Milne JC, Lambert PD, Schenk S, Carney DP, Smith JJ, Gagne DJ, Jin L, Boss O, Perni RB, Vu CB, Bemis JE, Xie R, Disch JS, Ng PY, Nunes JJ, Lynch AV, Yang H, Galonek H, Israelian K, Choy W, Iffland A, Lavu S, Medvedik O, Sinclair DA, Olefsky JM, Jirousek MR, Elliott PJ, Westphal CH (2007) Small molecule activators of SIRT1 as therapeutics for the treatment of type 2 diabetes. Nature 450:712–716
22. Barger J, Kayo T, Vann J, Arias E, Wang J, Hacker T, Wang Y, Raederstorff D, Morrow J, Leeuwenburgh C, Allison D, Saupe K, Cartee G, Weindruch R, Prolla T (2008) A low dose of dietary resveratrol partially mimics caloric restriction and retards aging parameters in mice. PLoS One 3(6):e2264
23. Price NL, Gomes AP, Ling AJ, Duarte FV, Martin-Montalvo A, North BJ, Agarwal B, Ye L, Ramadori G, Teodoro JS, Hubbard BP, Varela AT, Davis JG, Varamini B, Hafner A, Moaddel R, Rolo AP, Coppari R, Palmeira CM, de Cabo R, Baur JA, Sinclair DA (2012) SIRT1 is required for AMPK activation and the beneficial effects of resveratrol on mitochondrial function. Cell Metab 15:675–690
24. Yoshino J, Mills KF, Yoon MJ, Imai S (2011) Nicotinamide mononucleotide, a key NAD(+) intermediate, treats the pathophysiology of diet- and age-induced diabetes in mice. Cell Metab 14:528–536
25. Canto C, Houtkooper RH, Pirinen E, Youn DY, Oosterveer MH, Cen Y, Fernandez-Marcos PJ, Yamamoto H, Andreux PA, Cettour-Rose P, Gademann K, Rinsch C, Schoonjans K, Sauve AA, Auwerx J (2012) The NAD(+) precursor nicotinamide riboside enhances oxidative metabolism and protects against high-fat diet-induced obesity. Cell Metab 15:838–847
26. Kennedy BK, Austriaco NR Jr, Zhang J, Guarente L (1995) Mutation in the silencing gene SIR4 can delay aging in S. cerevisiae. Cell 80:485–496
27. Kennedy BK, Gotta M, Sinclair DA, Mills K, McNabb DS, Murthy M, Pak SM, Laroche T, Gasser SM, Guarente L (1997) Redistribution of silencing proteins from telomeres to the nucleolus is associated with extension of life span in S. cerevisiae. Cell 89:381–391
28. Kaeberlein M, McVey M, Guarente L (1999) The SIR2/3/4 complex and SIR2 alone promote longevity in Saccharomyces cerevisiae by two different mechanisms. Genes Dev 13:2570–2580
29. Sinclair DA, Guarente L (1997) Extrachromosomal rDNA circles—a cause of aging in yeast. Cell 91:1033–1042
30. Stumpferl SW, Brand SE, Jiang JC, Korona B, Tiwari A, Dai J, Seo JG, Jazwinski SM (2012) Natural genetic variation in yeast longevity. Genome Res 22:1963–1973
31. Tissenbaum H, Guarente L (2001) Increases dosage of a sir-2 gene extends lifespan in C. elegans. Nature 410:227–230
32. Viswanathan M, Guarente L (2011) Regulation of Caenorhabditis elegans lifespan by sir-2.1 transgenes. Nature 477:E1–E2
33. Viswanathan M, Kim SK, Berdichevsky A, Guarente L (2005) A role for SIR-2.1 regulation of ER stress response genes in determining C. elegans life span. Dev Cell 9:605–615
34. Rizki G, Iwata TN, Li J, Riedel CG, Picard CL, Jan M, Murphy CT, Lee SS (2011) The evolutionarily conserved longevity determinants HCF-1 and SIR-2.1/SIRT1 collaborate to regulate DAF-16/FOXO. PLoS Genet 7:e1002235
35. Kanfi Y, Naiman S, Amir G, Peshti V, Zinman G, Nahum L, Bar-Joseph Z, Cohen HY (2012) The sirtuin SIRT6 regulates lifespan in male mice. Nature 483:218–221
36. Baur JA, Pearson KJ, Price NL, Jamieson HA, Lerin C, Kalra A, Prabhu VV, Allard JS, Lopez-Lluch G, Lewis K, Pistell PJ, Poosala S, Becker KG, Boss O, Gwinn D, Wang M, Ramaswamy S,

Fishbein KW, Spencer RG, Lakatta EG, Le Couteur D, Shaw RJ, Navas P, Puigserver P, Ingram DK, de Cabo R, Sinclair DA (2006) Resveratrol improves health and survival of mice on a high-calorie diet. Nature 444:337–342
37. Lagouge M, Argmann C, Gerhart-Hines Z, Meziane H, Lerin C, Daussin F, Messadeq N, Milne J, Lambert P, Elliott P, Geny B, Laakso M, Puigserver P, Auwerx J (2006) Resveratrol improves mitochondrial function and protects against metabolic disease by activating SIRT1 and PGC-1alpha. Cell 127:1109–1122
38. Rodgers JT, Lerin C, Haas W, Gygi SP, Spiegelman BM, Puigserver P (2005) Nutrient control of glucose homeostasis through a complex of PGC-1alpha and SIRT1. Nature 434:113–118
39. Purushotham A, Schug TT, Xu Q, Surapureddi S, Guo X, Li X (2009) Hepatocyte-specific deletion of SIRT1 alters fatty acid metabolism and results in hepatic steatosis and inflammation. Cell Metab 9:327–338
40. Yang H, Zhang W, Pan H, Feldser HG, Lainez E, Miller C, Leung S, Zhong Z, Zhao H, Sweitzer S, Considine T, Riera T, Suri V, White B, Ellis JL, Vlasuk GP, Loh C (2012) SIRT1 activators suppress inflammatory responses through promotion of p65 deacetylation and inhibition of NF-kappaB activity. PLoS One 7:e46364
41. Walker AK, Yang F, Jiang K, Ji JY, Watts JL, Purushotham A, Boss O, Hirsch ML, Ribich S, Smith JJ, Israelian K, Westphal CH, Rodgers JT, Shioda T, Elson SL, Mulligan P, Najafi-Shoushtari H, Black JC, Thakur JK, Kadyk LC, Whetstine JR, Mostoslavsky R, Puigserver P, Li X, Dyson NJ, Hart AC, Naar AM (2010) Conserved role of SIRT1 orthologs in fasting-dependent inhibition of the lipid/cholesterol regulator SREBP. Genes Dev 24:1403–1417
42. Ponugoti B, Kim DH, Xiao Z, Smith Z, Miao J, Zang M, Wu SY, Chiang CM, Veenstra TD, Kemper JK (2010) SIRT1 deacetylates and inhibits SREBP-1C activity in regulation of hepatic lipid metabolism. J Biol Chem 285: 33959–33970
43. Donmez G, Wang D, Cohen DE, Guarente L (2010) SIRT1 suppresses beta-amyloid production by activating the alpha-secretase gene ADAM10. Cell 142:320–332
44. Donmez G, Arun A, Chung CY, McLean PJ, Lindquist S, Guarente L (2012) SIRT1 protects against alpha-synuclein aggregation by activating molecular chaperones. J Neurosci 32:124–132
45. Jeong H, Cohen DE, Cui L, Supinski A, Savas JN, Mazzulli JR, Yates JR 3rd, Bordone L, Guarente L, Krainc D (2011) Sirt1 mediates neuroprotection from mutant huntingtin by activation of the TORC1 and CREB transcriptional pathway. Nat Med 18:159–165
46. Jiang M, Wang J, Fu J, Du L, Jeong H, West T, Xiang L, Peng Q, Hou Z, Cai H, Seredenina T, Arbez N, Zhu S, Sommers K, Qian J, Zhang J, Mori S, Yang XW, Tamashiro KL, Aja S, Moran TH, Luthi-Carter R, Martin B, Maudsley S, Mattson MP, Cichewicz RH, Ross CA, Holtzman DM, Krainc D, Duan W (2011) Neuroprotective role of Sirt1 in mammalian models of Huntington's disease through activation of multiple Sirt1 targets. Nat Med 18:153–158
47. Kim D, Nguyen MD, Dobbin MM, Fischer A, Sananbenesi F, Rodgers JT, Delalle I, Baur JA, Sui G, Armour SM, Puigserver P, Sinclair DA, Tsai LH (2007) SIRT1 deacetylase protects against neurodegeneration in models for Alzheimer's disease and amyotrophic lateral sclerosis. EMBO J 26:3169–3179
48. Kawahara TL, Michishita E, Adler AS, Damian M, Berber E, Lin M, McCord RA, Ongaigui KC, Boxer LD, Chang HY, Chua KF (2009) SIRT6 links histone H3 lysine 9 deacetylation to NF-kappaB-dependent gene expression and organismal life span. Cell 136:62–74
49. Finley LW, Carracedo A, Lee J, Souza A, Egia A, Zhang J, Teruya-Feldstein J, Moreira PI, Cardoso SM, Clish CB, Pandolfi PP, Haigis MC (2011) SIRT3 opposes reprogramming of cancer cell metabolism through HIF1alpha destabilization. Cancer Cell 19:416–428
50. Bell EL, Emerling BM, Ricoult SJ, Guarente L (2011) SirT3 suppresses hypoxia inducible factor 1alpha and tumor growth by inhibiting mitochondrial ROS production. Oncogene 30:2986–2996
51. Sebastián C, Zwaans BM, Silberman DM, Gymrek M, Goren A, Zhong L, Ram O, Truelove J, Guimaraes AR, Toiber D, Cosentino C, Greenson JK, MacDonald AI, McGlynn L, Maxwell F, Edwards J, Giacosa S, Guccione E, Weissleder R, Bernstein BE, Regev A, Shiels PG, Lombard DB, Mostoslavsky R (2012) The histone deacetylase SIRT6 is a tumor suppressor that controls cancer metabolism. Cell 151:1185–1199. doi: 10.1016/j.cell.2012.10.0
52. Guarani V, Deflorian G, Franco CA, Kruger M, Phng LK, Bentley K, Toussaint L, Dequiedt F, Mostoslavsky R, Schmidt MH, Zimmermann B, Brandes RP, Mione M, Westphal CH, Braun T, Zeiher AM, Gerhardt H, Dimmeler S, Potente M (2011) Acetylation-dependent regulation of endothelial Notch signalling by the SIRT1 deacetylase. Nature 473:234–238
53. Mattagajasingh I, Kim CS, Naqvi A, Yamamori T, Hoffman TA, Jung SB, DeRicco J, Kasuno K, Irani K (2007) SIRT1 promotes

endothelium-dependent vascular relaxation by activating endothelial nitric oxide synthase. Proc Natl Acad Sci USA 104: 14855–14860

54. Csiszar A, Labinskyy N, Podlutsky A, Kaminski PM, Wolin MS, Zhang C, Mukhopadhyay P, Pacher P, Hu F, de Cabo R, Ballabh P, Ungvari Z (2008) Vasoprotective effects of resveratrol and SIRT1: attenuation of cigarette smoke-induced oxidative stress and proinflammatory phenotypic alterations. Am J Physiol Heart Circ Physiol 294:H2721–H2735

Chapter 2

Sirtuins in Yeast: Phenotypes and Tools

Scott Tsuchiyama, Elizabeth Kwan, Weiwei Dang, Antonio Bedalov, and Brian K. Kennedy

Abstract

Originally discovered as a transcriptional silencing protein, *SIR2* was later linked to yeast replicative aging and the rest was history. Sir2p is now known to be a member of a class of protein deacetylases with a unique enzymatic activity coupling the deacetylation event to NAD^+ hydrolysis. While still incompletely understood, the mechanism by which Sir2p modulates yeast aging is linked to inhibition of rDNA recombination. Here we describe phenotypes associated with yeast Sirtuins and assays used to monitor Sirtuin function in yeast, including the replicative aging assay.

Key words Sirtuins, Yeast replicative aging, rDNA, Extrachromosomal rDNA circles, Transcriptional regulation

1 Introduction

The *s*ilent *i*nformation *r*egulator 2 (*SIR2*) was identified as part of a protein complex required to silence expression of the silent mating loci in yeast [1, 2]. Sir2p, along with Sir3p and Sir4p, are recruited to silent mating type genes and other loci in the genome, including regions near telomeres [3]. In addition, Sir2p independently can be targeted to the rDNA locus, where it suppresses rDNA recombination and transcription of ncRNA transcription. Activity of Sir2p at these loci depends on overlapping sets of recruitment factors and several studies have indicated that competition for recruitment of Sir2 to different compartments can occur.

SIR2, along with four homologs, *HST1, HST2, HST3,* and *HST4*, are collectively known as the Sirtuins and all display histone deacetylase activity [4]. Sir2p and Hst1p have some overlapping activities, as do Hst3p and Hst4p. All Sirtuins are class III deacetylases and have an unusual mechanism of enzyme action that involves hydrolysis of NAD^+ to generate nicotinamide, an end-product enzymatic inhibitor. A range of assays to measure Sirtuin enzyme activity have been described in other chapters in this volume.

Matthew D. Hirschey (ed.), *Sirtuins: Methods and Protocols*, Methods in Molecular Biology, vol. 1077, DOI 10.1007/978-1-62703-637-5_2, © Springer Science+Business Media, LLC 2013

Here we focus on assays to measure replicative aging, as well as functions of Sir2p relevant to aging.

One of the current assays to measure aging in yeast is the yeast replicative lifespan assay, which determines the number of times a mother cell divides to produce daughters and, by definition, is measured in generations [5]. Interestingly, quantitative evidence has been generated demonstrating that there is significant overlap between the pathways modulating longevity in the yeast replicative lifespan assay and longevity in the nematode *C. elegans* [6], which substantiates that yeast is a useful model organism for studying aging.

In general, yeast replicative lifespan correlates with Sir2p levels; deletion of *SIR2* leads to reduced lifespan whereas overexpression extends lifespan. Whether *SIR2,* with or without other Sirtuins, is required for lifespan extension by dietary restriction remains to be fully resolved [7–10]. Part of the confusion surrounds the many ways to invoke dietary restriction in yeast, discussed elsewhere [11]. Furthermore, *hst1D*, *hst2D,* and *hst4D* strains do not have consistently altered lifespan relative to controls [9, 10]. The *hst3D* displays a reduced lifespan and, consistent with functional redundancy between Hst3p and Hst4p, the *hst3D hst4D* strain has dramatically reduced lifespan [10, 12]. Hst3p and Hst4p act to maintain genome stability through deacetylation of their target, histone 3 lysine 56 (H3K56) [12]. This activity is likely linked to their role as longevity assurance genes.

Two other recent methods have been devised to measure replicative lifespan in yeast and to study the behavior of aging cells mostly at the single cell level. One, the *m*other *e*nrichment *p*rogram utilizes a clever genetic technique to selectively and conditionally kill daughter cells [13]. Using this technique, old cells age in a liquid culture growing linearly. Old cells can then be isolated from their nondividing daughters by cell surface labeling methods, described below. Another recent method involves the use of microfluidics to trap dividing mother cells in a small area, where they can be exposed to circulating liquid medium and visualized microscopically [14, 15]. Readers should consult separate reports cited herein describing these methods.

Yeast cells lacking *SIR2* display a significantly reduced lifespan, which has been attributed to an increase in the frequency of recombination within the rDNA which when accumulated has been assumed to be toxic [16]. In contrast, strains lacking *FOB1*, identified by virtue of its ability to directionally impede replication fork progress in rDNA repeats [17], have reduced extrachromosomal rDNA formation (ERC) formation and enhanced lifespan [18]. The rDNA in yeast is a series of 100–200 repeats of a 35S RNA gene, followed by a non-transcribed spacer (NTS) region and then a 5S RNA gene. Sir2p is involved with the repression of mitotic recombination [19] and its activity as a part of the RENT complex localizes to the NTS region as well as the Pol I promoter in the rDNA [20].

This complex among other activities reduces the frequency with which origins of replication initiate in the rDNA locus [21].

Homologous recombination within the rDNA can lead to looping out of one or more rDNA repeat, forming ERCs. rDNA and thus also ERCs contain a replication origin [22]; however, they do not contain a CEN sequence which allows for them to be preferentially retained in the mother cell during cell division. An accumulation of these circles has been assumed to be toxic and this has been reported to be one of the ways that Sir2p and Fob1p regulate lifespan. Loss of *SIR2* promotes recombination, shortening lifespan, and loss of *FOB1* reduces recombination, lengthening it [16, 18, 19, 23]. A recent study has led to the hypothesis that origins of ERCs (and also rDNA repeats in the genome) compete for other origins in the rest of the genome in old mothers, conferring replication stress and limiting mother cell division potential [24].

Sir2p levels are reduced in aged cells, and this decrease in abundance correlates with an increase in H4K16 acetylation, a known target of Sir2p [25]. This age-related chromatin change is rescued by deletion of *SAS2*, the acetyltransferase responsible for acetylation of H4K16, which also results in replicative lifespan extension. Sir2-dependent silencing of genes near telomeres was originally found for auxotrophic markers inserted artificially in these regions. For instance, insertion of the *ADE2* gene in subtelomeric regions leads to colonies that sector red and white. The red state, which represents silencing of the *ADE2* gene, is lost in strains lacking *SIR2* [3]. More recently, an endogenous locus, *YFR057W*, was found to be regulated by Sir2p and expression of this gene can be assessed by quantitative PCR [26].

Together, Sirtuins are now one of the most studied pathways in aging research. Initial indications of a role for these deacetylases came from findings in yeast. While overexpression of *SIR2* orthologs have been linked to lifespan extension in other invertebrate organisms, a common mechanism underlying this phenotype has not been determined. As such, studies in yeast remain critical to identify all of the Sir2p activities linked to aging, generating hypotheses for testing in multicellular systems. Here we define methods for determination of replicative lifespan in yeast, also available in video format [27], as well as the study of Sir2-dependent functions which are important for lifespan.

2 Materials

2.1 Yeast Replicative Lifespan

1. Tetrad dissecting scope (with 160× magnification).
2. Computer with a spreadsheet program (optional).
3. YPD plates.
4. Blank spreadsheet.

2.2 Isolating Old Yeast Cells

1. Total of 10 L freshly made and filter-sterilized liquid rich media (synthetic complete or YPD) for four overnight aging periods, including 4 L for the first period, and 2 L for each of the following periods. Media for aging should be prepared on the day of usage. Additional media are required for startup cultures (*see* **Note 1**).
2. Phosphate-buffered saline (PBS), sterile, 1 L, cool to 4 °C.
3. EZ-Link Sulfo-NHS-LC-Biotin, 12 mg.
4. 15 mL 0.1 M Glycine-PBS (diluted in PBS from a 2.5 M Glycine stock solution just before use).
5. PBS + BE. PBS containing 1 mg/mL BSA and 2 mM EDTA. To make 80 mL, dissolve 80 mg BSA powder in 80 mL PBS, and then add 320 μL 0.5 M EDTA. Prepare just before use.
6. Dynabeads Biotin Binder, 9 mL.

2.3 Sir2-Dependent Modulation of Replicative Aging of ERC Formation in the rDNA

1. A *S. cerevisiae* strain that has *ADE2* integrated within the rDNA repeats.
2. Synthetic media lacking adenine.
3. Large (15 cm) dishes are preferred to allow a low plating density.

2.4 Direct Measurement of ERC Levels

2.4.1 Yeast Genomic Miniprep (Adapted from [28])

1. 100 mM $NaPO_4$ pH 7.2; for 50 mL, combine 3.4 mL 1 M NaH_2PO4, 1.6 mL 1 M Na_2HPO4, and 45 mL H_2O.
2. Lysis buffer: 10 mM $NaPO_4$, 10 mM EDTA pH 8.0, 0.1 M B-mercaptoethanol, 100 μg/mL 100 T zymolyase.
3. 20 mg/mL Proteinase K.
4. Lysing stock: 0.25 M EDTA, 0.5 M Tris–Hcl, 2.5 % SDS; 5 M potassium acetate.
5. TE pH 8.0; 6. 20 mg/mL RNase A.

2.4.2 Preparation of Yeast DNA Embedded in Agarose Plugs

This protocol was adapted from the Brewer/Raghuraman lab website: http://fangman-brewer.genetics.washington.edu/

1. 50 mM EDTA.
2. 1 % low melting temperature agarose (e.g., InCert or SeaPlaque) in 50 mM EDTA.
3. Spheroplasting solution: 40 mL 1 M Sorbitol (approximately 1 M final), 1.6 mL 0.5 M EDTA (20 mM final), 0.4 mL 1 M Tris–HCl pH 7.5 (10 mM final), 40 μL β-mercaptoethanol (14 mM final), 21 mg Zymolyase 20 T (0.5 mg/mL final).
4. 1× NDS solution pH 9.5: 0.5 M EDTA, 10 mM Tris base, 1 % Sarkosyl (Sarkosyl is also called "N-Lauroylsarcosine sodium salt").

5. LDS Solution: 1 % lithium dodecyl sulfate, 100 mM EDTA, 10 mM Tris–HCl pH 8.0.
6. Tris–EDTA (TE) pH 8.0.
7. Agarose plug molds (e.g., Bio-Rad catalog # 1703706).
8. Tissue culture plate, 6-well.
9. Metal spatula.

2.4.3 1D-ERC gel

1. 1× TBE: 90 mM Tris base, 90 mM Boric acid, 2 mM EDTA.
2. 0.7 % agarose in 1× TBE.
3. Standard loading buffer (e.g., blue juice).
4. DNA ladder.
5. 10 mg/mL Ethidium bromide.

2.4.4 2D-ERC gel (Adapted from [29])

1. 1× TBE.
2. 0.5 % agarose in 1× TBE (for first dimension).
3. 1 % agarose in 1× TBE (for second dimension).
4. DNA ladder.
5. 10 mg/mL Ethidium bromide.
6. Refrigerated gel electrophoresis apparatus (i.e., gel box in cold room) with a buffer recirculation pump.

2.4.5 Southern Blotting

1. Depurination solution: 0.25 N HCl.
2. Denaturation solution: 0.5 N NaOH, 1 M NaCl.
3. Neutralization solution pH 7.4: 0.5 M Tris–HCl, 3 M NaCl.
4. 20× SSC: 3 M NaCl, 300 mM sodium citrate.
5. Nylon membrane (e.g., GeneScreen hybridization transfer membrane, Hybond-N membrane).
6. Wicking papers (e.g., Whatman 3 mm chromatography paper).
7. Paper towels.
8. Sponge and tray.

2.4.6 ^{32}P Labeled Probe Synthesis

1. Fragment of interest (i.e., rDNA): PCR amplified and gel purified. Blots in Fig. 2 were probed with a 1 kb fragment from the NTS2 region of the rDNA, but any rDNA sequence can be used.

   NTS2 5′ primer: CTGGTAGATATGGCCGCAACC
   NTS2 3′ primer: GTCTTCAACTGCTTTCGCAT
2. Oligo mix for ^{32}P-dATP labeling: 12 μM dCTP (use 125 μM dATP if labeling with ^{32}P-dCTP), 125 μM dGTP, 125 μM dTTP, 1.25 M HEPES pH 6.6, 62.5 mM β-mercaptoethanol, 62.5 mM Tris–HCl, 0.4 mM EDTA pH 7.5, random hexamer oligos (32 OD units/mL).

3. Tris + $MgCl_2$ solution: 1 M Tris–HCl, 125 mM $MgCl_2$.
4. BSA (10 mg/mL).
5. ^{32}P-labeled nucleotides (e.g., dATP or dCTP, [α-^{32}P]- 800 Ci/mmol 10 mCi/mL EasyTide).
6. Klenow (exo-fragment) (e.g., NEB#M0212).
7. Optional, for probe purification to remove excess radionucleotides: Sephadex G-50, medium grade (Pharmacia #17.0043-01; Prepare by combining 10 g Sephadex G-50, 160 mL H_2O, 0.32 mL Tris–HCl pH8.0, 0.4 mL 0.2 M EDTA in a 250 mL bottle, autoclave, and let cool. Store at 4 °C.); 1 cc syringe; Glass wool; Carrier RNA: 10 mg/mL tRNA in TE pH 7.6 [30].

2.4.7 Southern Blot Hybridization (Adapted from [30])

1. Prehyb solution which consists of 30 mL 20× SSC *and* 10 mL 10× Denhardt's solution containing 1 % Polyvinylpyrrolidone-40; 1 % PEG-8000, 1 % BSA; 10 mg/mL sheared carrier DNA (e.g., herring testes DNA or salmon sperm DNA), 1 mL 10 % SDS, 4 mL 5 % sodium pyrophosphate, 54 mL H_2O, 10 mg sheared carrier DNA).
2. Wash solution A: 2× SSC, 0.1 % SDS (sodium dodecyl sulfate).
3. Wash solution B: 1× SSC, 0.1 % SDS.
4. Autoradiography film (e.g., GeneMate blue autoradiography film).
5. Phosphorimager (e.g., GE Healthcare Life Sciences Typhoon, Bio-Rad Personal Molecular Imager).

3 Methods

3.1 Yeast Replicative Lifespan

The standard yeast replicative lifespan measurement requires the physical separation of the mother cell from the daughter cell by use of a micromanipulator fitted with a fiber optic needle.

1. Pour YPD plates on a level surface (*see* **Note 2**).
2. For each test and control strains, from a single colony, streak out cells onto a new plate (*see* **Note 3**).
3. The next day, make a small patch of cells on another new plate.
4. Place plate onto tetrad dissecting scope and select coordinates on the plate to deposit single cells (*see* **Note 4**).
5. Move around 30–40 cells from the patch to the starting coordinate.
6. Pick ten of the smallest cells and separate them into a line of evenly spaced cells between the starting coordinate and the next coordinate that was marked. It is not essential that these cells be daughters at this stage.

7. Take remainder of cells and spread around ten out near the starting coordinate, take rest and place as far from starting coordinate as possible.
8. Move around 30–40 more cells newly isolated from patch to the second chosen coordinate, etc., until the desired number of cells for lifespan analysis are set out.
9. Incubate plate for 90 min.
10. Each of the cells that were lined out should have one bud, remove bud (smaller of two cells) and put into line. Discard larger bud. The small bud is the virgin mother cell that has now divided zero times.
11. Repeat for all cells. Do nothing for cells that have not divided.
12. Incubate plate for 90 min.
13. Use needle to separate buds from mothers, keep large cells, and count the smaller cells.
14. Use spreadsheet to keep track of how many buds have been produced by each mother.
15. As mother cells age, or produce more buds, the time between buddings will increase. As this happens, increase the incubation time. Ideally, the incubation time will be such that every time the plates are counted, the mothers have divided 1–2 times at most. Over incubation will result in a difficulty in determining mother from its daughters.
16. Continue until mothers have proceeded 8 h of incubation time without dividing; at this point the mother is considered to have entered senescence or "died," and is marked as such.
17. A typical lifespan experiment lasts around 2 weeks and varies due to the large variation in the number of buds produced by the mother cells (*see* **Notes 5** and **6**).
18. As a reference, the BY haploid parent strains used to create the deletion collection usually produce around 25–28 buds (Fig. 1) (*see* **Note 7**).
19. After all cells have "died," survival curves can be plotted by plotting fraction viable over age in number of divisions (*see* **Note 8**).

3.2 Isolating Old Yeast Cells

Large number of old cells sufficient for biochemical analysis can also be generated taking advantage of the fact that the yeast cell wall is retained in the mother cell during division. Thus, cell surface proteins can be labeled with biotin and the label is retained in the mother cell [31]. We have provided a thorough methodology for isolating old cells in sufficient numbers for most biochemical methods. Using this method, for instance, it was demonstrated that Sir2 protein levels decline with age [25].

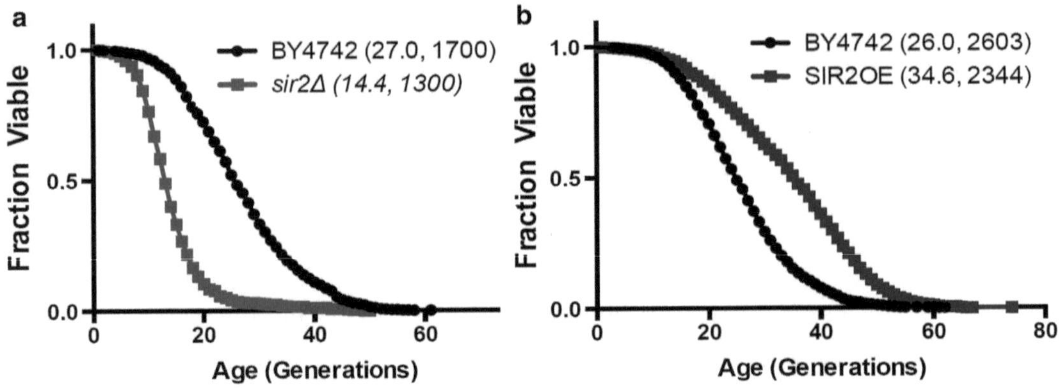

Fig. 1 Replicative lifespans of *SIR2* mutant strains. Lifespan analysis was performed in the BY4742 strain background. Note that genetic effects on replicative lifespan are strain-specific. (**a**) *sir2D* is short-lived. (**b**) *SIR2* overexpression (SIR2OE) results in enhanced means and maximum lifespan. Curves in both cases are compiled from a number of independently conducted lifespan experiments. In the legends, the numbers after each strain indicate (mean replicative lifespan, number of cells analyzed)

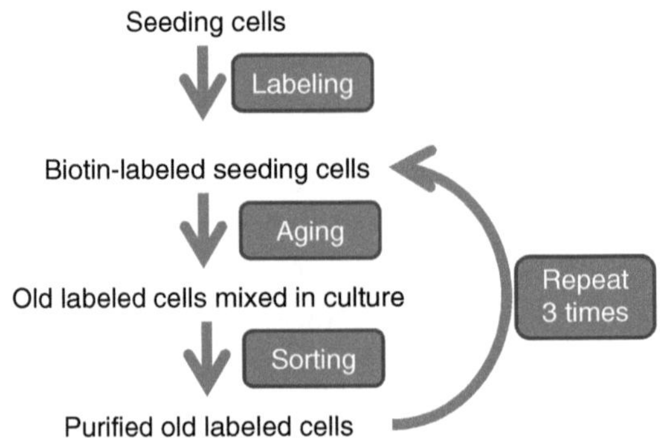

Fig. 2 Graphic of methodology for sorting of old cells. A protocol is described in the text detailing the process to generate large number of replicatively aged yeast cells

The objective of this protocol is to purify 150 million (1.5×10^8) yeast mother cells aged up to 80 % of their mean lifespan for biochemical analysis. Two strains can be sorted at the same time. This protocol only describes the procedure for one strain. The protocol requires a total of 6 days, of which four are devoted to one round per day of manual sorting (*see* Fig. 2 for graphic of protocol).

3.2.1 Starting Up Culture (Day 1)

1. Inoculate 5 mL media from a colony and grow at 30 °C overnight.

3.2.2 Cell Surface Labeling (Day 2)

1. Dilute the overnight culture to 100 mL and an optical density (OD_{600}) of 0.2. Shake at 30 °C until OD_{600} reaches 0.8–1.0.
2. Prepare 4 L fresh media and cool to 4 °C during this time.
3. Transfer the culture into two 50-mL conical tubes. Pellet cells in a tabletop centrifuge at 2,000 × *g*, 4 °C for 5 min.

Table 1
USC calculation table

	Strain 1
Count 1	
Count 2	
Count 3	
Count 4	
Average	
Counting dilution	100
Susp. density (/mL)	
Target cell no. for labeling	1.2×10^9
Target volume (mL)	
Target cell no. to save as USC	3.0×10^8
Target volume for USC (mL)	

4. While harvesting cells, take out the EZ-link Sulfo-NHS-LC-Biotin from the refrigerator to warm up to room temperature (RT).
5. Remove the supernatant, resuspend the cell pellet in each tube in 20 mL PBS, and combine into one 50-mL conical tube.
6. Pellet cells again as before.
7. Wash cells two more times in 40 mL PBS.
8. Resuspend the washed cell pellet in 20 mL PBS.
9. Dilute 1 μL cell suspension to 100 μL in PBS (100-fold dilution) and determine cell density by counting cells with a hemocytometer (*see* **Note 9**).
10. Calculate required cell suspension volume for biotin surface labeling and unsorted control (USC) samples in Table 1.
11. Transfer 1.2×10^9 cells in suspension to a 15-mL conical tube, centrifuge at $2{,}000 \times g$, 4 °C for 3 min, and remove the supernatant. These cells will be labeled with biotin.
12. Save an aliquot of 3×10^8 cells in a 1.5-mL screw-cap tube as a USC for analysis. Centrifuge at $2{,}000 \times g$, 4 °C for 1 min, remove the supernatant, and flash-freeze. Save another 50 μL USC sample for bud scar counting (BSC). Mix the BSC sample with 5 μL 37 % formaldehyde and store at 4 °C.
13. Make sure that EZ-link Sulfo-NHS-LC-Biotin is warmed up to RT before opening. Prepare a 1.2 mL 10 mg/mL (18 mM) solution and add to cells from previous step.
14. Rotate the tube at 30 °C for 15 min.

15. Prepare a 15 mL 0.1 M Glycine-PBS solution during incubation.
16. Wash biotin-labeled cells three times in 5 mL 0.1 M Glycine-PBS to quench and remove free biotin. Centrifuge at 2,000 × *g*, 4 °C for 3 min after each wash.
17. Resuspend cells in 2 mL PBS and aliquot into two 1.5-mL microcentrifuge tubes. Keep on ice until inoculation.
18. About 10 h before harvesting the cells, add each aliquot to 2 L cold (to slow down initial growth) media, saving 5 mL media before inoculation for blanking the spectrophotometer.
19. Shake at 30 °C.

3.2.3 Old Cell Sorting Round 1 (Day 2, Sort 1)

1. Harvest cells when OD600 is between 1.2 and 1.6. Collect culture in sterilized 500-mL centrifuge tubes and pellet cells at 2,000 × *g* at 4 °C for 20 min. Drain off media.
2. For each centrifuge tube, resuspend cells in 10 mL PBS, combine all cell suspensions into four 50-mL conical tubes. Pellet cells at 2,000 × *g* at 4 °C for 5 min. Wash two more times in 40 mL PBS (*see* **Note 10**).
3. Prepare 160 mL PBS + BE (40 mL for each conical tube) during wash.
4. Resuspend each cell pellet in 35 mL PBS + BE to get a final volume ~ 40 mL. Use a 10-mL serological pipet to help resuspend cell pellets.
5. Divide 3 mL (1.2×10^9 beads) Dynabeads Biotin Binder among four 1.5-mL microcentrifuge tubes and remove supernatants using a magnetic stand. For each tube, wash beads three times with 1 mL PBS + BE. Use a short and gentle vortex to completely resuspend beads during wash.
6. Process up to two conical tubes (2× 40-mL cell suspension) at a time using the following steps:
7. Add an aliquot of beads to each conical tube containing cell suspension and incubate at 4 °C for 1 h in a rotator (*see* **Note 11**).
8. In a cold room, transfer cell suspension in 3.3 mL aliquots to 5-mL round-bottom culture tubes with lids (12 tubes for each 40-mL suspension) in a magnetic stand. Wait 15 min for bead-bound old cells to move toward the magnet.
9. During the wait time, start another batch of binding reactions.
10. Remove young cells with a 5-mL serological pipet by slowly pipetting from the top of tube and descending without disturbing the strip of old cells/beads attached to the magnet.
11. Leave about 0.2 mL of cells at the bottom of tube to avoid removing old cells.
12. The supernatant is pooled into a 50-mL conical tube and put on the magnetic stand for 50 mL tubes to recover any

Dynabeads and bead-bound old cells pulled out with the supernatant.

13. Save the cells in the supernatant (young cells) temporarily as "FT" and keep on ice until later.
14. Wash each 5-mL tube twice with 3 mL PBS without removing tubes from the magnetic stand and wait 5 min. Remove free cells as before and save temporarily as "W1" and "W2."
15. The remaining sorting steps can be performed out of the cold room on ice.
16. Remove tubes from the magnetic stand, resuspend old cells/beads in all 12 tubes (per 40-mL cell suspension) with 1 mL PBS, and combine into a 1.5-mL microcentrifuge tube. Separate the old cells/beads in a magnetic stand for microcentrifuge tubes and save the supernatant as "W3."
17. Rinse the batch of twelve 5-mL tubes with 1 mL PBS and add to the microcentrifuge tube containing old cells/beads. Separate the old cells/beads in a magnetic stand and save the supernatant as "W4."
18. Continue to wash the old cells/beads four more times by resuspending in 1 mL PBS and separating in a magnetic stand. Save washes as "W5," "W6," "W7," and "W8." "W1" and "W2" can be spun down to recover lost beads and old cells during initial washing if they are visible. All wash samples may be discarded after yield is confirmed.
19. Repeat the above steps to process all 40-mL cell suspensions.
20. Resuspend and combine all purified bead-bound old cells in 2.5 mL PBS (total volume, expected density 4×10^8/mL) for the first round of sorting.
21. Dilute 1 μL suspended old cells to 200 μL in PBS (200-fold dilution) and determine cell density by counting cells with a hemocytometer. Calculate sorting yield using Table 2.
22. For both young (FT, S1Y) and old cells (S1O), save 10 μL (mixed with 1 μL 37 % formaldehyde) for BSC.
23. Do NOT save "sort 1" old cells for analysis. Save 1 mL young cells (20 mL culture worth of cells) in 1.5-mL screw-cap tubes for analysis. Centrifuge at $2{,}000 \times g$, 4 °C for 1 min, remove the supernatant, and flash-freeze.
24. Keep old cells on ice until inoculation. Prepare 2 L fresh media and cool to 4 °C.
25. About 12 h before next harvest, add old cells from "sort 1" to 2 L cold (to slow down initial growth) media, saving 5 mL media before inoculation for blanking the spectrophotometer.
26. Shake at 30 °C.

Table 2
Yield table

	Strain 1
Count 1	
Count 2	
Count 3	
Count 4	
Average	
Counting dilution	200
Susp. density (/mL)	
Susp. volume (mL)	2.5
Total sorted cells	
Cells before sorting	1.2×10^9
Yield	

3.2.4 Old Cell Sorting Round 2 (Day 4, Sort 2)

1. Harvest cells when OD600 is between 1.2 and 1.6. Collect culture in sterilized 500-mL centrifuge tubes and pellet cells at 3,220×*g* at 4 °C for 20 min. Drain off media (*see* **Note 12**).
2. For each centrifuge tube, resuspend cells in 10 mL PBS, combine all cell suspensions into two 50-mL conical tubes. Pellet cells at 2,000×*g* at 4 °C for 5 min. Wash two more times in 40 mL PBS.
3. Prepare 80 mL PBS+BE (40 mL for each conical tube) during wash.
4. To remove used beads (lost binding capacity, free from cells), resuspend cell pellets in 35 mL PBS and make 3.3 mL aliquots in 5-mL round-bottom culture tubes (12 tubes for each 40-mL suspension) in a magnetic stand. Keep cells in a cold room or cold box and wait 5 min for free beads to move toward the magnet.
5. Remove supernatant cells gently with a 5-mL pipet without disturbing the strip of beads attracted to the magnet. Try to remove as much cells as you can without taking significant amount of beads. All supernatants from a batch of 12 tubes are pooled into a 50-mL conical tube as "FB-FT."
6. Wash each 5-mL tube with 3 mL cold PBS without removing tubes from the magnetic stand and wait for 5 min.
7. Remove supernatant as before and save as "FB-W1."
8. Resuspend beads with 1 mL PBS for each group of 12 tubes and combine into a 1.5-mL microcentrifuge tube.

9. Separate the beads on a magnetic stand and save supernatant as "FB-W2."
10. Wash with 1 mL PBS two more times and save as "FB-W3" and "FB-W4."
11. Centrifuge "FB-FT," "FB-W1-W4" at 2,000 × *g* for 3 min.
12. Resuspend and combine "FB-FT," "FB-W1-W4" with 35 mL PBS + BE to get a final volume ~ 40 mL. Use a 10-mL serological pipet to help resuspend cell pellets.
13. Divide 3 mL (1.2×10^9 beads) Dynabeads Biotin Binder between two 1.5-mL microcentrifuge tubes and remove supernatants using a magnetic stand.
14. For each tube, wash beads three times with 1 mL PBS + BE. Use a short and gentle vortex to completely resuspend beads during wash.
15. Add an aliquot of beads to each conical tube containing cell suspension and incubate at 4 °C for 1 h in a rotator (*see* **Note 13**).
16. During the wait time, resuspend the used beads with 1 mL PBS for each microcentrifuge tube and check under the microscope to make sure no significant amount of cells are pulled down.
17. In a cold room, transfer cell suspension in 3.3 mL aliquots to 5-mL round-bottom culture tubes with lids (12 tubes for each 40-mL suspension) in a magnetic stand. Wait 15 min for bead-bound old cells to move toward the magnet.
18. Remove young cells with a 5-mL serological pipet by slowly pipetting from the top of tube and descending without disturbing the strip of old cells/beads attached to the magnet (*see* **Note 14**).
19. Leave about 0.2 mL of cells at the bottom of tube to avoid removing old cells.
20. Pool the supernatant into a 50-mL conical tube and put on the magnetic stand for 50 mL tubes to recover any Dynabeads and bead-bound old cells pulled out with the supernatant.
21. Save the cells in the supernatant (young cells) temporarily as "FT" and keep on ice until later.
22. Wash each 5-mL tube twice with 3 mL PBS without removing tubes from the magnetic stand and wait 5 min.
23. Remove free cells as before and save temporarily as "W1" and "W2."
24. The remaining sorting steps can be performed out of the cold room on ice.
25. Remove tubes from the magnetic stand, resuspend old cells/beads in all 12 tubes (per 40-mL cell suspension) with 1 mL PBS, and combine into a 1.5-mL microcentrifuge tube.

26. Separate the old cells/beads in a magnetic stand for microcentrifuge tubes and save the supernatant as “W3.”
27. Rinse the batch of 12 5-mL tubes with 1 mL PBS and add to the microcentrifuge tube containing old cells/beads.
28. Separate the old cells/beads in a magnetic stand and save the supernatant as “W4.”
29. Continue to wash the old cells/beads four more times by resuspending in 1 mL PBS and separating in a magnetic stand.
30. Save washes as “W5,” “W6,” “W7,” and “W8.” “W1” and “W2” can be spun down to recover lost beads and old cells during initial washing if they are visible. All wash samples may be discarded after yield is confirmed.
31. Resuspend and combine all purified bead-bound old cells in 2 mL PBS (total volume, expected density 4×10^8/mL) for the first round of sorting.
32. Dilute 1 μL suspended old cells to 200 μL in PBS (200-fold dilution), and determine cell density by counting cells with a hemocytometer. Count cells in triplicates or quadruplicates to ensure accuracy. Calculate sorting yield using Table 3.

Table 3
Sorting yield table

	Strain 1
Count 1	
Count 2	
Count 3	
Count 4	
Average	
Counting dilution	200
Susp. density (/mL)	
Susp. volume (mL)	2
Total sorted cells	
Cells before sorting	
Yield	
Overall yield	
Cells to save for analysis	1.5×10^8
Volume to save for WD (mL)	
Cells left to sort	

Table 4
Complete sorting table

	Sort 1	Sort 2	Sort 3	Sort 4
No. of cells to start with	1.2×10^9	1.0×10^9	6.5×10^8	4×10^8
Volume of media to inoculate	4 L (cold)	2 L (cold)	2 L (RT)	2 L (pre-warmed)
Targeting OD_{600}	1.2–1.6	1.2–1.6	1.2–1.6	>0.8, <1.6
Culturing time	10 h	12 h	14 h	No more than 14 h
No. of conical tube to harvest in	4	2	2	2
Beads to use	1.2×10^9 (3 mL)	1.2×10^9 (3 mL)	8×10^8 (2 mL)	4×10^8 (1 mL)
Expected to recover	1.0×10^9	8×10^8	4×10^8	1.5×10^8
Saving cells for analysis	–	1.5×10^8	–	1.5×10^8
Remaining cells for next round	1.0×10^9	6.5×10^8	4×10^8	–

33. For both young (FT, S2Y) and old cells (S2O), save 10 μL of cells (mixed with 1 μL 37 % formaldehyde) for BSC.
34. Save up to 1.5×10^8 sorted old cells and 1 mL young cells (20 mL culture worth of cells) in a 1.5-mL screw-cap tube for analysis.
35. Centrifuge at $2{,}000 \times g$, 4 °C for 1 min, remove the supernatant, and flash-freeze.
36. Use the remaining old cells for the next round of sorting.
37. Keep old cells on ice until inoculation. Prepare 2 L fresh media and cool to 4 °C.

3.2.5 Old Cell Sorting Round 3 and 4 (Day 5 and 6, Sort 3 and 4)

Subsequent sortings are carried out the same way as sort 2, except for the differences noted in Table 4 (*see* **Note 15**).

3.3 rDNA Sectoring

It is possible to measure Sir2-dependent suppression of rDNA recombination within rDNA repeats by a half-sectoring assay (*see* **Note 16**). In this assay, the *ADE2* gene is inserted into one rDNA repeat. When a recombination event excises the rDNA repeat from the genome, *ADE2* is removed, and is analogous to an *ADE2* deletion. Daughter cells from the *ADE2* deletion will accumulate adenine precursors which turn the cells a red color. Thus, the recombination event leads to a half-sectored red/not-red (white) colony. When cells are plated at low density, colonies will arise from single cells, and the ratio of red/white cells can be used as a direct read-out of recombination rate in a single colony. The rate can be measured by the frequency of half-sectored colonies over total colonies. Normally homologous recombination between repeats of the rDNA locus typically occurs once in every 1,000 mitotic divisions. A protocol for this assay is provided below.

1. Obtain a mid-log phase culture of sectoring strain grown in synthetic complete media lacking adenine to ensure that all yeast cells retain the *ADE2* gene.
2. Count density of cells using a hemocytometer.
3. Plate 1,000 cells per dish, spread evenly and allow plates to fully dry.
4. Incubate at 30 °C for 3–5 days (*see* **Note 17**).
5. At end of incubation, red colonies should appear. Count the number of colonies on the plate and the number of colonies that are half red and half white (*see* **Note 18**).
6. The rate of sectoring is the number of half red half white colonies over the total number of colonies, typical publications contain data from around 20,000 colonies (*see* **Note 19**).

3.4 Direct Measurement of ERC Levels

3.4.1 Yeast Genomic DNA Miniprep

1. Inoculate 5 mL media from a colony and grow at 30 °C overnight.
2. Centrifuge cells in a 1.5 mL microcentrifuge tube for 1 min.
3. Aspirate off liquid from pelleted cells using a vacuum manifold. Drain residual liquid on kimwipes.
4. Resuspend cells in 0.5 mL lysis buffer.
5. Incubate at 37 °C for 1 h. Every 20 min, gently invert tubes to mix. Do not vortex tubes as it will break the DNA.
6. Add 5 μL 20 mg/mL Proteinase K to each sample. Then add 0.1 mL lysing stock.
7. Gently invert to mix.
8. Incubate at 65 °C for 1 h.
9. Cool samples to room temperature.
10. Add 0.1 mL 5 M potassium acetate and mix immediately by inversion.
11. Chill samples on ice water for 10 min.
12. Centrifuge samples for 20 min in refrigerated centrifuge (4 °C).
13. Transfer the supernatant to a fresh tube containing 0.5 mL 100 % ethanol.
14. Invert a few times and centrifuge for 20 s.
15. Aspirate off ethanol, leaving DNA pellet at the bottom of the tube. Centrifuge tube briefly to collect residual ethanol and aspirate ethanol.
16. Air dry the sample in a warm room or hood for 15 min.
17. Dissolve pellet in 0.5 mL TE + 10 μg/mL RNaseA.

18. Incubate at 37 °C for 30 min.
19. Add 0.5 mL isopropanol to sample, invert several times, and centrifuge for 10 s.
20. Aspirate off supernatant from pelleted DNA.
21. Air dry in hood for 15–30 min.
22. Dissolve DNA in 100 μL TE pH 8.0.
23. Quantify DNA concentration of sample by spectrophotometry and by gel electrophoresis.
24. Because a significant amount of Proteinase K may remain with DNA pellet and may interfere with subsequent enzymatic reactions, an optional phenol–chloroform clean-up procedure is provided (**steps 23–29**).
25. Add 200 μL 25:24:1 phenol:chloroform:isoamyl alcohol to sample. Invert several times to mix.
26. Centrifuge for 5 min.
27. Transfer aqueous (top) layer to fresh 1.5 mL microcentrifuge tube, keeping clear of the cell debris at the phenol interface.
28. Add 200 μL chloroform to sample. Invert several times to mix.
29. Centrifuge for 2 min.
30. Transfer aqueous (top) layer to fresh 1.5 mL microcentrifuge tube.
31. Quantify DNA concentration of sample by spectrophotometry and by gel electrophoresis.

3.4.2 Preparation of Yeast DNA Embedded in Agarose Plugs

1. Melt 1 % low melting temperature agarose and place in 45 °C water bath to maintain agarose temperature at 45 °C.
2. Transfer yeast cells into a 1.5 mL microcentrifuge tube (want 10^8 cells per agarose plug). Centrifuge at 21130×*g* for 1 min to pellet cells.
3. Aspirate off supernatant.
4. Wash the cell pellet twice in 1 mL 50 mM EDTA, centrifuging cells and aspirating the supernatant as in **step 1**.
5. Resuspend the cells in 45 μL 50 mM EDTA per agarose plug. Warm the cell suspension to 45 °C.
6. Add an equal volume of 1 % low melting temperature agarose, quickly pipet-mix the cells with the agarose, and transfer the cell-agarose mixture into the plug molds (*see* **Note 20**).
7. Allow plugs to set for 30 min at 4 °C.
8. Extrude each plug from the plug mold into a 6-well plate. Each well can hold three plugs. Keep plugs of different cell types separate.
9. To each well, add 6 mL of freshly prepared spheroplasting solution.

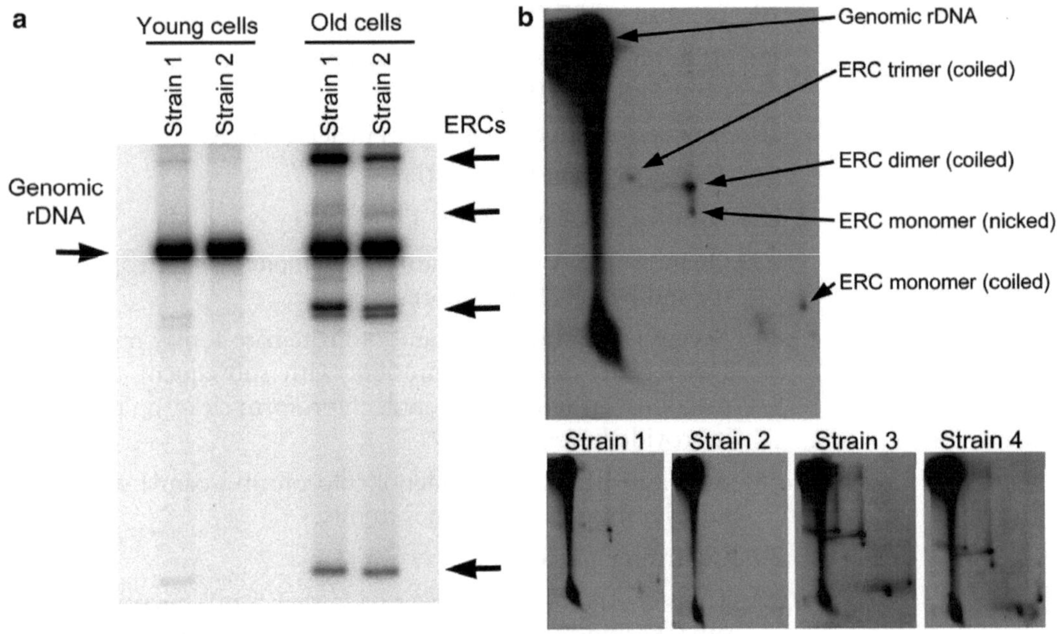

Fig. 3 Expected results from 1D and 2D gels to measure extrachromosomal rDNA circles. (**a**) 1D gel to measure ERCs showing an increase in ERC levels with aging. Note that ERCs can exist as monomers, dimers, trimers, and more complicated species, justifying multiple migrating bands. (**b**) 2D gel to measure ERCs on top, detailing the identification of multiple ERC species. On bottom, strains are shown with different ERC levels

10. Incubate at 37 °C for 2–4 h with gentle shaking (*see* **Note 21**).
11. Aspirate off the spheroplasting solution from each well. Add 6 mL of the LDS solution.
12. Incubate with gentle shaking for 15 min at 37 °C.
13. Aspirate off LDS and replace with 6 mL fresh LDS.
14. Incubate with gentle shaking overnight at 37 °C.
15. Aspirate off the LDS and wash the plugs 3× 30 min in 6 mL 0.2× NDS solution with gentle shaking at room temperature.
16. Wash the plugs with 6 mL TE pH8.0 for 5× 30 min with gentle shaking at room temperature.
17. Plugs may be stored at 4 °C in TE pH8.0.
18. Since small molecular weight circular DNA (such as ERCs) will diffuse out of the agarose plugs, it is recommended to run the samples as soon as possible.

3.4.3 1D-ERC Gel (See Fig. 3)

1. Pour a large 0.7 % agarose gel in 1× TBE without ethidium bromide (*see* **Note 22**).
2. Place gel into gel electrophoresis apparatus with 1× TBE.
3. Run 1 μg genomic DNA isolated from yeast genomic mini-prep. For each sample, mix DNA with loading buffer and load into wells (Fig. 4).

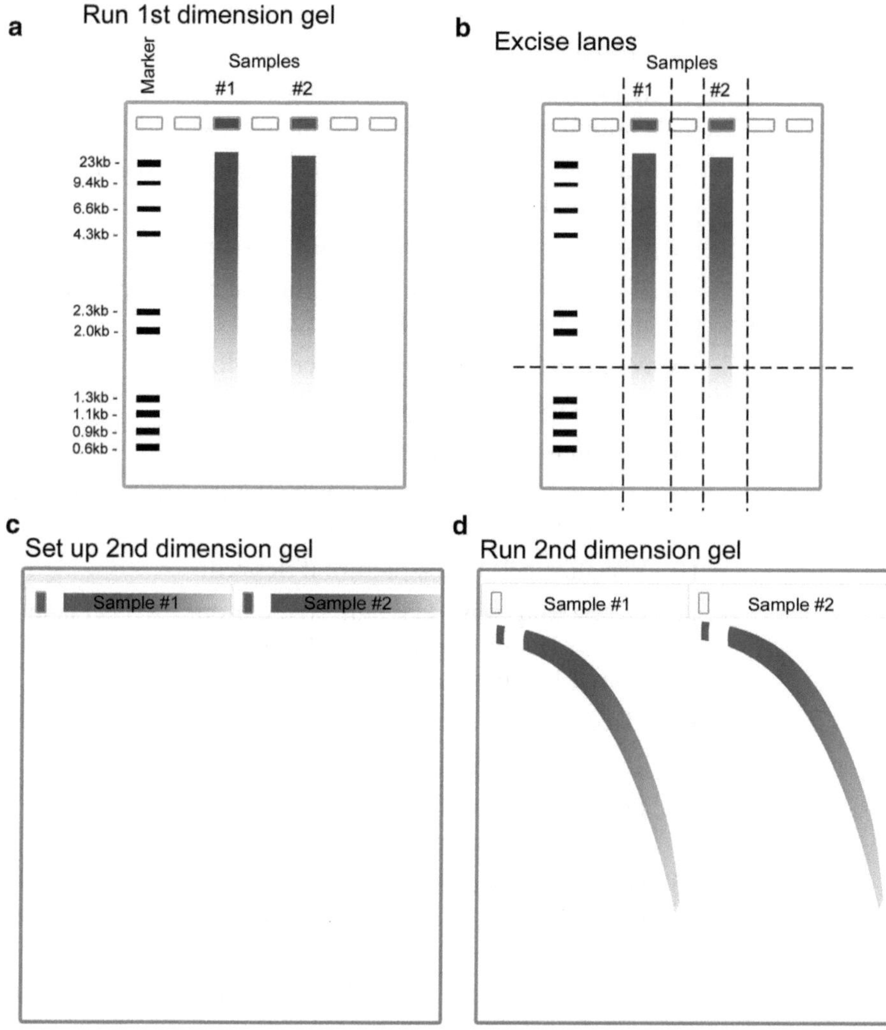

Fig. 4 Schematic of 2D-gel. (**a**) Example of layout of 1D gel; (**b**) Schematic of how to cut out lanes from 1D gel to use in 2D gel; (**c**) Sample set up of 2D gel; (**d**) Example of a 2D gel run

4. Load DNA ladder on a side well to gauge size of bands.
5. Run gel for 36 h at 1 V/cm.
6. To visualize DNA, stain gel in 1× TBE with 0.3 μg/mL ethidium bromide for 30 min.
7. Transfer DNA from gel onto membrane using a Southern blot protocol, such as the one outlined below.

3.4.4 2D-ERC Gel

This gel can be run with yeast DNA from a yeast genomic miniprep (Subheading 3.4.1) or DNA embedded in agarose plugs (Subheading 3.4.2). Shown in Fig. 1b is a 2D ERC gel run with yeast DNA embedded in agarose plugs, which is a more gentle DNA extraction method but may impale ERCs on agarose polymers.

1. Pour a large 0.5 % agarose gel in 1× TBE without ethidium bromide.
2. If using DNA in agarose plugs, use a comb with wells wide enough to accommodate a full agarose plug.
3. After gel is set and before putting it into 1× TBE, slide plug into the bottom of each well.
4. Cover rest of the well with 0.5 % agarose.
5. If using DNA from a yeast genomic miniprep, load gel as normal: place gel in 1× TBE buffer.
6. Mix DNA with loading buffer and load into gel.
7. Load DNA ladder to gauge size.
8. Run first dimension for 21 h at 1 V/cm.
9. Stain gel in 1× TBE buffer with 0.3 μg/mL ethidium bromide for 30 min with gentle shaking.
10. Visualize gel, take a picture if desired.
11. Over a long wave UV lightbox, cut out each gel lane from the well to 10 cm below the well (should contain DNA fragments from 2 kb and above, Fig. 4B).
12. Place each gel slice horizontally in new casting tray (Fig. 4C).
13. Tape ends of casting tray.
14. Pipet a small amount of 1 % agarose gel behind each gel slice, which creates a brace for the slice.
15. When gel brace is solidified (~5 min), fill the rest of the casting tray with 1 % agarose in 1× TBE with 0.3 μg/mL ethidium bromide (*see* **Note 23**).
16. Make 3 L of 1× TBE with 0.3 μg/mL ethidium bromide.
17. Cool gel electrophoresis apparatus to cool to 4 °C.
18. Turn on buffer recirculation pump to maintain even distribution of ethidium bromide when gel is running.
19. Place gel in gel electrophoresis apparatus (with the tape removed).
20. Run at 275 V (~5 V/cm) for 4 h in 4 °C 1× TBE.
21. Transfer DNA from gel onto membrane using a Southern blot protocol, such as the one outlined below.

3.4.5 Southern Blotting

1. Place gel in large dish (i.e., Pyrex dish).
2. Wash gel in Depurination solution for 2×10 min (20 min total) with gentle shaking (*see* **Notes 24** and **25**).
3. Wash gel in Denaturation solution for 2×15 min (30 min total) with gentle shaking (*see* **Note 26**).

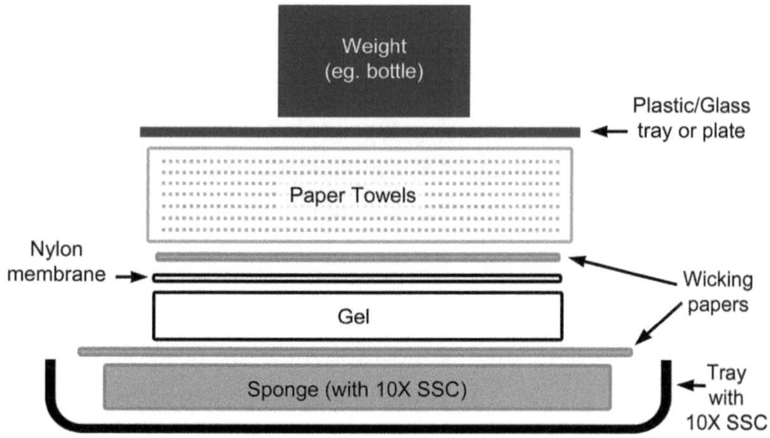

Fig. 5 Schematic of transfer stack setup for 2D ERC gel

4. Wash gel in Neutralization solution for 2×15 min (30 min total) with gentle shaking (*see* **Notes 27** and **28**).
5. Cut 2 wicking papers and nylon membrane to size of gel.
6. Label the membrane using a pencil.
7. Float membrane on ddH_2O for 15 min to allow diffusion of water across membrane thickness (*see* **Note 29**).
8. Soak sponge in10× SSC (i.e., 250 mL 20× SSC + 250 mL H_2O) and place in a new tray.
9. Make 5× SSC (100 mL 20× SSC + 300 mL H_2O) and place in a new tray.
10. Assemble transfer stack (Fig. 5).
11. Soak one wicking paper in 5× SSC and place on sponge. Remove bubbles by rolling a serological pipet across the wicking paper.
12. Place gel upside-down (face down) on soaked wicking paper/sponge. Remove bubbles.
13. Briefly soak the membrane in 5× SSC. Place face down on top of gel. Remove bubbles by rolling a serological pipet across the membrane.
14. Soak the other wicking paper in 5× SSC and place on top of the membrane. Remove bubbles.
15. Place parafilm around the gel to prevent "short circuiting" of the upward transfer.
16. The DNA is carried from the gel onto the membrane via capillary action. If the buffer does not travel upward through the gel, the DNA will not travel.
17. Place 2 in. of paper towels on top of the wicking paper.

18. Place a flat object on top of stack (such as a casting tray or glass plate). Place a weighted object (such as a bottle of water) on top, which creates even pressure across and the stack flat.
19. Let transfer for at least 16 h.
20. Dissemble stack and remove membrane.
21. Cross-link the DNA to the membrane (*see* **Note 30**).
22. Membrane is now ready for probe hybridization (*see* **Note 31**).

3.4.6 ^{32}P Labeled Probe Synthesis

1. Transfer 25–50 ng of DNA fragment of interest to a microcentrifuge tube and bring volume to 10.5 μL with H_2O.
2. Boil for 10 min and then cool for 30 s on ice.
3. Spin down for 10 s at maximum speed on a tabletop centrifuge.
4. For ^{32}P oligo labeling, place 25–50 ng DNA fragment in microcentrifuge tube and bring the volume up to 15 μL with H_2O.
5. Boil for 10 min.
6. Spin down for 1 min. Keep at room temperature.
7. Add 4 μL oligo mix, 1 μL Tris + $MgCl_2$ solution, and 1 μL BSA.
8. Add 50 μCi ^{32}P-labeled nucleotides (~2.5 μL).
9. Add 1 μL Klenow (exo-) (~5 units).
10. Incubate at room temperature for 3 h (*see* **Note 32**).
11. Purify ^{32}P labeled probe (optional):
    a. Prepare Sephadex bead column: Place some glass wool at the bottom of a 1 mL syringe. Fill column with Sephadex bead slurry. Place syringe in 15 mL round bottom Falcon tube. Spin column in centrifuge to pack Sephadex beads and to remove liquid from bead slurry. Discard liquid. Repeat as necessary to obtain 0.8 mL packed Sephadex beads in the syringe column.
    b. Add an equal volume of 10 mg/mL tRNA to ^{32}P labeling reaction. Transfer to Sephadex column in 15 mL round-bottom Falcon tube. Centrifuge column in tube to elute labeled probe.
    c. Add 50 μL 5mg/mL tRNA to Sephadex column. Centrifuge syringe in tube again to elute residual labeled probe.
    d. Discard column in radioactive waste.
12. Add 8 mL prehyb solution to labeled probe in 15 mL round-bottom tube.

3.4.7 Southern Blot Hybridization

1. Add 8 mL of prehyb solution to blot membrane in hybridization tube and cap tube.
2. Place in 65 °C hybridization oven (with roller) and incubate rolling for 15–30 min.

3. While membrane is incubating in prehyb solution, denature the labeled probe by boiling for 10 min.
4. Loosen the 15 mL tube cap to allow for steam exit.
5. After incubation, discard the prehyb from the hybridization tube, quickly add the freshly boiled probe to the hybridization tube, cap tube securely, and replace in 65 °C hybridization oven.
6. Hybridize probe to membrane, rolling, at 65 °C for at least 12 h.
7. Pour labeled probe back into 15 mL round-bottom tube.
8. The probe can be reused for 1–2 weeks as long as you boil it before each hybridization.
9. Rinse the membrane in the tube with ~25 mL 2× SSC.
10. Discard wash into radioactive wash.
11. Transfer membrane to a glass tray with 1 L Wash Solution A (room temperature).
12. Wash with gentle shaking for 20 min.
13. Discard wash in sink.
14. Wash blot in tray with 1 L Wash Solution B (heated to 37 °C) with gentle shaking for 20 min.
15. Discard wash in sink.
16. Wash blot in tray with 1 L Wash Solution B (heated to 65 °C) with gentle shaking for 20 min.
17. Discard wash in sink.
18. Briefly dry the membrane to remove excess buffer.
19. Wrap labeled membrane with plastic wrap.
20. Expose labeled membrane to film or phosphorimager screen.
21. Quantive relative ERC levels by measuring the intensity of each ERC spot in comparison to the intensity of the genomic rDNA band.

4 Notes

1. Filter-sterilized synthetic complete media are recommended for this protocol because they produce more consistent results (culturing time and yield). When YPD media are used, a 10× (20 %) Dextrose solution should be prepared separately and filter-sterilized, instead of autoclave, to prevent carbonization. The YP portion should be autoclaved and cooled before use. Add the 10× Dextrose solution to the YP portion just before inoculation.
2. Pour plates within a few days prior to beginning of lifespan to avoid issues with desiccation. Thickness of the plate is important so that the surface is accessible to the dissecting needle.

3. Avoid placing too many mother cells on one plate because that will result in prolonged time out of the incubator and can result in plate drying. We usually use 60 cells per plate or less. Also, the genetics of yeast replicative longevity are highly strain-specific and researchers should carefully note strain background when performing lifespan studies.
4. We mark coordinates on the plate by poking holes in the surface of the media with the needle.
5. It is generally advisable to keep plates at 30 °C for as long as possible; however, bringing the plates down to 4 °C every night is the standard procedure. For most strains overnight residence at 4 °C does not impact lifespan, but we have recently discovered a small set of mutant strains that are dramatically impacted by this treatment (Kennedy lab, unpublished data).
6. Parafilm wrapping of plates when not being counted is important in avoiding desiccation.
7. BY diploid strains have slightly faster doubling times than the haploids; therefore, 60 min between rounds is advisable at the beginning of an experiment.
8. Wilcoxon rank sum test is the generally accepted test to determine if two lifespan curves are statistically different.
9. Count cells in triplicates or quadruplicates to ensure accuracy.
10. It is important to keep cells cold during the sorting process. This is because warming up will accelerate the rate cells degrade Streptavidin on the beads, hence causing the Dynabeads to lose biotin binding capacity faster.
11. Do not incubate for more than 2 h since Dynabeads lose biotin binding capability over time in the presence of living cells.
12. Culturing times are affected by a number of factors, including strain genetic backgrounds and media types, and hence should be empirically determined. The culturing times in this protocol are used for BY4741/BY4742 or S288c culturing in synthetic complete media.
13. Do not incubate for more than 2 h since Dynabeads lose biotin binding capability over time in the presence of living cells.
14. Extra care and good aseptic techniques should be used in all steps prior to the last sorting. During all bead separation steps, tube caps (5-mL round-bottom tubes and microcentrifuge tubes) should be kept closed and only open during the short time of pipetting. For washing steps performed outside the cold room or cold box, it is recommended to carry out in an ice bucket near a flame. Avoid coughing and sneezing over samples.

15. The average age for each sort can be estimated by counting bud scars using the saved BSC samples. Bud scars can be stained with WGA Alexa Fluor 488 and visualized by fluorescence microscopy.
16. In addition, Sir2 can silence transcription of Pol II-dependent genes within or near repeats and this assay is frequently used.
17. Wrap parafilm on the plates to avoid excessive drying.
18. If no red pigment has accumulated after 4 or 5 days, incubation at 4 °C may help the cells accumulate pigment. Incubation at 4 °C typically takes 5 or more days.
19. It may be useful to use an *sir2D* strain as a positive control in this assay, it has been previously observed that in the W303 strain background, the rate of recombination is around three- to fivefold higher in the *sir2D* [19]. Recombination is reduced in an *fob1D* background [23].
20. As an example, the Bio-Rad plug molds hold approximately 80 μL. This procedure will give you 0.5 % agarose plugs, which are fairly fragile, but are better if further manipulations, like *in gelo* digestion, are needed.
21. Longer incubation times are better.
22. Recommended dimensions: at least 20 cm long.
23. The ethidium bromide intercalation into DNA will pop out DNA structures, which will allow for 2D separation of DNA replication intermediates by shape.
24. Check that there is enough solution to cover the gel during each wash.
25. Do not overwash, as the DNA will have excess depurination.
26. When switching to the Denaturation wash, it is recommended to briefly rinse the gel with water to remove excess Depurination solution.
27. When switching to the Neutralization wash, it is recommended to briefly rinse the gel with Neutralization solution to remove excess Denaturation solution and to maintain the high salt concentration.
28. It is fine to leave gel in Neutralization solution for a while (~1 h) if need be.
29. Pre-wetting both sides may trap bubbles in membrane, which may impede the Southern blot transfer.
30. We use the "AUTO CROSS LINK" function on the Stratagene UV Stratalinker 1800.
31. Can store in hybridization tube until ready.
32. Incubation can be extended to overnight.

References

1. Shore D, Squire M, Nasmyth KA (1984) Characterization of two genes required for the position-effect control of yeast mating-type genes. EMBO J 3(12):2817–2823
2. Ivy JM, Klar AJ, Hicks JB (1986) Cloning and characterization of four SIR genes of Saccharomyces cerevisiae. Mol Cell Biol 6(2): 688–702
3. Aparicio OM, Billington BL, Gottschling DE (1991) Modifiers of position effect are shared between telomeric and silent mating-type loci in S. cerevisiae. Cell 66(6):1279–1287
4. Imai S, Armstrong CM, Kaeberlein M, Guarente L (2000) Transcriptional silencing and longevity protein Sir2 is an NAD- dependent histone deacetylase. Nature 403(6771):795–800
5. Mortimer RK, Johnston JR (1959) Life span of individual yeast cells. Nature 183: 1751–1752
6. Smith ED, Tsuchiya M, Fox LA, Dang N, Hu D, Kerr EO, Johnston ED, Tchao BN, Pak DN, Welton KL, Promislow DE, Thomas JH, Kaeberlein M, Kennedy BK (2008) Quantitative evidence for conserved longevity pathways between divergent eukaryotic species. Genome Res 18(4):564–570. doi:10.1101/gr.074724.107
7. Lin SJ, Defossez PA, Guarente L (2000) Requirement of NAD and SIR2 for life-span extension by calorie restriction in Saccharomyces cerevisiae. Science 289(5487): 2126–2128
8. Kaeberlein M, Kirkland KT, Fields S, Kennedy BK (2004) Sir2-independent life span extension by calorie restriction in yeast. PLoS Biol 2:1381–1387
9. Lamming DW, Latorre-Esteves M, Medvedik O, Wong SN, Tsang FA, Wang C, Lin SJ, Sinclair DA (2005) *HST2* mediates *SIR2-independent* life-span extension by calorie restriction. Science 309:1861–1864
10. Tsuchiya M, Dang N, Kerr EO, Hu D, Steffen KK, Oakes JA, Kennedy BK, Kaeberlein M (2006) Sirtuin-independent effects of nicotinamide on lifespan extension from calorie restriction in yeast. Aging Cell 5(6):505–514
11. Schleit J, Wasko BM, Kaeberlein M (2012) Yeast as a model to understand the interaction between genotype and the response to calorie restriction. FEBS Lett 586(18):2868–2873. doi:10.1016/j.febslet.2012.07.038
12. Hachinohe M, Hanaoka F, Masumoto H (2011) Hst3 and Hst4 histone deacetylases regulate replicative lifespan by preventing genome instability in Saccharomyces cerevisiae. Genes Cells 16(4):467–477. doi:10.1111/j.1365-2443.2011.01493.x
13. Lindstrom DL, Gottschling DE (2009) The mother enrichment program: a genetic system for facile replicative life span analysis in Saccharomyces cerevisiae. Genetics 183(2): 413–422. doi:10.1534/genetics.109. 106229, 411SI-413SI
14. Lee SS, Vizcarra IA, Huberts DH, Lee LP, Heinemann M (2012) Whole lifespan microscopic observation of budding yeast aging through a microfluidic dissection platform. Proc Natl Acad Sci U S A 109(13):4916–4920. doi:10.1073/pnas.1113505109
15. Xie Z, Zhang Y, Zou K, Brandman O, Luo C, Ouyang Q, Li H (2012) Molecular phenotyping of aging in single yeast cells using a novel microfluidic device. Aging Cell. doi:10.1111 /j.1474-9726.2012.00821.x
16. Kaeberlein M, McVey M, Guarente L (1999) The *SIR2/3/4* complex and *SIR2* alone promote longevity in *Saccharomyces cerevisiae* by two different mechanisms. Genes Dev 13: 2570–2580
17. Kobayashi T, Horiuchi T (1996) A yeast gene product, Fob1 protein, required for both replication fork blocking and recombinational hotspot activities. Genes Cells 1(5):465–474
18. Defossez PA, Prusty R, Kaeberlein M, Lin SJ, Ferrigno P, Silver PA, Keil RL, Guarente L (1999) Elimination of replication block protein Fob1 extends the life span of yeast mother cells. Mol Cell 3(4):447–455
19. Gottlieb S, Esposito RE (1989) A new role for a yeast transcriptional silencer gene, SIR2, in regulation of recombination in ribosomal DNA. Cell 56(5):771–776
20. Huang J, Moazed D (2003) Association of the RENT complex with nontranscribed and coding regions of rDNA and a regional requirement for the replication fork block protein Fob1 in rDNA silencing. Genes Dev 17(17):2162–2176
21. Pasero P, Bensimon A, Schwob E (2002) Single-molecule analysis reveals clustering and epigenetic regulation of replication origins at the yeast rDNA locus. Genes Dev 16(19):2479–2484. doi:10.1101/gad.232902
22. Skryabin KG, Eldarov MA, Larionov VL, Bayev AA, Klootwijk J, de Regt VC, Veldman GM, Planta RJ, Georgiev OI, Hadjiolov AA (1984) Structure and function of the nontranscribed spacer regions of yeast rDNA. Nucleic Acids Res 12(6):2955–2968
23. Kobayashi T, Heck DJ, Nomura M, Horiuchi T (1998) Expansion and contraction of ribosomal DNA repeats in Saccharomyces cerevisiae: requirement of replication fork blocking (Fob1) protein and the role of RNA polymerase I. Genes Dev 12(24):3821–3830

24. Kwan EX, Foss EJ, Tsuchiyama S, Alvino GM, Kruglyak L, Kaeberlein M, Raghuraman MK, Brewer BJ, Kennedy BK, Bedalov A (2013) A natural polymorphism in rDNA replication origins links origin activation with calorie restriction and lifespan. PLoS Genet 9(3):e1003329. doi:10.1371/journal.pgen.1003329
25. Dang W, Steffen KK, Perry R, Dorsey JA, Johnson FB, Shilatifard A, Kaeberlein M, Kennedy BK, Berger SL (2009) Histone H4 lysine 16 acetylation regulates cellular lifespan. Nature 459(7248):802–807. doi:10.1038/nature08085
26. Vega-Palas MA, Martin-Figueroa E, Florencio FJ (2000) Telomeric silencing of a natural subtelomeric gene. Mol Gen Genet 263(2): 287–291
27. Steffen KK, Kennedy BK, Kaeberlein M (2009) Measuring replicative lifespan in budding yeast. J Vis Exp pii:1209
28. Cao L, Alani E, Kleckner N (1990) A pathway for generation and processing of double-strand breaks during meiotic recombination in S. cerevisiae. Cell 61(6):1089–1101
29. Brewer BJ, Fangman WL (1987) The localization of replication origins on ARS plasmids in *S. cerevisiae*. Cell 51:463–471
30. Sambrook J, Fritsch EF, Maniatis T (1989) Molecular cloning: a laboratory manual. Cold Spring Harbor Laboratory Press, Cold Spring Harbor, NY
31. Smeal T, Claus J, Kennedy B, Cole F, Guarente L (1996) Loss of transcriptional silencing causes sterility in old mother cells of S. cerevisiae. Cell 84(4):633–642

Chapter 3

C. elegans Sirtuins

Mohan Viswanathan and Heidi A. Tissenbaum

Abstract

The nematode *Caenorhabditis elegans (C. elegans)* has four Sir2 paralogs, *sir-2.1, sir-2.2, sir-2.3*, and *sir-2.4*. Thus far, most of the research tools to study worm sirtuins have been developed for *sir-2.1*, due to its homology to yeast SIR2 and human SIRT1. Here, we have compiled a listing of the currently available strains (including both loss-of-function alleles and transgenics), antibodies, and RNAi constructs relevant to studies on all *C. elegans* sirtuin family members. We also describe the methods used in the analysis of *C. elegans* sirtuin function, including life span analysis, various stress-resistance assays, and fat content analysis and provide an overview of all phenotypic data relevant to *C. elegans sir-2.1*.

Key words *C. elegans*, sirtuin, *sir-2.1*, *sir-2.2*, *sir-2.3*, *sir-2.4*, Resveratrol

1 Introduction

The nematode *Caenorhabditis elegans* (*C. elegans*) has been one of the premiere model organisms in aging research due to its short, reproducible life span, invariant cell lineage and a plethora of genetic, molecular and genomic tools that make it easy to identify new modifiers of life span. Work on sirtuins in *Caenorhabditis elegans* was prompted by two research findings: the discovery of the NAD+-dependent deacetylase activity of mammalian SirT1 [1] and that overexpression of Sir2 in the yeast *Saccharomyces cerevisiae* delayed the onset of replicative senescence extending organismal mean replicative life span by as much as 50 % [2]. In *C. elegans*, there are four *C. elegans* Sir2 paralogs; *sir-2.1*, *sir-2.2*, *sir-2.3*, and *sir-2.4*, each encoding proteins with regions of homology to the sirtuin core domain [3, 4]. When compared with other sirtuin family members, there is little to be gleaned from the protein sequences outside the conserved catalytic core with respect to specific conserved domains or to homology with other sirtuin family members.

Matthew D. Hirschey (ed.), *Sirtuins: Methods and Protocols*, Methods in Molecular Biology, vol. 1077,
DOI 10.1007/978-1-62703-637-5_3, © Springer Science+Business Media, LLC 2013

Phylogenetically, *sir-2.1* is most closely related to both yeast SIR2, the founding member of the sirtuin family of proteins, and mammalian SIRT1. It is also the most extensively studied gene among the worm sirtuin homologs. Like its yeast and mammalian homologs, *C. elegans* SIR-2.1 protein demonstrates robust protein deacetylase activity in in vitro peptide-based deacetylase assays (unpublished observation, Viswanathan and Guarente); however, to date, no endogenous worm protein substrate has been identified, as biochemical studies on *C. elegans* sirtuins have been limited. *C. elegans* sirtuin studies have largely focused on phenotypic analysis of gene dosage effects on life span; including the analysis of loss-of-function mutants, RNAi, duplications, and transgenic overexpression strains.

Here, we have compiled a listing of the currently available strains (including both loss-of-function alleles and transgenics), antibodies, and RNAi constructs relevant to studies on all *C. elegans* sirtuin family members. We also describe the methods used in the analysis of *C. elegans* sirtuin function, including life span analysis, various stress-resistance assays, and fat content analysis and provide an overview of all phenotypic data relevant to *C. elegans sir-2.1*.

2 Materials

2.1 Sirtuin-Related C. elegans Strains (See Fig. 1, Tables 1–4)

1. All strains are handled according to standard procedures, as outlined in Brenner 1974 [5].
2. Strains are grown at different temperatures ranging from 15 to 25 °C depending on the experiment.
3. Strains are usually available from the lab that created the strain or from the *Caenorhabditis* Genetics Center (CGC, http://www.cbs.umn.edu/cgc).
4. Tables 1, 2, 3, and 4 contain a list of available strains relevant to studying *C. elegans* sirtuins.

2.2 Sirtuin-Related Molecular and Biochemical Reagents

1. SIR-2.1 antibody, rabbit, raised against recombinant His6-SIR-2.1 (full length); this antibody is good for Western, IP, and IF [6].
2. SIR-2.1 antibody, rabbit, raised against recombinant His6-SIR-2.1 (full length); this antibody is good for Western, IP, and IF, although the serum requires antibody immunopurification before use [7].
3. SIR-2.1 antibody, rabbit, raised against recombinant SIR-2.1 (Novus Biologicals) [8].
4. SIR-2.4 antibody, rabbit, raised against recombinant His6-SIR-2.4 (full length), unpublished reagent available from Viswanathan and Guarente; this antibody is good for Western.

a

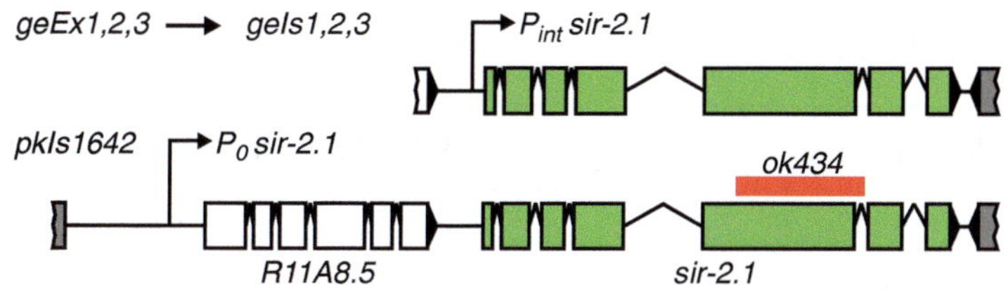

b

	geEx1,2,3 geIs1,2,3	geIs3	geEx1,2	pkIs1642	sir-2.1(ok434)
Life Span	Increase	Increase	Increase	Increase	Decrease
Stress Resistance					
Heat		Increase		Increase	Decrease
Oxidative		Increase		Increase	Decrease
UV					Decrease
Dauer Formation		Increase	Increase	Increase	
Fat Storage		Reduce			Increase
PolyQ Toxicity		Suppress			

Fig. 1 (**a**) Graphical representation of genomic sequences used to generate the various *sir-2.1* transgenic alleles. P_0 is the operonic promoter that drives expression of *sir-2.1* in muscle cells and the intestine, while P_{int} is a cryptic intergenic promoter that drives *sir-2.1* expression in neurons and the hypodermis. Lines designated *geEx* were generated by injection of the *sir-2.1* transgene (*upper graphical construct*) along with the dominant roller gene *rol-6* (*su10006*) carried on the plasmid pRF4. *pkIs1642* was generated by bombardment of *unc-119*(*ed3*) bearing worms with the full operon construct (*lower graphical construct*) along with pRP2510, containing a copy of wild type *unc-119*, which rescues the UNC phenotype of *unc-119*(*ed3*) bearing animals. *geIs1*, *2*, and *3* lines are the result of gamma-irradiation induced random integration of *geEx3* into the wild-type *C. elegans* genome. It has been demonstrated that the original integrated line LG100 bearing *geIs3* contained a secondary dye-filling mutation, which alone extends worm life span. It should be noted that some researchers do not find an extension with *geIs3* or *pkIs42* when backcrossed into their respective control background [17], while other laboratories still see significant and reproducible extension with both transgenic alleles [8, 18] following backcross. Two independent groups have also shown that RNAi specific to *sir-2.1* abolishes the life span increase of the *sir-2.1* transgenic worms [8, 18]. For *sir-2.1* mutants, the *sir-2.1*(*ok434*) strain is shown. The *red box* indicates the region of *sir-2.1* deleted in this strain. (**b**) Summary of various aging and stress-related assays [6–8, 10, 18–22] performed on different *C. elegans* strains harboring either the *sir-2.1*(*ok434*) loss-of-function allele or various *sir-2.1* transgenes. Dauer assays results with *pkIs1642* are unpublished results of Viswanathan and Guarente

2.3 Nematode Growth Media: Solid Media for Worm Growth and Propagation

1. Combine 3.0 g NaCl, 17.0 g Bacto™Agar (BD Biosciences), and 2.5 g Bacto™ Peptone (BD Biosciences) in a 2 L bottle with a stir bar. Add 975 mL of deionized water, loosely cap, and autoclave for 30 min.
2. Remove bottle from autoclave and allow media to cool with gentle stirring to 55–65 °C.
3. Add the following sterile solutions using sterile technique: 1 mL 1 M $CaCl_2$, 1 mL 5 mg/mL cholesterol (Sigma-Aldrich) dissolved in ethanol, 1 mL 1 M $MgSO_4$, and 25 mL 1 M KPO_4 pH 6.0 (*see* **Note 4**).

Table 1
Strains relevant to *sir-2.1* (*see* Note 1)

Strain	Gene (Allele)	Outcross to N2	Available from	References
VC199	*sir-2.1(ok434)IV*	0x	CGC	[10]
LG339	*sir-2.1(ok434)IV*	8x	LG	[10]
NL3518	*sir-2.2(pk1571)IV*	0x	LG	Unpublished
NL3904	*sir-2.1(pk1571)IV*	3x	LG	Unpublished
NL4250	*mut-7(pk204)III; sir-2.1(pk1640::Tc1)IV*	0x	LG	[23]
NL4251	*sir-2.1(pk1640::Tc1)*IV	4x	LG	[10]
LG444	*geEx1*[P_{int} *sir-2.1(+) rol-6(su1006)*]	0x	LG/HT	[21]
LG445	*geEx2*[P_{int} *sir-2.1(+) rol-6(su1006)*]	0x	LG/HT	[21]
LG446	*geEx3*[P_{int} *sir-2.1(+) rol-6(su1006)*]	0x	LG/HT	[21]
HT122	*geIs1*[P_{int} *sir-2.1(+) rol-6(su1006)*]	0x	HT	[21]
HT123	*geIs2*[P_{int} *sir-2.1(+) rol-6(su1006)*]	0x	HT	[21]
LG269	*geIs3*[P_{int} *sir-2.1(+) rol-6(su1006)*]; *dyf(?)*	0x	LG/HT	[18]
LG391	*geIs3*[P_{int} *sir-2.1(+) rol-6(su1006)*]	8x	LG	[18]
NL3909	*pkIs1642[unc-119(+) P_O R11A8.5(+) sir-2.1(+)] unc-119(ed3)*	0x	CGC	[7, 10]
NL3912	*pkIs1645[unc-119(+) P_O R11A8.5(+) sir-2.1(+)] unc-119(ed3)*	0x	LG	Unpublished
HT809	*lpIs15* [P_{int} *sir-2.1::GFP rol-6(su1006)*]		HT	[30]
HT810	*lpIs16* [P_{int} *sir-2.1::GFP rol-6(su1006)*]		HT	[30]
UL3294	*leEx3294* [*sir-2.1::mCherry+ R11A8.5::GFP+ rol-6(su1006)*]	0x	CGC	[24]
DR1786	*dpy-13(e184) unc-24(e138) IV; mDp4[unc-17(e245)] (IV;?)*	0x	CGC	[21]

4. Using a 50 mL sterile serological pipette or peristaltic pump (*see* **Note 5**), 25 mL of molten media can be pipetted into 100 mm plastic petri dishes (large plates), 12 mL can be pipetted into 60 mm petri dishes (medium), and 5 mL can be pipetted into 35 mm petri dishes (small) (*see* **Note 6**).
5. Plates should be left at room temperature for 2–3 days to dry before seeding with bacterial food source. Plates can be stored unseeded at 4 °C in closed boxes.
6. Seeding plates with *C. elegans* bacterial food source OP50 (*see* **Note 7**).

Table 2
Strains relevant to *sir-2.2 and sir-2.3* (*see* Note 2)

Strain	Gene (Allele)	Outcross to N2	Available from	References
MT18069	*sir-2.2(n5136)X*	6xox	MT	Unpublished
FX2673	*sir-2.2(tm2673)X*	0xox	FX	Unpublished
FX2648	*sir-2.2(tm2648)X*	0xox	FX	Unpublished
RB654	*sir-2.3(ok444)X*	0xox	CGC	Unpublished
LG340	*sir-2.3(ok444)X*	6xox	LG	Unpublished
TH188	*(pRedFlp-Hgr)::unc-119-Nat (sir-2.2[33311]::2XTY1-eGFP (C.ele)-3XFLAG)dFRT*	0x	TH	[25]
SP75	*mnDp25(X;I)/+ I; unc-3(e151)X*	0x	CGC	[21]
SP116	*mnDp9(X;I); unc-3(e151)X*	0x	CGC	[21]
SP117	*mnDp10(X;I); unc-3(e151)X*	0x	CGC	[21]

Table 3
Strains relevant to *sir-2.4* (*see* Note 3)

Strain	Gene (allele)	Outcross to N2	Available from	References
FX2817	*sir-2.4(tm2817)I/+*	0x	FX	Unpublished
LG383	*sir-2.4(tm2817I)/hT2 (I;III)::qIs48[myo-2::gfp; pes-10::gfp; ges-1::gfp bli-4(e937) let-?(q782)]; him-8 (e1469)*	0x	LG	Unpublished
MT18068	*sir-2.4(n5137)I*	6x	MT	Unpublished
EQ 137	*iqEx47 [sir-2.4p::sir-2.4::gfp rol-6(su10006)]*	0x	EQ	[26]
EQ158	*iqEx50 [sir-2.4p::sir-2.4 myo-3p::rfp]*	0x	EQ	[26]
KR1293	*dpy-5(e61) dpy-14(e188) rec-1(s180)? I; hDp12(I;f)*	0x	CGC	[21]
KR1725	*dpy-14(e188) unc-13(e51)I; hDp44(I;X;f)*	0x	CGC	[21]
KR1722	*dpy-14(e188) unc-13(e51)I; hDp52(I;X;f)*	0x	CGC	[21]
KR1704	*dpy-14(e188) unc-13(e51)I; hDp68(I;X;f)*	0x	CGC	[21]
KR1732	*dpy-14(e188) unc-13(e450) I; hDp48(I;X;f)*	0x	CGC	[21]
KR1108	*unc-11(e47) dpy-5(e61)I; hDp8(I;f)*	0x	CGC	[21]
KR1110	*unc-11(e47) dpy-5(e61)I; hDp11(I;f)*	0x	CGC	[21]
KR1112	*unc-11(e47) dpy-5(e61)I; hDp9(I;f)*	0x	CGC	[21]
KR1236	*unc-74(x19) dpy-5(e61)I; hDp2(I;f)*	0x	CGC	[21]

(continued)

Table 3
(continued)

Strain	Gene (allele)	Outcross to N2	Available from	References
KR1284	*dpy-5(e61) dpy-14(e188) rec-1(s180)? I; hDp15(I;f)*	0x	CGC	[21]
KR1280	*dpy-5(e61) dpy-14(e188) rec-1(s180)? I; hDp13(I;f)*	0x	CGC	[21]
KR1548	*dpy-5(e61) dpy-14(e188) rec-1(s180)? I; hDp22(I;f)*	0x	CGC	[21]
KR1282	*dpy-5(e61) dpy-14(e188) rec-1(s180)? I; hDp16(I;f)*	0x	CGC	[21]
KR1815	*dpy-5(e61) dpy-14(e188) rec-1(s180)? I; hDp30(I;f)*	0x	CGC	[21]

Table 4
RNAi

Sirtuin	Strain	Clone	References	Phenotype noted
sir-2.1	HT115(DE3) pL4440-*sir-2.1*	*sir-2.1* Exons 4–5	[7]	Reduced life span
sir-2.2	Ahringer RNAi Library	F46G10.7	[27, 28]	Mutator
sir-2.3	Ahringer RNAi Library	F46G10.3	[28, 29]	Enhanced tau-induced UNC phenotype
sir-2.4	HT115(DE3) pL4440-*sir-2.4*	1–467 *sir-2.4* cDNA	[26]	Prevents stress-induced *daf-16* nuclear localization

7. Streak OP50 bacteria from frozen stock onto LB agar plates to achieve growth of single colonies.
8. Inoculate a bottle containing 100 mL liquid LB media with a single OP50 colony and incubate as a standing culture overnight at 37 °C (*see* **Note 8**).
9. Using a repeater pipettor (Cat# 022,260,201, Eppendorf, eppendorf.com) with an appropriately sized sterile barrel pipette, 35 mm nematode growth media (NGM) + FuDR plates are seeded with 50 μL OP50, 60 mm plates seeded with 200 μL OP50, and large plates are seeded with 200 μL OP50 and the culture is spread around the plate using a glass spreader.

2.4 NGM + FuDR Plates (See Note 9): Life Span Assay Plates

1. Dissolve 0.125 g 2′-Deoxy-5-fluorouridine (FuDR) in 10 mL of deionized water and pass it through a 0.2 μM syringe filter to sterilize. This will be 1,000× stock solution at 12.5 mg/mL FuDR.

2. Sufficient molten NGM media is prepared as described above and supplemented with FuDR for a final FuDR concentration of 12.5 μg/mL. 500 mL of NGM supplemented with 500 μL of 12.5 mg/mL FuDR is enough for approximately one hundred 30 mm plates, as each plate can be filled with 5 mL of molten media.
3. Plates can be left at room temperature for 2–3 days or dried overnight at 37 °C before seeding with 50 μL of OP50 bacteria. Dry plates allow for the rapid absorption of the LB containing OP50 bacteria leading to the formation of perfect circular lawns (*see* **Note 10**).
4. Plates are typically poured fresh for each life span experiment to prevent desiccation associated with long-term storage of plates. However, plates can be stored for up to 2 weeks at 4 °C in a sealed container.
5. To test for effects of small molecules upon life span, the small molecule and DMSO are added into the molten media of both the NGM plates used to produce the progeny assayed for life span and the NGM + FuDR life span plates (*see* **Note 11**).

3 Methods

3.1 Transferring Worms and Maintenance

1. Worm transfer can be accomplished by picking individual worms or by chunking a large number of worms from a donor plate to a recipient plate.
2. Worm picks are typically made from platinum, platinum/iridium, or chromium wire (*see* **Note 12**), as they rapidly cool following sterilization by flaming. The most common type of platinum pick is made by placing a 0.2 mm diameter platinum wire into the open end of a Pasteur pipette and melting the glass around the wire with a Bunsen burner flame. The other end of the wire is then bent into a loop and closed tightly using a flat plier in order to close any gap in the looped region creating a closed off hook.
3. Individual worms can be picked up on the end of the wire loop using a small amount of *E. coli* from the seeded plate as a "glue" to gently gather up one or more worms. Worms can then be deposited on the agar surface of the recipient NGM plate by brushing off the worms from the *E. coli*.
4. An alternative transfer method involves cutting a small piece of agar from one plate and inverting it onto the surface of the recipient plate. A sterile metal weighing spatula or scalpel that has been immersed in alcohol and flamed is used to cut a small square or similarly shaped piece of agar from the donor plate and then used to hoist the piece of agar out and flip it upside

down onto the donor plate allowing the worms in contact with the new seeded NGM media.

5. *C. elegans* can be stored at a number of temperatures for maintenance and growth purposes, although care should be taken when choosing incubation temperature as growth and storage temperature can greatly affect assay conditions. It is standard for labs to have constant temperature incubators set at 25, 20, 15, and 10 °C for *C. elegans* maintenance. Life span assays can be done at any temperature. Most strains show a change in life span dependent on the temperature. In addition, many classic life span mutants are temperature-sensitive. In such cases, worms are usually propagated at 15 °C and then transferred to 20 or 25 °C as an L4 or young adult for the assay. For wild type and non-temperature-sensitive mutants, to increase growth rates, worms can be stored at 25 °C or room temperature and to slow down growth rates they can be kept at 15 °C or for much longer-term storage (up to several months) at 10 °C. Worms can survive storage at 4 °C for several hours to overnight to essentially halt growth and development, but this is typically not needed as 10 °C is sufficient to significantly slow growth. Typically, strains are stored at 15 °C (temperature-sensitive strains) or 20 °C. Care should be taken to grow all strains at the same temperature prior to life span analysis.
6. *C. elegans* can be maintained for long-term storage at −80 °C or in liquid Nitrogen by slow freezing in S-buffer with 30 % glycerol. When freezing worms, a plate with many freshly starved L1-stage worms is washed with S-Freezing buffer #1 (*see* **Note 13**) recovering approximately 0.5 mL of worms. An equal volume of S-Freezing buffer #2 containing 60 % glycerol (*see* **Note 13**) is added and the worms are aliquoted into five cryogenic freezing vials (200 μl/vial) and allowed to slow freeze slowly at −80 °C (*see* **Notes 14** and **15**).
7. To recover frozen animals, cryovials containing worms are rapidly thawed and the vials inverted over seeded NGM plates. Plates are left open to air dry on the bench top and then incubated at 15 or 20 °C. The following day 10–20 surviving larval stage animals are moved by picking to a new plate and incubated at 15 or 20 °C.

3.2 Standard Life Span Assays

1. Eight to ten L4-stage or young adult worm animals are serially passaged to medium seeded NGM plates for at least two generations at 20 °C to ensure (1) animals have not been starved, (2) animals have been generated under constant temperature conditions, and (3) lack of fungal or bacterial contamination prior to the start of life span analysis. Worms should be grown at 20 °C unless using a temperature-sensitive strain in which case an alternative temperature can be utilized (*see* **Note 16**).

2. To generate worms used for life span analysis, eight to ten worms for each strain being tested are moved onto three to five 60 mm seeded NGM plates and incubated for 3–4 days at 20 °C in order to generate sufficient numbers of L4-stage animals that can be selected for the start of life span analysis (*see* **Note 17** for an alternative method).
3. Animals between L4-stage and young adult stage of development are used for life span analysis. Typically, for each strain being tested, 200–300 staged animals are selected by picking and are moved to a 60 mm medium-size seeded NGM plate for each strain being analyzed. Moving to a new plate prevents any eggs or young larval stage worms from being moved to life span plates. Strains are picked at room temperature; however, whenever possible keep strains in the incubator since it can take many hours to set up life span assays.
4. Life spans are performed with 100–250 animals (*see* **Note 18**). Once all the worms have been stage selected onto seeded NGM plates the animals are then moved to 30 mm NGM plates containing FuDR seeded with OP50 on the same day, day 0 of the life span. Approximately 25–30 worms are placed onto each plate. The number of plates used depends on the total number of worms in the experiment (*see* **Note 19**). For a life span of 150 worms, seven plates are utilized, with the seventh plate being a spare plate used to replace worms that might crawl off the plates during the life span (*see* **Note 20**). This seventh plate is optional; many labs do not replace the missing/lost worms.
5. Life span plates are incubated for 6 days at 20 °C without perturbation. On day 6, the animals are moved to new 30 mm NGM plates containing FuDR, 25 worms per plate. Moving the animals to new plates gets the animals away from the many dead eggs laid on the plate and makes finding worms for subsequent counting easier. Any extra animals left on the original plates are moved to a new seventh plate, which is again labeled as spare. (This spare plate is optional.)
6. Counting of live and dead animals is performed every other day. On each count day, first the number of total worms is counted to verify the same number is on the plate as last time. If a crawl off or vulval bursting has taken place, an age-matched live worm is put onto the plate from the spare. Animals are monitored for movement. Any worm found not moving can be checked for life by gentle prodding with the wire pick, or if needed, can be moved by pick to check for response. Care should be taken to limit physical perturbation of the animals, which may affect life span (*see* **Note 21** for an alternative method). If any animal is injured by picking it can be replaced with an age-matched animal from the spare plate. If no move-

ment is seen, the worm is counted in the death toll for the day and removed from the plate.

7. Mean day of survival for both control and experimental strains are calculated as a weighted mean (m), where $m = \sum$ ((day number) × (# of animals dead on given day)/(total number of animals assayed (n))). Statistical significance of differences between survival curves are determined by Kaplan–Meyer-based log-rank statistical analysis using survival curve plotting tools within GraphPad Prism or JMP software (*see* **Note 22**).

3.3 Fat Assays

3.3.1 Determining Fat Content from an Asynchronous Population (See Note 23)

1. From a plate of growing worms, rinse off the plates with phosphate-buffered saline (PBS) (*see* **Note 24**) and then rinse the worms three times to remove bacteria.
2. Fix worms by placing the non-starved animals in 1 % paraformaldehyde (Cat# 158,127, Sigma Aldrich) in PBS and place tubes at −70 °C.
3. Sequentially dehydrated samples by washing thawed fixed worms in 25, 50, and 70 % ethanol.
4. Fixed animals are incubated overnight in 250 μL of Saturated Sudan Black solution (Cat# 199,664, Sigma-Aldrich) dissolved in 70 % ethanol, on a shaker at room temperature or incubated overnight in Nile Red (Cat# 72,485, Sigma Aldrich) dissolved in acetone and then diluted in water to a final concentration of 100 ng/mL.
5. Rinse worms 2–3 times in PBS and then multiple worms are placed on a small drop of 50 mM Sodium Azide (Cat #S2002 Sigma-Aldrich) onto a glass slide (Cat# 12-550-343, Fisher) and then a cover slip (Cat# 2,890-18, Corning) is gently placed over the worm for visualization. Care is taken to avoid air bubbles.
6. ImageJ software is used to measure the average pixel intensity for a 84-pixel radius below the pharynx of each animal in the anterior intestine area. In addition, an 84-pixel radius of the background is measured and subtracted from the values obtained for the staining. At least ten animals are measured for each strain and significance is determined by Student's t-test (*see* **Note 25**).

3.3.2 Determining Fat Content from a Synchronous Population

1. Grow worms on medium or large plates until there are many gravid adult worms. Plates should not be starved.
2. Wash worms off the plate into a 15 mL conical vials with sterile PBS.
3. Wash twice with PBS centrifuging at 2,000 × g for 30 s after each wash. After the second wash, add 5 mL of a 1N sodium hydroxide and 3 % sodium hypochlorite solution. Vortex tube every 30 s for 5 min to dissolve the adult and release eggs.

4. Wash eggs twice with sterile PBS centrifuging at 2,000 × *g* for 30 s after each wash PBS. Aspirate eggs with a glass pipette or low retention sterile pipette tips into fresh 15 mL conical tubes with 5–6 mL M9 or PBS.
5. Incubate eggs overnight on a shaker at the desired incubator temperature (12-h 25 °C; 16–18-h 20 °C, 24-h 15 °C).
6. After desired incubation, spin down worms and remove supernatant. Hatched L1 animals can be aliquoted onto fresh plates with a glass pipette.
7. Worms should grow on plates to the L4/young adult stage and then be extensively washed for fixation. Follow steps outlined in Subheading 3.3.1.

3.3.3 Imaging Conditions for Dye-Labeled Assays

1. Fixed worms are mounted on slides and visualized using a Zeiss Axioscope 2+ microscope.
2. ImageJ software is used to quantify total fluorescence from pictures obtained from the microscope.
3. Exposure times should be kept constant within each trial. Nile Red and Sudan Black images can be obtained using an ORCA ER CCD camera (Hamamatsu Photonics, Japan). ORO images are obtained using an Axiocam HRc CCD camera (Carl Zeiss, Thornwood, NY). Fluorescent dyes are excited using a 100 W HBO mercury lamp in conjunction with an FITC (480ex/535em) or TRITC (540ex/605em) filter set (Chroma Technology, Bellows Falls, VT) for Nile Red.

3.4 Stress Assays (See Note 26)

3.4.1 Oxidative Stress (H_2O_2)

1. 50–60 young adult worms are transferred to a 96-well plates (with 1–2 worms/well) containing 40 mL of 10 mM hydrogen peroxide solution at room temperature.
2. Animals are tapped every 20–30 min and scored as dead when they do not respond to the platinum wire pick.
3. Assay is continued until all animals are dead. Plotting percent survival over time generates survivorship curves.

3.4.2 Heat Stress

4. Thirty young adults are picked to NGM plates and kept at 35 °C and scored for survival.
5. One can either set up one plate with 30 worms and then pick off the dead worms a each time point or set up separate plates for each time point and discarding the plate after scoring.
6. Plates should be prepared for a 1 and 10 h time point and then for every subsequent hour until strains are no longer alive. For wild type and *sir-2(ok434)* animals, enough plates (10 per strain) should be prepared to score up to 18 h [6].

3.4.3 UV Irradiation

1. 40–50 young adults are picked onto NGM plates without *E. coli*.
2. Worms are irradiated with 2,000 J/m^2 of UV with a UV Stratalinker (Stratagene).
3. Irradiated worms are then transferred to NGM plates with *E. coli* and placed at 20 °C.
4. Worms are scored every day to determine the number of dead animals until all animals are dead. Dead animals are counted and removed from the plates.

3.5 Brood Size Measurements

1. Five L4 worms are picked onto five individual plates and allowed to lay eggs at temperature of interest (15–25 °C).
2. Every 12 h worms are transferred to a new plate.
3. After the parental worms are picked off the plate and transferred to a new individual plate, the total number of eggs laid is counted.
4. The number of adult progeny from these eggs is scored again after 2–5 days, depending on the temperature. The percentage of hatched eggs is calculated as the number of adult progeny over the number of eggs laid (*see* **Note 27**).
5. Average brood size is the total number of fully developed adult animals divided by the number of original parental animals.
6. Several such measurements can be made for each strain tested and averaged, significance of any differences between strains can be determined by Student's *t*-test using standard statistical software such as JMP.

4 Notes

1. *sir-2.1* is located on chromosome IV, *sir-2.1* is part of the CEOP4372 operon containing R11A8.5, a putative glutathione transferase. *sir-2.1* is the downstream gene in the operon (Fig. 1a).
2. *sir-2.2* and *sir-2.3* are adjacently located on chromosome X and likely arose as a consequence of a gene duplication event. *sir-2.2* is part of the CEOPX178 operon containing F46G10.2, a nematode-specific gene of unknown function. *sir-2.2* is the downstream gene in the operon. *sir-2.3* is not in an operon.
3. *sir-2.4* is located on chromosome I, *sir-2.4* is part of the CEOP1976 operon with the upstream gene C06A5.3, a gene of unknown function containing a PWWP domain an essential part of the hepatoma-derived growth factor (HDGF) family of proteins and other transcription factors and chromatin remodeling proteins [9].

4. Combine 35.6 g K_2HPO_4 and 108.3 g KH_2PO_4 and dissolve in 1 L deionized water, check pH and adjust to pH = 6.0 with KOH or HCl as needed. Autoclave or filter sterilize with 0.22 μm sterile filter.
5. Alternatively, a large number of plates can be poured using a volumetric peristaltic pump, such as a Wheaton Omnispence (Cat# 13-687-434, Fisher Scientific, www.fishersci.com), using sterilized plastic tubing to dispense a fixed volume of media into plates.
6. It is important that plates be filled with molten NGM volumetrically. The height of the agar should be similar for all plates to prevent the need for repeated adjustment of the plane of focus when viewing animals with a stereo dissection scope.
7. OP50 is a streptomycin resistance strain of *E. coli* B containing a uracil auxotrophy, which limits its growth on NGM.
8. OP50 cultures are sterilely transferred to 50 mL conical vials and can be used for several months when kept at 4 °C.
9. Life span assays are performed using 30 mm petri dishes containing NGM supplemented with FuDR. A nucleotide analog, FuDR prevents progeny production by interfering with DNA replication. Since adult worms are purely post-mitotic, FuDR has no effect upon worm adult life span, but prevents the development of fertilized oocytes.
10. For standard worm propagation and for standard life span assays, plates are seeded with the bacterial strain OP50 available from the *C. elegans* stock center. Care should be taken not to move the plates after seeding to prevent the culture from spreading to the edge, which can cause worms to crawl off the plates with higher frequency.
11. The worm cuticle is a substantial physical barrier to many small molecules; to surmount this barrier, dimethyl sulfoxide (DMSO) is routinely added to achieve a final concentration of 1 %. DMSO is a water miscible polar aprotic solvent that dissolves both polar and nonpolar compounds and allows for penetration of compounds through the *C. elegans* cuticle, as it does human skin. In the case of resveratrol, whose life span extending effects are fully *sir-2.1*-dependent, DMSO is required. Resveratrol (Cat# R5010, Sigma-Aldrich) is poorly solvated in aqueous solutions, so the presence of DMSO aids compound solubility in the aqueous plate media and drug delivery into worms. Media containing resveratrol dissolved in ethanol, but lacking DMSO fails to show any life span extension [10, 11]. DMSO is routinely added in *C. elegans* drug screens to aid in compound solubility and delivery; however, caution should be taken to insure that DMSO alone be added as a vehicle control as at least one group found evidence that

DMSO alone can affect life span [12], though that has not been found in other studies using 1 % DMSO [10, 13, 14].

12. 0.22–0.5 mm diameter platinum wire (Cat# 357,359, Sigma-Aldrich) is commonly used for worm picks. Alternatively, premade picks can also be obtained (Cat# 59-X30P6, Genesee Scientific, www.geneseesci.com) in a variety of metal types and thicknesses.
13. S-Freezing Buffer #1: Combine 5.85 g NaCl, 50 mL 1 M KPO_4 pH 6.0, and add deionized water to 1 L. S-Freezing Buffer #2: Combine 5.85 g NaCl, 50 mL 1 M KPO_4 pH 6.0, 600 mL 100 % glycerol, and add deionized water to 1 L. Filter sterilize or autoclave both solutions can then be aliquoted into smaller volumes.
14. For best yields from frozen vials: (a) Freeze worms in a larger volume (ca. 1 mL). (b) Use freshly starved large (150 mm) plates containing many L1-stage animals, or bleach gravid adults and hatch eggs overnight in S-Freezing Buffer #1 in the absence of food producing L1-arrested worms. (c) Mutants that affect the worm cuticle (such as *rol-6*) do not survive freezing as well; therefore, care should be taken to maximize results.
15. Nalgene Cryo 1 °C Freezing Containers (Cat# 5,100-0,001, Nalgene, nalgenelabware.com) can be used to freeze worms to −80 °C, alternatively, worms can be placed in a Styrofoam box and placed in the −80 °C for slow freezing. Two days later, thaw one tube of frozen worms to check that the freezing worked. At this time, frozen worms can be transferred to a cardboard box at −80 °C.
16. Strains being analyzed for life span must be treated to the same growth conditions prior to analysis. For example, *daf-2* mutant strains, such as *daf-2(e1370)* are temperature-sensitive for dauer entry at temperatures above 15 °C. *daf-2(e1370)* mutants must be raised at 15 °C until they pass the dauer decision stage (typically the L4 stage) at which time they can be transferred to life span plates and placed at 20 or 25 °C. Therefore, all strains in this assay must be grown with the same conditions. Without similar growth conditions, strains cannot be compared for life span. While other published protocols for life span analysis require inducing dauer formation in the parents of tested animals [15], we strongly disagree with this methodology as it is likely to introduce variable influences upon the life span of animals being tested and cannot be used for dauer-defective animals such as *daf-16* mutant animals.
17. Alternatively, progeny animals can be developmentally semi-synchronized, reducing the need to stage worms when setting up life span plates, by allowing 20–30 gravid adults to lay eggs overnight and then picking off the adult worms the next day.

In this case, it is not necessary to move stage-selected worms to a seeded NGM plate prior to moving to NGM+FuDR plates, as worms are all the same stage and there are no eggs or young larval stage worms that can be accidently picked; worms can be directly transfered to the NGM+FuDR plates.

18. The choice on the number of animals depends upon the expected difference in life span. The smaller the expected difference in mean survival, the greater the number of animals needed in order to see statistical differences. As a rule of thumb, for expected differences in mean life span between 20 and 50 % 100 animals is sufficient, for difference between 10 and 20 % 150–200 animals is sufficient. Two hundred worms is a good starting point when the life span effect is not known. When learning to perform life span analysis, it is easier to place fewer worms on the plates and simply increase the number of plates used.
19. Life spans can also be performed in the absence of FuDR. In such cases, worms must be moved to a new plate every other day until such time as the animals cease progeny production, which is typically within 6–8 days for reaching adulthood. After this period, the animals can be maintained on NGM media without perturbation. Some labs have reported life span changes associated with the use of FuDR; however, we have found no such affects in our studies on *sir-2.1* mutant or transgenic animals.
20. Crawl-off events, where worms crawl onto the side of the petri dish typically happen within the first 6 days, when the worms are most active, these animals are not included in the life span. Animals that crawl off the plate after day 6 are replaced with age-matched worms from the spare plate. Similarly, any animals that demonstrate vulval bursting, where the reproductive tract exteriorizes through the vulva, are removed and replaced with an age-match live animal from the spare plate. Some labs include the animals that die due to vulval bursting. One should record these events and note this in the results of any analysis. Otherwise, animals are not censored or replaced for any other reason, including protruding vulva, as these animals have been found to live as long as other animals.
21. As an alternative to testing movement by pick prodding. Plates can be briefly and rapidly shaken by placing the plate bottom on a vortexer, which causes most live worms to move. Any worm that fails to move following vortexing can be checked for life by gentle stroking with a sterile platinum wire to check for any response. Finally, if no response is elicited the worm is picked up and moved to a new part of the plate to check for movement.

22. Life span assays need to be repeated at least two independent times. It is also helpful if more than one individual in the lab repeats the life span assay or if the identity of the strains being analyzed are blinded.
23. Fat assays can be performed on asynchronous populations of growing worms as well as synchronous populations. Since fat content is affected by the developmental stage of the worm [16] as well as the bacterial food source, when comparing different strains, these conditions should be similar for accurate comparison.
24. 1× PBS: Dissolve 8 g NaCl, 0.2 g KCl, 1.44 g Na_2HPO_4, 0.24 g KH_2PO_4 in 800 mL water, adjust pH to 7.4, and bring to 1 L with water. Autoclave to sterilize.
25. Assays should be done multiple independent times and many worms need to be examined to achieve accurate results.
26. Any given stress assays should be repeated at least three times. Statistical differences can be determined using Student's *t*-test or by plotting percent survival over time generating survivorship curves that can be subject to Kaplan–Meyer analysis using GraphPad Prism or JMP software.
27. It is important to compare the number of progeny that reach adulthood with the number of eggs laid in order to determine if there is any embryonic lethality or larval arrest phenotype, which can be reported separately (Fig. 1b).

References

1. Imai S, Armstrong CM, Kaeberlein M, Guarente L (2000) Transcriptional silencing and longevity protein Sir2 is an NAD-dependent histone deacetylase. Nature 403(6771):795–800
2. Kaeberlein M, McVey M, Guarente L (1999) The SIR2/3/4 complex and SIR2 alone promote longevity in Saccharomyces cerevisiae by two different mechanisms. Genes Dev 13(19):2570–2580
3. Greiss S, Gartner A (2009) Sirtuin/Sir2 phylogeny, evolutionary considerations and structural conservation. Mol Cells 28(5):407–415. doi:10.1007/s10059-009-0169-x
4. Frye RA (2000) Phylogenetic classification of prokaryotic and eukaryotic Sir2-like proteins. Biochem Biophys Res Commun 273(2): 793–798. doi:10.1006/bbrc.2000.3000, pii: S0006-291X(00)93000-6
5. Brenner S (1974) The genetics of Caenorhabditis elegans. Genetics 77(1):71–94
6. Wang Y, Oh SW, Deplancke B, Luo J, Walhout AJ, Tissenbaum HA (2006) C. elegans 14-3-3 proteins regulate life span and interact with SIR-2.1 and DAF-16/FOXO. Mech Ageing Dev 127(9):741–747. doi:10.1016/j.mad.2006.05.005, pii: S0047-6374(06)00144-8
7. Berdichevsky A, Viswanathan M, Horvitz HR, Guarente L (2006) C. elegans SIR-2.1 interacts with 14-3-3 proteins to activate DAF-16 and extend life span. Cell 125(6): 1165–1177
8. Rizki G, Iwata TN, Li J, Riedel CG, Picard CL, Jan M, Murphy CT, Lee SS (2011) The evolutionarily conserved longevity determinants HCF-1 and SIR-2.1/SIRT1 collaborate to regulate DAF-16/FOXO. PLoS Genet 7(9):e1002235. doi:10.1371/journal.pgen.1002235, pii:PGENETICS-D-11-00207
9. Alvarez-Venegas R, Avramova Z (2012) Evolution of the PWWP-domain encoding genes in the plant and animal lineages. BMC Evol Biol 12:101. doi:10.1186/1471-2148-12-101, pii: 1471-2148-12-101
10. Viswanathan M, Kim SK, Berdichevsky A, Guarente L (2005) A role for SIR-2.1 regula-

tion of ER stress response genes in determining C. elegans life span. Dev Cell 9(5):605–615

11. Bass TM, Weinkove D, Houthoofd K, Gems D, Partridge L (2007) Effects of resveratrol on lifespan in Drosophila melanogaster and Caenorhabditis elegans. Mech Ageing Dev 128(10):546–552. doi:10.1016/j.mad.2007.07.007, pii: S0047-6374(07)00121-2
12. Wang X, Li L, Wang D (2010) Lifespan extension in Caenorhabditis elegans by DMSO is dependent on sir-2.1 and daf-16. Biochem Biophys Res Commun 400(4):613–618. doi:10.1016/j.bbrc.2010.08.113, pii: S0006-291X(10)01634-7
13. Cai WJ, Huang JH, Zhang SQ, Wu B, Kapahi P, Zhang XM, Shen ZY (2011) Icariin and its derivative icariside II extend healthspan via insulin/IGF-1 pathway in C. elegans. PLoS One 6(12):e28835. doi:10.1371/journal.pone.0028835, pii: PONE-D-11-20117
14. Wood JG, Rogina B, Lavu S, Howitz K, Helfand SL, Tatar M, Sinclair D (2004) Sirtuin activators mimic caloric restriction and delay ageing in metazoans. Nature 430(7000):686–689. doi:10.1038/nature02789, pii: nature02789
15. Sutphin GL, Kaeberlein M (2009) Measuring Caenorhabditis elegans life span on solid media. J Vis Exp (27). doi: 10.3791/1152. pii: 1152
16. Yen K, Le TT, Bansal A, Narasimhan SD, Cheng JX, Tissenbaum HA (2010) A comparative study of fat storage quantitation in nematode *Caenorhabditis elegans* using label and label-free methods. PLoS One 5(9):e12810. doi: 10.1371/journal.pone.0012810
17. Burnett C, Valentini S, Cabreiro F, Goss M, Somogyvari M, Piper MD, Hoddinott M, Sutphin GL, Leko V, McElwee JJ, Vazquez-Manrique RP, Orfila AM, Ackerman D, Au C, Vinti G, Riesen M, Howard K, Neri C, Bedalov A, Kaeberlein M, Soti C, Partridge L, Gems D (2011) Absence of effects of Sir2 overexpression on lifespan in C. elegans and Drosophila. Nature 477(7365):482–485. doi:10.1038/nature10296, pii: nature10296
18. Viswanathan M, Guarente L (2011) Regulation of Caenorhabditis elegans lifespan by sir-2.1 transgenes. Nature 477(7365):E1–E2. doi:10.1038/nature10440, pii: nature10440
19. Berdichevsky A, Nedelcu S, Boulias K, Bishop NA, Guarente L, Horvitz HR (2010) 3-Ketoacyl thiolase delays aging of Caenorhabditis elegans and is required for lifespan extension mediated by sir-2.1. Proc Natl Acad Sci U S A 107(44):18927–18932. doi:10.1073/pnas.1013854107, pii: 1013854107
20. Parker JA, Vazquez-Manrique RP, Tourette C, Farina F, Offner N, Mukhopadhyay A, Orfila AM, Darbois A, Menet S, Tissenbaum HA, Neri C (2012) Integration of beta-catenin, sirtuin, and FOXO signaling protects from mutant huntingtin toxicity. J Neurosci 32(36):12630–12640. doi:10.1523/JNEUROSCI.0277-12.2012, pii: 32/36/12630
21. Tissenbaum HA, Guarente L (2001) Increased dosage of a sir-2 gene extends lifespan in Caenorhabditis elegans. Nature 410(6825): 227–230. doi:10.1038/35065638, pii: 350656 38
22. Walker AK, Yang F, Jiang K, Ji JY, Watts JL, Purushotham A, Boss O, Hirsch ML, Ribich S, Smith JJ, Israelian K, Westphal CH, Rodgers JT, Shioda T, Elson SL, Mulligan P, Najafi-Shoushtari H, Black JC, Thakur JK, Kadyk LC, Whetstine JR, Mostoslavsky R, Puigserver P, Li X, Dyson NJ, Hart AC, Naar AM (2010) Conserved role of SIRT1 orthologs in fasting-dependent inhibition of the lipid/cholesterol regulator SREBP. Genes Dev 24(13):1403–1417. doi:10.1101/gad.1901210, pii: 24/13/1403
23. van der Linden AM, Plasterk RH (2004) Shotgun cloning of transposon insertions in the genome of Caenorhabditis elegans. Comp Funct Genomics 5(3):225–229. doi:10.1002/cfg.392
24. Bamps S, Wirtz J, Savory FR, Lake D, Hope IA (2009) The Caenorhabditis elegans sirtuin gene, sir-2.1, is widely expressed and induced upon caloric restriction. Mech Ageing Dev 130(11–12):762–770. doi:10.1016/j.mad.2009.10.001, pii: S0047-6374(09)00146-8
25. Sarov M, Murray JI, Schanze K, Pozniakovski A, Niu W, Angermann K, Hasse S, Rupprecht M, Vinis E, Tinney M, Preston E, Zinke A, Enst S, Teichgraber T, Janette J, Reis K, Janosch S, Schloissnig S, Ejsmont RK, Slightam C, Xu X, Kim SK, Reinke V, Stewart AF, Snyder M, Waterston RH, Hyman AA (2012) A genome-scale resource for in vivo tag-based protein function exploration in C. elegans. Cell 150(4):855–866. doi:10.1016/j.cell.2012.08.001, pii: S0092-8674(12)00946-4
26. Chiang WC, Tishkoff DX, Yang B, Wilson-Grady J, Yu X, Mazer T, Eckersdorff M, Gygi SP, Lombard DB, Hsu AL (2012) C. elegans SIRT6/7 homolog SIR-2.4 promotes DAF-16 relocalization and function during stress. PLoS Genet 8(9):e1002948. doi:10.1371/journal.pgen.1002948,pii: PGENETICS-D-11-01071
27. Pothof J, van Haaften G, Thijssen K, Kamath RS, Fraser AG, Ahringer J, Plasterk RH, Tijsterman M (2003) Identification of genes

that protect the C. elegans genome against mutations by genome-wide RNAi. Genes Dev 17(4):443–448. doi:10.1101/gad.1060703

28. Kamath RS, Martinez-Campos M, Zipperlen P, Fraser AG, Ahringer J (2001) Effectiveness of specific RNA-mediated interference through ingested double-stranded RNA in Caenorhabditis elegans. Genome Biol 2(1): RESEARCH0002

29. Kraemer BC, Burgess JK, Chen JH, Thomas JH, Schellenberg GD (2006) Molecular pathways that influence human tau-induced pathology in Caenorhabditis elegans. Hum Mol Genet 15(9):1483–1496. doi:10.1093/hmg/ddl067, pii: ddl067

30. Wang Y, Tissenbaum HA (2006) Overlapping and distinct functions for a *Caenorhabditis elegans* SIR2 and DAF-16/FOXO. Mech Ageing Dev 127:48–56

Chapter 4

Genetic and Biochemical Tools for Investigating Sirtuin Function in *Drosophila melanogaster*

Jason G. Wood, Rachel Whitaker, and Stephen L. Helfand

Abstract

Drosophila melanogaster is one of the most widely used genetic model systems in biology. The ease of working in an invertebrate model system allows the design and execution of many experiments that would be infeasible in a vertebrate model. Although the strength of the fly as a model system lies primarily in the ease of genetic manipulation, it is flexible enough that biochemical and proteomic approaches can also be used to build a more comprehensive study. Here we present a pair of complementary protocols that we have used to examine sirtuin biology in *Drosophila*. First, we describe our protocol for measuring lifespan in flies expressing a gene of interest under the inducible control of the Gene-Switch system. Finally, we describe a method for performing chromatin immunoprecipitation on adult flies, including some of the difficulties associated with using this technique in chitinous tissue.

Key words *Drosophila melanogaster*, Lifespan, Demography, Chromatin immunoprecipitation

1 Introduction

The common fruit fly *Drosophila melanogaster* is an excellent model system in which to investigate many aspects of human biology, including sirtuin biology. The fly has highly developed nervous, endocrine, and muscular systems, and exhibits complex behaviors, while still retaining the advantages of invertebrates such as low cost of husbandry and maintenance, and manageable lifespans (50–80 days) [1–3]. *Drosophila* has well-developed molecular genetic tools for tight temporal and spatial regulation of gene expression, including a binary driver/target system (GAL4/UAS) and a drug-inducible counterpart (the Gene-Switch system) [4]. In addition, over the last decade, tools for gene targeting have been developed and refined, allowing straightforward generation of true null alleles as well as in situ epitope tagging of native genes [5–7]. Although the fly has been used historically primarily for genetic experiments, the organism is also well suited for biochemical approaches. Methods designed for mammalian tissues or tissue

Matthew D. Hirschey (ed.), *Sirtuins: Methods and Protocols*, Methods in Molecular Biology, vol. 1077,
DOI 10.1007/978-1-62703-637-5_4,

culture can typically be adapted for use with embryonic or adult flies, or a *Drosophila* tissue culture system. This combination of possible approaches allows a fly researcher great flexibility in designing experiments, and allows for projects that may be too time-consuming or costly to perform in a mammalian system, such as multifactorial lifespan assays.

In this chapter, we present two complementary methods that can be used in the *Drosophila* model system to investigate sirtuin biology. First, we describe the approach we use for assaying lifespan of a population of flies overexpressing a gene of interest. Second, we present a method for performing chromatin immunoprecipitation (ChIP) experiments on a population of adult flies or fly tissues, which present some challenges compared to the more commonly used tissue culture or soft tissue systems. Combining methods such as these can be especially powerful when studying genes such as sirtuins, which are known to have effects on longevity as well as chromatin structure.

First, overexpression of sirtuins and other genes in *Drosophila* is accomplished using the UAS-Gal4 system [8]. However, when using this system, genetic background and appropriate controls are crucial to correctly interpret the effects of overexpression. A method of overexpression has been developed that allows for sirtuin overexpressing flies to be genetically identical to their controls. This method, known as the Gene-Switch or P{Switch} system, utilizes a chimeric Gal4 protein that is only able to bind UAS sequences in the presence of the antiprogestin known as RU486 [9, 10]. Using this system, the experimental flies are given RU486 in the food, while the control flies are genetically identical to the treatment flies and come from the same cohort, but are given only the diluent without RU486. Importantly, this system allows for normal expression levels of the gene of interest during development, since RU486 can be added to the food only during the adult phase of life.

Specifically, lifespan analysis in *Drosophila* involves scoring the deaths of a population of flies and comparing the survivorship of the treatment group with the control group. When evaluating the effects of sirtuin overexpression on lifespan, it is crucial that several different levels of overexpression are tested. Just as more than two different caloric levels need to be tested when evaluating whether a gene is involved in caloric restriction [11], more than two levels of overexpression need to be examined before determining the effect of the gene on lifespan. Importantly, the Gene-Switch system is responsive to changes in the levels of RU486 in the food when overexpressing dSir2, the *Drosophila* Sir2/SIRT1 homolog [12]. Since giving a higher dose of RU486 in the food results in higher levels of expression of the gene, this system can be used to overexpress sirtuins in a dosage gradient. We recommend at least three different concentrations of RU486, 100, 200, and 500 μM, in addition to the control condition without RU486 in the food to obtain a dosage response curve for examining the effect of overexpressing sirtuins on lifespan in *Drosophila*.

Secondly, most ChIP protocols are designed for either tissue culture cells, or soft tissue (e.g., liver, fly embryos), which are easily dissociated and homogenized and have uniform nuclei. Tissue chromatin preps involve a tradeoff—you need sufficient detergent for good sonication which can be much trickier than cultured cells, but too much will interfere with IP conditions later on. Yields will generally be lower and more variable than with cultured cells. This protocol uses a ground glass homogenizer (can also use bead mill/tissuelyser) for homogenization, formaldehyde cross-linking, a high SDS wash (to help disrupt cells), and sonication in 0.1 % SDS. Good homogenization is critical. Here we present a protocol for adult fly heads, but this protocol also works for whole flies or other tissues (e.g., thorax). We generally use heads as the tissue is relatively homogeneous (primarily nervous tissue) and they are easy to isolate in large quantities by the freezing/vortexing/sieving approach detailed in the protocol. This protocol was adapted from published protocols from the Kuroda group [13, 14] and the Cavalli group [15], with helpful input from Art Alekseyenko (personal communication).

2 Materials

2.1 Lifespan Assays in Drosophila melanogaster

1. Normal Fly Food for brood vials: 30 g/L Autolyzed Yeast, 120 g/L sucrose, 50 g/L cornmeal, 10 g/L agar, poured ~1 in. thick in 28 mm shell vials.
2. 1.5 N EtOH Lifespan Food: 150 g/L Autolyzed Yeast, 150 g/L sucrose, 50 g/L cornmeal, 20 g/L agar, 20 mL/L ethanol.
3. 1.5 N 100 μM RU486 Lifespan Food: 150 g/L Autolyzed Yeast, 150 g/L sucrose, 50 g/L cornmeal, 20 g/L agar, 20 mL/L 5 mM RU486 in EtOH.
4. 1.5 N 200 μM RU486 Lifespan Food: 150 g/L Autolyzed Yeast, 150 g/L sucrose, 50 g/L cornmeal, 20 g/L agar, 20 mL/L 10 mM RU486 in EtOH.
5. 1.5 N 500 μM RU486 Lifespan Food: 150 g/L Autolyzed Yeast, 150 g/L sucrose, 50 g/L cornmeal, 20 g/L agar, 20 mL/L 25 mM RU486 in EtOH.
6. Fly line containing UAS-driven gene of interest (e.g., Bloomington Stock Number 24859, also known as EP2300, is a fly line containing an UAS in the endogenous promoter of dSir2).
7. Fly line containing Gene-Switch Gal4 driver for tissue(s) of interest (e.g., Bloomington Stock Number 9431 is a fly line that expresses the Gene-Switch Gal4 driver ubiquitously using an Actin5C promoter).

2.2 Chromatin Preparation and Immunoprecipitation in Adult *Drosophila melanogaster*

1. Buffer A basic: 60 mM KCl, 15 mM NaCl, 4 mM $MgCl_2$, 15 mM HEPES (pH7.6). Can store long term. For buffer A complete add fresh: 0.25 % Triton X-100, 0.5 mM DTT, 1× PLAAC (protease inhibitor mix).
2. Buffer B basic: 140 mM NaCl, 15 mM HEPES pH 7.6, 1 mM EDTA, 0.5 mM EGTA. Can store long term. For buffer B complete add fresh: 0.01 % Triton X-100, 1× PLAAC.
3. TE+0.01 % SDS: 10 mM Tris pH 7.4, 1 mM EDTA, 0.01 % SDS, 1× PLAAC. Make fresh.
4. Sonication buffer: 10 mM Tris pH 7.4, 1 mM EDTA, 0.1 % SDS, 1× PLAAC. Make fresh.
5. IP Buffer: 10 mM Tris pH 7.4, 1 mM EDTA, 0.1 % SDS, 0.1 % sodium deoxycholate, 1 % Triton X-100, 140 mM NaCl, 1× PLAAC.
6. PBS: 8 g NaCl, 0.2 g KCl, 1.44 g Na_2HPO_4, 0.24 g KH_2PO_4, to pH 7.4 in 1 L water. Store long term. Add 1× PLAAC fresh before use.
7. 2.5 M glycine.
8. 1,000× PLAAC protease inhibitor mixture: pepstatin A, leupeptin, antipain, aprotinin, chymostatin. 2.5 mg/mL each in DMSO. Aliquot and store at −20 °C. Dilute 1:1000 for final 1× working concentration in buffers.
9. Brass sieve if using fly heads. US Standard Sieve Series No. 25 (710 μm) retains female bodies, No. 30 (600 μm) retains male bodies, and No. 45 (355 μm) retains heads of both sexes.
10. 37 % formaldehyde solution.
11. Ground glass tissue grinder, size #23 (Kimberly Kontes #885500-0023, available from Fisher Scientific).
12. Dounce homogenizer, 15 mL (Wheaton #357544, available from Fisher Scientific).
13. Sonicator. We use Fisher FB505 (same as Qsonica Q500) 500 W sonicator with a 1/8″ microtip.
14. Proteinase K. Make stock to 20 mg/mL with 15 % glycerol and store at −20 °C.
15. Agencourt Ampure XP beads (Beckman Coulter) or Qiagen MinElute PCR purification columns for cleaning up DNA. Ampure XP beads are generally easier to use and have significantly less sample loss in our hands than columns.
16. Qubit (Invitrogen) and accompanying DNA-HS dye kit for quantifying low levels of DNA.
17. Dynabeads Protein A and Dynabeads Protein G (Invitrogen).
18. IP buffer: same as above. Add 1× PLAAC before use.

19. IP buffer + NaCl: IP buffer, with NaCl supplemented to 300 mM. Add 1× PLAAC before use.
20. TE: 10 mM Tris pH 7.4, 1 mM EDTA. No PLAAC.
21. Bicarbonate elution buffer: 1 % SDS, 0.1 M $NaHCO_3$.
22. Glycogen, 5 mg/mL (Ambion).

3 Methods

3.1 Lifespan Assays in Drosophila melanogaster

1. To create the brood vials used to obtain adult flies of the genotype of interest, cross 8–10 virgin driver females with 5–6 males containing your overexpression construct of interest. Also include a genetic background control (*see* **Note 1**). Place flies on "normal" food, sprinkled with dry live yeast. Pass flies every 24 h onto new food, keeping the old vials (*see* **Note 2**).
2. Keep brood vials at 25 ° C in a humidity-controlled area with a 12-h on/off light cycle, and adult flies for your lifespans should eclose approximately 10 days after brood vials are set up, if they are kept at 25 ° Celsius. Twenty-four hours before collecting adult flies for lifespans, clear the brood vials of any adults that have eclosed early.
3. At collection time, briefly anesthetize flies under CO_2, sort flies, and place 25 males and 25 females in each "lifespan" food vial (*see* **Notes 3**, **4**, and **5**).
4. Collect additional flies for qPCR or Western blot validation of the doses of the gene of interest (*see* **Note 6**).
5. Pass flies onto new food every day or every other day, remaining consistent for the duration of the experiments (*see* **Note 7**).
6. When each vial is passed, examine it under a dissecting microscope and record the number of dead females and males. Also take care to note whether any dead animals have transferred to the new vial, which will need to be noted and subtracted out of the next day's total "dead" in that vial (*see* **Note 8**).
7. Record the number of dead flies per vial with age, and calculate the survivorship of animals for each condition and sex.

3.2 Chromatin Preparation and Immunoprecipitation (ChIP) in Adult Drosophila melanogaster

3.2.1 Chromatin Preparation (Day 1)

1. Prep all working solutions. Per sample (*see* **Note 9**), make up the following: 15 mL PBS, 5 mL buffer A complete, 5 mL buffer B complete, 20 mL TE + 0.01 % SDS, 1 mL sonication buffer. Ensure all buffers have PLAAC protease inhibitors in them. Keep all solutions on ice once prepared.
2. Start with 200–250 mg tissue (fly heads, bodies, etc., *see* **Note 10**). Heads can be shaken off by freezing in liquid nitrogen in a 50 mL conical tube, vortexing (3× 20s, LN_2 freezing between rounds), and quickly sifting through a brass sieve,

shaking vigorously for 30 s. For heads, start with 2–3 g flies to get enough material. Weigh tissue before homogenization for yield calculation. Set aside some material for RNA preps if desired (100–150 heads in an Eppendorf tube). Return bodies to −80 °C freezer for later processing if desired (*see* **Note 11**).

3. Aliquot 10 mL PBS for each sample (from amount prepared above) and add 520 μL formaldehyde (37 % solution, to 1.8 % total) to each aliquot.
4. Dump fly heads into ground glass homogenizer, add aliquot of PBS + formaldehyde, start timer for 20 m. Homogenize well until no fly parts are visible (approx. 1–2 m of grinding, *see* **Note 12**).
5. Transfer to 15 mL Dounce homogenizer and dounce 5× with loose pestle then 5× with tight pestle. Decant into 15 mL falcon tube and place on rotator at room temperature. Total time exposed to formaldehyde should be 20 m.
6. Quench cross-linking by adding 1/10 volume (1 mL) 2.5 M glycine. Put tube on ice for 5 m.
7. Spin samples in refrigerated tabletop centrifuge, max speed (4,000 rpm = 3,000 g) for 5 m at 4 °C.
8. Decant supernatant into formaldehyde liquid waste container and dispose of properly. Wash pellet by resuspending in 5 mL PBS. Add 2 mL, resuspending by pipetting well with 1 mL pipetter, then add remaining 3 mL and mix by inversion. Spin 5 m at 4 °C as in **step 7** for each wash.
9. Wash pellet as above 1× with 5 mL buffer A. Prepare Protein A/G beads during washes if not already blocked (*see* **Note 13**).
10. Wash pellet 1× with 5 mL buffer B.
11. Wash pellet 1× with 5 mL TE + 0.01 % SDS. Low concentration of SDS is to keep sample off tube walls.
12. Resuspend pellet in 5 mL TE + 0.01 % SDS. Bring to 1 % SDS total by adding 1/9 volume (550 μL) of 10 % SDS. Mix quickly by inverting 4–5 times and immediately centrifuge as above.
13. Wash pellet 2× in 5 mL TE + 0.01 % SDS with 5 m spin.
14. Resuspend pellet in 750 μL sonication buffer for a total volume of ~1 mL. Split into two 500 μL aliquots in microcentrifuge tubes. Sonicate using 1/8″ microtip, constant duty cycle, amplitude 20 %. Sonicate for 5× 20s bursts with 3 m on ice in between, for total sonication time of 1 min 40 s. Sonication conditions will need to be determined empirically for each instrument (*see* **Note 14**).
15. Recombine aliquots and spin at max speed (12,000 × *g*) 10 m in refrigerated centrifuge. Remove supernatant containing soluble chromatin to new tube and discard pellet.

16. Repeat centrifugation as in **step 15** to pellet any remaining debris and remove supernatant into new tube. Bring final volume to 900 μL in sonication buffer.
17. Make up complete IP buffer by adding sequentially: 10 % Triton X-100 to 1 % final (100 μL), 10 % sodium deoxycholate to 0.1 % final (10 μL), and 5 M NaCl to 140 mM final (28 μL). Mix for 1 m between steps.
18. Remove 50 μL aliquot for QC (quality control) into PCR tube. Add 1 μL Proteinase K (20 mg/mL) to aliquot and put in PCR machine overnight to reverse cross-links. Program: 12 h at 65 °C, 20 m at 72 °C (to inactivate proteinase K).
19. Add 100 μL blocked Protein A/G beads to remainder of sample to preclear chromatin and place on rocker in cold room overnight.

3.2.2 Chromatin Preparation (Day 2)

20. For QC samples, add 1 μL RNAse A, incubate at 37 °C for 30 m. Purify DNA using Ampure XP beads (or Qiagen MinElute column) and run sample on 1 % agarose gel with ethidium bromide. Chromatin should be visible as a smear, ideally with a size distribution centered around 300–500 bp.
21. Remove beads from chromatin samples with magnet and quantify DNA using Qubit DNA-HS dye. Yields vary, but 30–50 μg chromatin is expected from 200 mg original starting material.
22. Divide chromatin into 10 μg aliquots and freeze in low-retention screw-top tubes at −80 °C. Set aside 50 μL aliquot for input. Aliquots do not need to be exactly 10 μg but should be close. If volume is significantly less than 250 μL, add IP buffer to bring volume up to 250. Samples can be stored at −80 °C for months.

3.2.3 Immuno-precipitation (Day 3)

1. Thaw chromatin aliquot. Add antibody and label tube. Amount of antibody has to be empirically determined, but we generally use 4–5 μg of histone antibodies, or 10 μL of concentrated monoclonal supernatant. Put on rocker in cold room overnight (*see* **Note 15**).

3.2.4 Immuno-precipitation (Day 4)

2. Add 50 μL Protein A/G beads (washed and blocked) and incubate 3–4 h.
3. Make wash buffers up during incubation. For each sample, make: 3 mL IP buffer (*see* Subheading 2), 1 mL IP buffer with NaCl supplemented to 300 mM, 2 mL TE. The first two wash buffers have PLAAC protease inhibitors, the last one (TE) does not.
4. Capture beads on magnet and wash IPs. Each wash is 1 mL, for 5 m on rocker at 4 °C. Six washes total as follows (keep buffers on ice): Washes 1–3: IP buffer. Wash 4: IP buffer + NaCl. Washes 5–6: TE (no PLAAC, *see* **Note 16**). Between washes 5 and 6,

transfer samples to new microcentrifuge tubes (to minimize contaminants that have bound to the tube).

5. Elute in 125 μL bicarbonate elution buffer for 15 m at 65 °C. Capture beads, remove eluate to new low retention tube, and repeat elution with 125 μL, combining eluates.
6. Add 16 μL 5 M NaCl to IPs (*see* **Note 17**).
7. Thaw input sample if necessary. To 50 μL input sample add: 40 μL IP buffer, 10 μL 10 % SDS (to 1 % total), and 5 μL 5 M NaCl.
8. Incubate all samples at 65 °C overnight to reverse cross-links. We split into PCR tubes and do incubation in the thermal cycler to prevent condensation.

3.2.5 Immunoprecipitation (Day 5)

9. Move samples to low retention microcentrifuge tubes and add 150 μL TE (IP samples) or 300 μL TE (inputs) to bring total volume to 400 μL. Add 1 μL RNAse A to inputs and incubate at 37°C for 30 m.
10. Add 20 μL 1 M Tris 7.0, 10 μL 0.5 M EDTA, and 3 μL Proteinase K (20 mg/mL). Incubate at 42 °C for 2–3 h.
11. Extract with phenol/chloroform: Add 500 μL P/C, vortex well, spin maximum speed (12,000 × *g*) 5 m, keep upper layer and move into new tube. Discard phenol into appropriate hazardous waste.
12. Extract with 500 μL chloroform as above to get rid of residual phenol.
13. Add 40 μL 3 M sodium acetate, 10 μL glycogen, and 2–2.5 volumes EtOH.
14. Precipitate overnight at −20 °C.

3.2.6 Immunoprecipitation (Day 6)

15. Spin samples 30 m at maximum speed (12,000 × *g*) in 4 °C centrifuge to pellet DNA.
16. Wash once with 500 μL cold 70 % EtOH. Spin 10 m max speed.
17. Remove EtOH and let dry with lid open for 10 m to evaporate remaining EtOH.
18. Resuspend pellet in 11 μL nuclease-free water (samples) or 100 μL water (inputs) and quantitate 1 μL in Qubit DNA-HS assay. Yields will depend on efficiency of IP and antibody used (*see* **Note 18**).
19. Dilute ChIP samples to 1 ng/μL for convenience (or leave as is if lower concentration). Freeze and store DNA, or proceed directly to library prep. Use 20 ng for Illumina ChIP-seq library, or 1 ng for qPCR.

4 Notes

1. When designing your experiment, be sure to include a genetic background control for all of your RU486 conditions. In other words, cross flies of the same genetic background as those that contain your UAS-overexpression construct to the Gene-Switch flies and include the progeny in your lifespans.
2. It is best to pass the brood vials for at least 2–4 days. This way, if there is a small discrepancy in eclosion timing between different fly lines from which you want to collect, there is a window of several days to work with for collection. However, always collect adults for your lifespans over a 24 h period or shorter.
3. Use a moderate to high calorie food to avoid any potential complications between your gene of interest and caloric restriction.
4. Collect at least 12 vials (300 flies/sex) for each condition (experimental and control) for best lifespan results. Be sure to evenly distribute your collected flies among the EtOH control foods and the RU486-containing foods.
5. Be sure to select healthy looking flies with well-formed wings and monitor their waking after anesthesia to ensure that flies are healthy.
6. Collect all flies for validation of levels at the same age, we recommend 10 days.
7. It is important to remain consistent during the lifespan with regard to how often the flies are passed. If they are passed and counted every day, this must continue for the duration of the experiment. Passing the flies into new vials every day tends to extend the lifespan of the flies in comparison to every other day passing.
8. Practice is helpful in identifying dead males and females. Especially with every other day passing, dead females whose tissues have become necrotic can turn black in color and can easily be mistaken for males. Carefully note the sex of each fly, verifying the sex using the sex combs first, and then confirming with genitalia/abdominal markings.
9. Since there can be significant batch effects, it is best to process a batch of samples all at once.
10. Flies can be collected in advance and frozen at −80 °C until preparation.
11. Work quickly once the flies come out of the liquid nitrogen. They will start to thaw almost immediately and once thawed and wet they will not sift well.

12. Proper homogenization is critical to good yields. Other options include mortar and pestle grinding under liquid nitrogen, or a bead mill/tissuelyser. In our hands, the yield from mortar and pestle was noticeably lower due to residue sticking to the mortar. We have also seen success using a bead mill that works well but the packed pulverized pellet can be difficult to resuspend, so we generally prefer hand homogenization using the ground glass surface of the Kontes tissue grinder. If another homogenization procedure is used, the homogenate should still be passed through the Dounce homogenizer.
13. Start with equal amount of Protein A and Protein G Dynabeads (i.e., mix 1:1). Need 100 μL per sample for preclear, plus 50 μL per IP. Block beads in 1 mg/mL BSA in IP buffer with 1× PLAAC protease inhibitors. Do at least 3–4 washes in this buffer, overall equilibration time at least 1 h. Resuspend at end in same volume you started with. Note that Dynabeads literature indicates BSA blocking may not be very effective or necessary with these beads as they are hydrophilic, but we have generally continued to use the BSA blocking.
14. Sonication times are highly dependent on instrument, as well as concentration of the sample. It is a good idea to try several different conditions until you find one that consistently gives you a smear of chromatin centered in the 300–500 bp range. If the chromatin is insufficiently sonicated, the resolution of your ChIP will be impacted negatively. Although instruments such as the Bioruptor from Diagenode are popular for tissue culture chromatin preps, we and others have found that they provide insufficient power transfer to disrupt dense tissues, and better results are obtained from a traditional sonicator.
15. Ideally it is good to start with 10 μg chromatin for each IP, but we have gone down to 4–5 μg for samples with low yield. Usually there is enough from each prep for 3 IPs plus an input sample. Starting amount will depend on efficiency of IP and relative abundance of the target. For abundant histone targets, less starting material is required. For less abundant transcription factors with very inefficient IPs, it may be desirable to begin with 25 μg or more of chromatin, and pooling of multiple IPs may be necessary.
16. NaCl concentration can be increased during the high salt wash if background of IP is too high.
17. 200 mM NaCl keeps SDS in solution and helps reverse cross-links—volume is based on 400 μL final volume per **step 9**.
18. For histone antibodies we usually obtain at least 30–50 ng DNA and sometimes up to 500+ ng depending on antibody and relative abundance of mark.

References

1. Chippindale AK, Leroi M, Kim SB et al (1993) Phenotypic plasticity and selection in Drosophila life-history evolution. I. Nutrition and the cost of reproduction. J Evol Biol 6:171–193
2. Helfand SL, Rogina B (2003) From genes to aging in Drosophila. Adv Genet 49:67–109
3. Tatar M (2011) The plate half-full: status of research on the mechanisms of dietary restriction in Drosophila melanogaster. Exp Gerontol 46:363–368
4. Duffy JB (2002) GAL4 system in Drosophila: a fly geneticist's Swiss army knife. Genesis 34:1–15
5. Xie HB, Golic KG (2004) Gene deletions by ends-in targeting in Drosophila melanogaster. Genetics 168:1477–1489
6. Gong WJ, Golic KG (2003) Ends-out, or replacement, gene targeting in Drosophila. Proc Natl Acad Sci USA 100:2556–2561
7. Gong WJ, Golic KG (2004) Genomic deletions of the Drosophila melanogaster Hsp70 genes. Genetics 168:1467–1476
8. Brand AH, Perrimon N (1993) Targeted gene expression as a means of altering cell fates and generating dominant phenotypes. Development 118:401–415
9. Roman G, Endo K, Zong L et al (2001) P[Switch], a system for spatial and temporal control of gene expression in Drosophila melanogaster. Proc Natl Acad Sci USA 98: 12602–12607
10. Osterwalder T, Yoon KS, White BH et al (2001) A conditional tissue-specific transgene expression system using inducible GAL4. Proc Natl Acad Sci USA 98:12596–12601
11. Tatar M (2007) Diet restriction in Drosophila melanogaster. Design and analysis. Interdiscip Top Gerontol 35:115–136
12. Bauer JH, Nur S, Morris S et al (2009) dSir2 and Dmp53 interact to mediate aspects of CR—dependent life span extension in D. melanogaster. Aging 1:1–11
13. Alekseyenko AA, Larschan E, Lai WR et al (2006) High-resolution ChIP-chip analysis reveals that the Drosophila MSL complex selectively identifies active genes on the male X chromosome. Genes Dev 20:848–857
14. Larschan E, Alekseyenko AA, Gortchakov AA et al (2007) MSL complex is attracted to genes marked by H3K36 trimethylation using a sequence-independent mechanism. Mol Cell 28:121–133
15. Negre N, Lavrov S, Hennetin J et al (2006) Mapping the distribution of chromatin proteins by ChIP on chip. Methods Enzymol 410:316–341

Chapter 5

Generating Mammalian Sirtuin Tools for Protein-Interaction Analysis

Kathleen A. Hershberger, Jonathan Motley, Matthew D. Hirschey, and Kristin A. Anderson

Abstract

The sirtuins are a family of NAD$^+$-dependent deacylases with important effects on aging, cancer, and metabolism. Sirtuins exert their biological effects by catalyzing deacetylation and/or deacylation reactions in which Acyl groups are removed from lysine residues of specific proteins. A current challenge is to identify specific sirtuin target proteins against the high background of acetylated proteins recently identified by proteomic surveys. New evidence indicates that *bona fide* sirtuin substrate proteins form stable physical associations with their sirtuin regulator. Therefore, identification of sirtuin interacting proteins could be a useful aid in focusing the search for substrates. Described here is a method for identifying sirtuin protein interactors. Employing basic techniques of molecular cloning and immunochemistry, the method describes the generation of mammalian sirtuin protein expression plasmids and their use to overexpress and immunoprecipitate sirtuins with their interacting partners. Also described is the use of the Database for Annotation, Visualization, and Integrated Discovery for interpreting the sirtuin protein-interaction data obtained.

Key words Sirtuin, SIRT3, Deacetylation, Protein–protein interaction, Immunoprecipitation, DAVID

1 Introduction

Sirtuins are a family of NAD$^+$-dependent deacylases that regulate protein function by removing acetyl and/or acyl modifications from lysine residues of proteins. This activity has critical regulatory roles in aging, cancer, and metabolism. Originally identified as an activity occurring uniquely with histones, reversible protein acetylation is now considered a widespread regulatory mechanism that controls the function of several nonhistone proteins. For example, recent proteomics studies reveal widespread acetylation of mitochondrial proteins with greater than 30 % of all mitochondrial proteins acetylated on at least one lysine residue [1]. Although the mechanism of mitochondrial protein acetylation remains unclear, deacetylation is well characterized and identification of sirtuin substrates is a common approach for understanding how reversible

Matthew D. Hirschey (ed.), *Sirtuins: Methods and Protocols*, Methods in Molecular Biology, vol. 1077,
DOI 10.1007/978-1-62703-637-5_5, © Springer Science+Business Media, LLC 2013

acetylation modulates mitochondrial function. Although hundreds of acetyllysine residues have been identified, in only a few cases have proteins targeted by SIRT3, the most studied mitochondrial sirtuin, been identified; acetyl-CoA synthetase (AceCS2) long-chain acyl-CoA dehydrogenase (LCAD), 3-hydroxy,3-methyl-glutaryl-CoA synthase (HMGCS2), and glutamate dehydrogenase (GDH) are examples (for a review, *see* ref. 1). These studies show that these validated SIRT3 substrates, as well as several other substrates identified to date, physically associate with SIRT3. Therefore, identification of sirtuin interacting partners represents one approach to focus the search for substrates. Described here is a method for identifying sirtuin interacting partners and for interpreting this information. Employing basic techniques of molecular cloning and immunochemistry [2, 3], the method describes the generation of mammalian sirtuin expression plasmids and their use to overexpress and immunoprecipitate sirtuins with their interacting partners. The interacting proteins are identified by liquid chromatography-tandem mass spectrometry (LC-MS/MS) to create a list of potential sirtuin targets. The list is then analyzed with bioinformatics approaches such as DAVID to map the identified interactors to an associated biological annotation such as a pathway or disease [4, 5]. Presented is a list of SIRT3 interactors identified from HEK293T cells using this method and analyzed with DAVID. We anticipate that by varying the sirtuin, the cell type used for overexpression and the cell culture conditions, unique data sets can be obtained and novel sirtuin substrates discovered. In this way, the method can be used to expand the scope of sirtuin substrate identification and provide a deeper understanding into sirtuin biology.

2 Materials

2.1 Generating Sirtuin Mammalian Expression Plasmids

1. Sirtuin cDNA (Addgene, Cambridge, MA, USA or Open Biosystems, Huntsville, AL, USA).
2. Custom-made PCR primers.
3. Taq DNA Polymerase, such as Platinum High Fidelity.
4. pcDNA3.1 plasmid (Invitrogen Life Technologies).
5. Plasmid Maxi Kit.

2.2 Transfection

1. HEK293T cells.
2. Cell culture medium: Dulbecco's Modified Eagle's Medium (DMEM) with 4,500 mg/L glucose, L-glutamine, and sodium bicarbonate, containing 10 % fetal bovine serum.
3. Lipofectamine LTX Plus.
4. Cell culture PBS without calcium, without magnesium.
5. NP-40 Lysis buffer: 25 mM Tris–HCl, pH 7.5, 50 mM NaCl, 1 % NP-40, 2 mM EGTA, cocktail of protease inhibitors.

6. Bicinchoninic acid (BCA) Kit for determining protein concentration.
7. Monoclonal ANTI-FLAG M2 antibody.
8. Tris-buffered saline (TBS): 150 mM NaCl, 3 mM KCl, 25 mM Tris base, pH adjusted to 8.0 with HCl.

2.3 Immunoprecipitation

1. Anti-FLAG M2 Affinity Gel.
2. TBS: 150 mM NaCl, 3 mM KCl, 25 mM Tris base, pH adjusted to 8.0 with HCl.
3. 3× FLAG Peptide.

3 Methods

3.1 Prepare Sirtuin Mammalian Expression Plasmids

1. Obtain sirtuin cDNA that encodes the entire sirtuin coding region (*see* **Note 1**).
2. Design primers for introducing an in-frame FLAG affinity tag (peptide sequence DYKDDDKD) at the sirtuin N- or C-terminus by PCR (*see* **Note 2**). Design these primers so they also introduce appropriate restriction enzyme sites at each end of the amplified PCR product for subsequent subcloning into the mammalian expression plasmid. This process uses primers with additional 5′ restriction enzyme or FLAG sequence not complementary to the target sirtuin. The extensions are added to the new product sequence during PCR. This process is illustrated in Fig. 1 with primers used to introduce a FLAG tag at the C-terminus of human *SIRT3*.
3. Perform PCR using the sirtuin cDNA (Subheading 3.1, **step 1**) as template, the mutagenic primers (Subheading 3.1, **step 2**), and Platinum Taq DNA Polymerase High Fidelity following the cycling recommendations of the manufacturer (*see* **Note 3**).
4. Check the PCR fragment size on an agarose gel and sequence the entire PCR fragment to confirm that correct changes have been introduced and that unintended mutations are not present.
5. Follow standard molecular biology techniques for cloning the restriction enzyme-digested PCR product into the pcDNA3.1 mammalian expression vector (*see* **Note 4**).
6. Purify the sirtuin-pcDNA3.1 plasmid on a large scale with Qiagen Plasmid Maxi Kit following the manufacturer's instructions. In parallel purify empty pcDNA3.1 plasmid to use as a control. The resulting DNA is of suitable quality for transfection into cell lines (*see* **Note 5**).

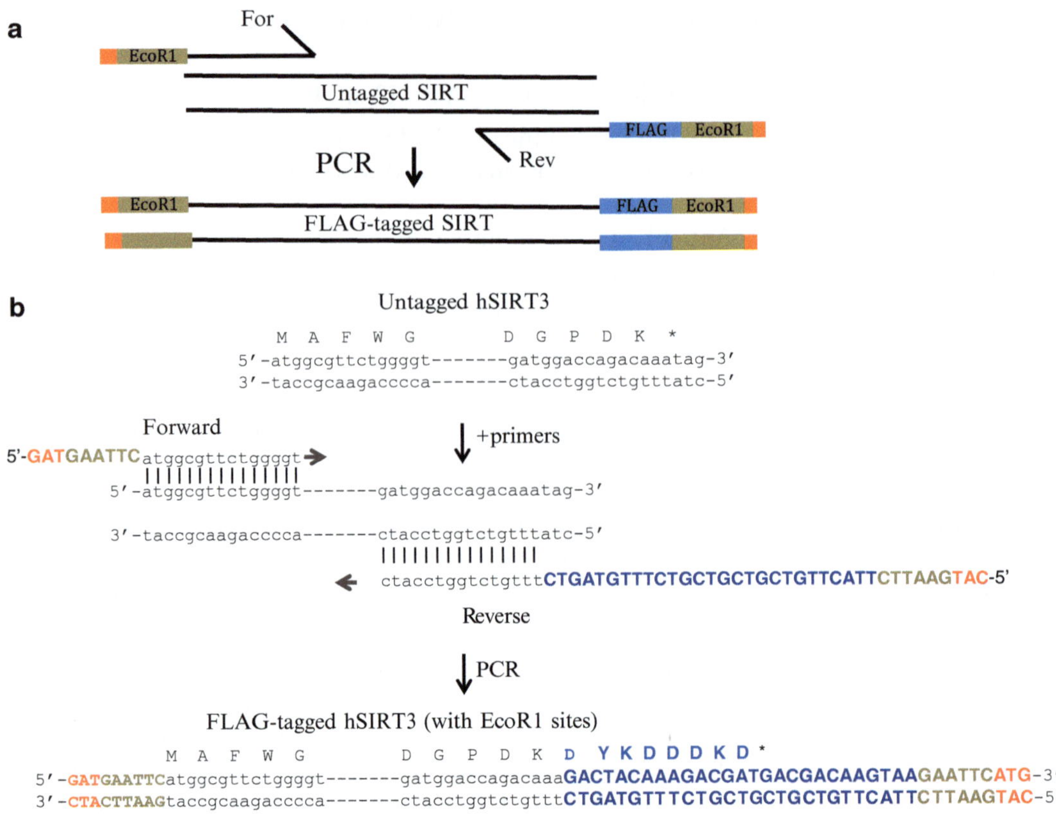

Fig. 1 Addition of restriction enzyme and FLAG sequences to sirtuin by PCR. (**a**). General schematic of PCR reaction. The template is untagged sirtuin cDNA. Custom primers are designed that are complementary to the target sirtuin with additional 5′ restriction enzyme or FLAG sequence extensions. The extensions are added to the new product sequence during PCR. FLAG-encoding, restriction enzyme (RE), and restriction enzyme flanking sequences (*see* **Note 14**) are color coded *blue, yellow*, and *red*, respectively. (**b**). Primer sequence for addition of EcoR1 and FLAG sequences to hSIRT3. The relevant sequence of untagged hSIRT3 template cDNA is shown (sequence encoding the first five amino acids of the protein, "MAFWG," and the C-terminal 5 amino acids, "DGPDK," plus the STOP codon). The forward primer contains an EcoR1 restriction site (gaattc, shown in *yellow*) and a region complementary to the antisense strand of SIRT3. The reverse primer has a region complementary to the sense strand of SIRT3 and encodes the FLAG peptide (DYKDDDKD, shown in *blue*) followed by a STOP codon. After the STOP codon is an EcoR1 site, shown in *yellow*. In *red* are restriction enzyme flanking nucleotides (*see* **Note 14**)

3.2 Transfect Mammalian Cells

1. Culture HEK293T cells in cell culture medium (*see* **Note 6**). In preparation for transfection grow the cells to near-confluence.
2. Transfect HEK293T cells with the purified sirtuin-pcDNA3.1 plasmid using Lipofectamin LTX Plus reagent following the manufacturer's instructions. Optimize the concentrations of Lipofectamin reagent and DNA as described by the manufacturer since the most effective concentrations will vary depending on the HEK293T cell passage number and history. Perform the optimization step on a small-scale in 6-well dishes plated with near-confluent HEK293T cells.
3. Control cells receive Lipofectamin LTX Plus reagent and empty pcDNA3.1 plasmid.

4. Culture cells at 37 °C for 24 h.
5. All remaining steps are completed with the sirtuin-pcDNA3.1 and empty pcDNA3.1 plasmid samples performed in parallel.
6. Prepare HEK293T cell extract and verify expression of FLAG-sirtuin protein using the following steps below.
7. Remove culture media from 6-well plate.
8. Wash cells 2× with ice-cold cell culture PBS.
9. Mechanically scrape attached cells into 200 μL of ice-cold NP-40 lysis buffer (*see* **Note** 7).
10. Transfer to a chilled 1.5 mL Eppendorf tube and incubate 30 min on ice.
11. Vortex, then homogenize with glass Dounce homogenizer (five slow passes up and down).
12. Centrifuge 14,000 × *g*, 10 min at 4 °C.
13. Discard pellet.
14. Quantify the protein concentration in the supernatant (cell extract) with the BCA assay.
15. Run 50 μg protein on a 10 % SDS polyacrylamide gel.
16. Before dye front runs off the gel stop electrophoresis and transfer proteins to membrane for immunoblotting with Sigma M2 FLAG antibody at 1:1,000.

3.3 Immunoprecipitate FLAG-Sirtuin with Interactors

1. Using transfection conditions from Subheading 3.2 that yield a FLAG-sirtuin protein signal by immunoblotting, scale-up the transfection reaction.
2. Calculate the amount of LF reagent and DNA to scale-up from the 6-well to a p150 dish of cells (a 15-fold increase).
3. Transfect four p150 dishes with the scaled-up amounts of LF reagent and FLAG-sirtuin or empty plasmid DNA.
4. Incubate 24 h at 37 °C.
5. Prepare cell extracts by removing culture medium from each p150 dish of cells.
6. Wash each dish 2× with 10 mL ice-cold PBS.
7. Mechanically scrape cells from each dish into 2 mL ice-cold NP-40 lysis buffer.
8. Pool scraped cells from the 4 dishes into one 15 mL conical tube and incubate on ice 30 min.
9. Vortex and then homogenize with glass Dounce homogenizer (five slow passes up and down).
10. Centrifuge 7,000 × *g* for 20 min at 4 °C.
11. Discard pellet.
12. Retain supernatant (cell extract) for immunoprecipitation.

13. Prepare FLAG resin for immunoprecipitation. For each set of 4 p150 dishes worth of cell extract, remove 160 μL of FLAG resin from the 50 % glycerol stock bottle.
14. Pellet the resin by centrifugation for 2 min at 2,000 × *g*, discard supernatant.
15. Resuspend resin in 1 mL TBS.
16. Centrifuge for 2 min at 2,000 × *g* and discard supernatant.
17. Repeat wash step one additional time (*see* **Note 8**).
18. Set up immunoprecipitation. Add washed, TBS-equilibrated resin to pooled cell extract in 15 mL conical tube and rotate end-to-end at 4 °C overnight.
19. Wash FLAG resin with NP-40 lysis buffer (*see* **Note 9**). After overnight incubation, pellet resin by centrifugation at 4 °C for 2 min at 2,000 × *g*.
20. Discard supernatant and wash resin by rotating tube in 10 mL NP-40 lysis buffer for 10 min.
21. Pellet resin as before and discard supernatant.
22. Repeat washing step with lysis buffer 2× more.
23. Final FLAG resin wash. To the FLAG resin pellet from above, add 10 mL ice-cold TBS to 15 mL conical tube.
24. Rotate end-to-end at 4 °C for 10 min.
25. Remove 1 mL for elution step (*see* **Note 10**).
26. Pellet resin from remaining 9 mL and discard supernatant.
27. Add TBS to barely submerge resin pellet.
28. Store at −20 °C until digestion with trypsin.
29. Elute FLAG-sirtuin protein from FLAG resin to confirm success of the FLAG-sirtuin immunoprecipitation.

    Pellet the 1 mL FLAG resin reserved from above and discard supernatant.

    Resuspend in 100 μL 3× FLAG peptide at 300 μg/mL.

    Incubate at room temperature for 15 min, occasionally flicking tube to mix.

    Pellet resin at 2,000 × *g* for 2 min.

    Remove supernatant with 1 mL syringe fitted with 31 gauge needle to avoid picking up resin (*see* **Note 11**).

    Repeat elution step 1× and pool.

    Run 20 μL on 10 % SDS polyacrylamide gel.

    Before dye front runs off transfer protein to membrane and immunoblot with FLAG M2 antibody diluted 1:1,000; Fig. 2 shows eluted hSIRT3-FLAG protein that had been overexpressed and immunoprecipitated from HEK293T cells.

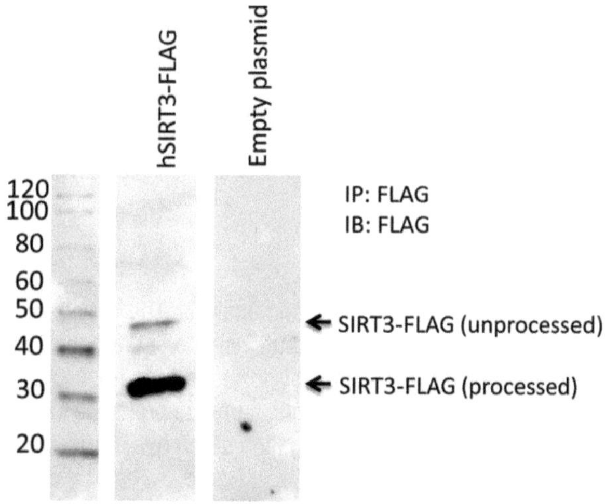

Fig. 2 Immunoprecipitation of hSIRT3-FLAG from HEK293T cells. Cells were transfected with hSIRT3-FLAG or with empty plasmid. Cell extracts were made, incubated with FLAG resin, and the FLAG proteins eluted with FLAG peptide for analysis by western blotting with FLAG antibody. The major band observed matches the predicted molecular weight (32 kDa) of the mitochondrial processed hSIRT3. A lesser amount of the cytosolic unprocessed form (MW 45 kDa) is also observed

3.4 Identify Immunoprecipitated Sirtuin Interactors by Mass Spectrometry

1. Transfer the FLAG resin from above, containing the immunoprecipitated FLAG-sirtuin or empty vector, to a mass spectrometry lab for trypsin digestion and identification of sirtuin interacting proteins by mass spectrometry analysis.

3.5 Filter Sirtuin Interactor Data in Preparation for Bioinformatics Analysis

1. Remove background proteins from list of interactors. Protein hits that are not significant as determined by mass spectrometry analysis in Subheading 3.4 and proteins used as controls for mass spectrometry should not be used for subsequent bioinformatics analysis (*see* Fig. 3).
2. Filter out non-mitochondrial proteins (*see* **Note 12**). Utilizing the Human MitoCarta dataset published by the Broad Institute [6], remove proteins that are not localized to the mitochondria.

3.6 Bioinformatics Analysis

1. Go to the DAVID website (http://david.abcc.ncifcrf.gov) for pathway analysis (*see* **Note 13**).
2. Upload the gene names of the proteins identified in Subheading 3.5 to DAVID and select "Official Gene Symbol" as an identifier. Select *Homo sapiens* as the species for both the gene list and background.
3. Examine the gene ontology terms (cellular component, biological process, and molecular function) to verify that the gene list is specific for protein products found in the mitochondria.

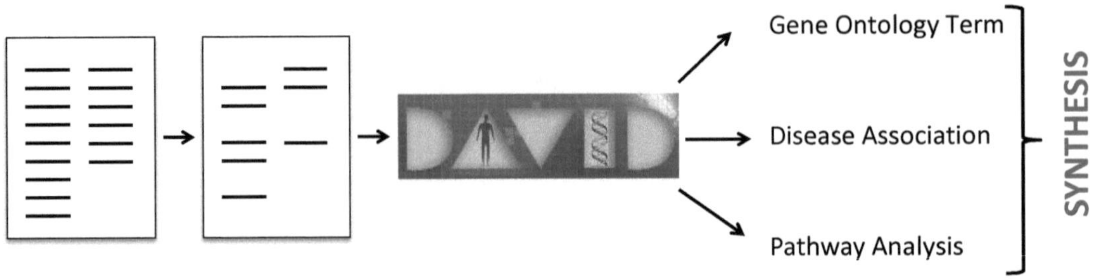

Fig. 3 Flowchart description of bioinformatic analysis. The list of sirtuin interactors is filtered to remove background, controls, and non-mitochondrial proteins. The list of genes associated with these proteins is uploaded into DAVID and analyzed by gene ontology terms, disease association, and pathway identification. Based on information gathered from multiple analyzes, a unique protein is identified as being regulated by a sirtuin and a hypothesis is generated for further investigation of the role of reversible deacetylation of the identified substrate

4. Identify diseases associated with the genes listed by accessing the OMIM (Online Mendelian Inheritance in Man) database.
5. Identify pathways that include genes from the uploaded interactor data. Kegg pathway analysis provides a list of pathways that contain a number of genes from your list that are statistically significant.
6. Synthesize a list of proteins of interest identified by DAVID analyzes and follow up with studies to verify interaction with the sirtuin and investigate the regulatory role of reversible deacetylation of the substrate.

4 Notes

1. The more commonly studied human and mouse sirtuin cDNAs can be obtained through plasmid repository services such as Addgene and Open Biosystems.
2. The FLAG tag should be introduced at the C-terminus of the mitochondrial sirtuins, SIRT3-5. Like many mitochondrial proteins, SIRT3-5 are synthesized as larger polypeptides carrying N-terminal mitochondrial localization pre-sequences that are cleaved by peptidases once the precursor has reached its final destination within the mitochondria. An affinity tag positioned at the N-terminus will thus be removed at this cleavage step. The affinity tag should therefore be placed at the C-terminus. Alternatively, affinity tags can be introduced at the N-terminus of truncated versions of the mitochondrial sirtuins that lack the N-terminal pre-sequence. Nuclear and cytosolic sirtuins may have affinity tags introduced at either the N- or C-terminus.

3. A high fidelity polymerase with proofreading activity, such as Platinum Taq DNA Polymerase High Fidelity should be used to avoid introduction of unwanted mutations within the sirtuin coding sequence during PCR.
4. The vector pcDNA3.1 works well for high-level sirtuin expression in mammalian cells. However, other commercially available mammalian expression vectors work well also.
5. DNA plasmid prepared by some methods is toxic to mammalian cells and does not transfect well. Methods using phenol at any step should be avoided.
6. HEK293T are human cells and recommended because they grow quickly, are easy to care for, and transfect well. They can be purchased from ATCC who provides detailed instructions for maintaining the cells in culture. However, other cell lines, challenged in various ways, may be desirable depending on the specific research goals and can be substituted in this step.
7. The use of strong detergents such as SDS and sodium deoxycholate denature the FLAG antibody during the immunoprecipitation step and should be avoided. A full list of chemicals incompatible with FLAG resin is available from the manufacturer.
8. Avoid centrifugation of FLAG resin above 2,000 × *g*. Higher forcc may damage the resin and FLAG antibody.
9. The detergent and salt present in the NP-40 lysis buffer make it a good buffer for washing FLAG resin.
10. FLAG-sirtuin protein is eluted from a small aliquot of resin for analysis by immunoblotting. This is performed to verify that immunoprecipitation has worked.
11. It is necessary to avoid contamination of eluted protein with FLAG resin. The FLAG antibody attached to the resin is released during preparation of sample for immunoblotting and produces high background signal through interaction with the anti-mouse secondary antibody.
12. Other criteria can be used to filter the interactor data for subsequent bioinformatic analysis according to your specific research goals.
13. DAVID (database for annotation, visualization, and integrated discovery) is a free and publically available database developed at the National Institute of Allergy and Infectious Diseases, NIH. The methods included here are particularly relevant to our analysis. Please refer to references by Huang et al. for an introduction to DAVID analysis and a step-by-step procedure for using DAVID [4, 5].
14. Many restriction enzymes require at least three nucleotides flanking either end of their recognition sequence in order to cut efficiently.

References

1. Anderson KA, Hirschey MD (2012) Mitochondrial protein acetylation regulates metabolism. Essays Biochem 52:23–35. doi:10.1042/bse0520023
2. Sambrook J, Frirsch EF, Maniatis T (1989) Molecular cloning, a laboratory manual. Cold Spring Harbor Laboratory Press, Cold Spring Harbor, NY
3. Harlow E, Lane D (1988) Antibodies, a laboratory manual. Cold Spring Harbor Laboratory, Cold Spring Harbor, NY
4. da Huang W, Sherman BT, Lempicki RA (2009) Systematic and integrative analysis of large gene lists using DAVID bioinformatics resources. Nat Protoc 4(1):44–57. doi:10.1038/nprot.2008.211
5. da Huang W, Sherman BT, Lempicki RA (2009) Bioinformatics enrichment tools: paths toward the comprehensive functional analysis of large gene lists. Nucleic Acids Res 37(1):1–13. doi:10.1093/nar/gkn923
6. Pagliarini DJ, Calvo SE, Chang B, Sheth SA, Vafai SB, Ong SE, Walford GA, Sugiana C, Boneh A, Chen WK, Hill DE, Vidal M, Evans JG, Thorburn DR, Carr SA, Mootha VK (2008) A mitochondrial protein compendium elucidates complex I disease biology. Cell 134(1):112–123. doi:10.1016/j.cell.2008.06.016

Part II

Methods to Identify Sirtuin Substrates

Chapter 6

Mass Spectrometry-Based Detection of Protein Acetylation

Yu Li, Jeffrey C. Silva, Mary E. Skinner, and David B. Lombard

Abstract

Improved sample preparation techniques and increasingly sensitive mass spectrometry (MS) analysis have revolutionized the study of protein post-translational modifications (PTMs) (Rush et al., Nat Biotechnol 23:94–101, 2005). Here, we describe a general approach for immunopurification and MS-based identification of acetylated proteins in biological samples. This approach is useful to characterize changes in the acetylome in response to biological interventions (Schwer et al., Aging Cell 8:604–606, 2009).

Key words Post-translational modification, Sirtuin, Immunopurification, Immunoprecipitation, Immunoblotting

1 Introduction

Reversible acetylation on the ε-amino group of lysines (acetylation; acK) has emerged as a PTM with a crucial role in regulating the biology of its target proteins, analogous to phosphorylation [2]. This modification is the target of many sirtuins, NAD^+-dependent deacetylases/deacylases/ADP-ribosyltransferases that regulate key physiologic processes involved in healthspan in mammals [3]. There is great current interest in whole-proteome, MS-based approaches to identifying changes in acetylation occurring in vivo in response to modulation of sirtuin activity or that of other deacetylases, changes in diet, and other biological interventions. Many groups have published such characterizations (e.g., [1, 4–7]). Here we described a general method for generation and immunoaffinity purification (IAP) of acetylated peptides from complex biological samples for MS, analysis of the MS data, and validation of the results via immunoprecipitation (IP)/immunoblotting (IB). Although we describe this method specifically in the context of protein acetylation, the approach is generally useful for identification of a variety of different PTMs [11]. Included among these methods are a series of

*Correspondence: davidlom@umich.edu

Matthew D. Hirschey (ed.), *Sirtuins: Methods and Protocols*, Methods in Molecular Biology, vol. 1077, DOI 10.1007/978-1-62703-637-5_6, © Springer Science+Business Media, LLC 2013

immunoaffinity techniques using motif antibodies to enrich for a specific subset of post-translational modifications (PTMs) (acetylation or others such as serine-, threonine-, and tyrosine-phosphorylation, methylation, and ubiquitination) that can be utilized to probe specific areas of signal transduction or protein modification (http://www.cellsignal.com/services/discovery_overview.html). In addition, other immunoaffinity applications have been developed using site-specific antibodies to target particular signaling pathways (http://www.cellsignal.com/services/direct_overview.html). If the reader wishes simply to use IP/IB to confirm hits from previous MS or other studies, she/he can skip to Subheading 3.12.

2 Materials

2.1 Stock Solutions for Peptide Preparation and IAP

Prepare solutions for cell lysis, Sep-Pak purification, IAP enrichment, and IP/IB with Milli-Q or equivalent grade water. Prepare MS solutions with HPLC grade water (e.g., Burdick and Jackson Water, Honeywell, AH365-4).

1. 200 mM HEPES/NaOH, pH 8.0: Dissolve 23.8 g HEPES in approximately 450 mL water, adjust pH with 5 M NaOH to 8.0, and bring to a final volume of 500 mL. Filter through a 0.22 μM filter, use for up to 6 months.
2. Sodium pyrophosphate: to make 50× stock, add 1.1 g/20 mL. Store at 4 °C, use for up to 1 month.
3. β-glycerophosphate: to make 1,000× stock, add 2.2 g/10 mL. Divide into 0.10 mL aliquots and store at −20 °C.
4. Sodium orthovanadate (Na_3VO_4): to make 100× stock, add 1.84 g/100 mL. Sodium orthovanadate must be depolymerized (activated) according to the following protocol. For a 100 mL solution, fill up with water to approximately 90 mL. Adjust the pH to 10.0 using 1 M NaOH with stirring. At this pH, the solution will be yellow. Boil the solution until it turns colorless and cool to room temperature (put on ice for cooling). Readjust the pH to 10.0 and repeat boiling until the solution remains colorless and the pH stabilizes at 10.0 (usually it takes two rounds). Adjust the final volume with water. Store the activated sodium orthovanadate in 1 mL aliquots at −20 °C. Thaw one aliquot for each experiment; do not refreeze thawed vial.
5. Trypsin-TPCK: Store dry powder for up to 2 years at −80 °C. Parafilm cap of trypsin container to avoid collecting moisture, which can lead to degradation of the reagent. Prepare 1 mg/mL stock in 1 mM HCl. Divide into 1 mL aliquots, store at −80 °C for up to 1 year.
6. Urea Lysis Buffer: 20 mM HEPES pH 8.0, 9 M urea, 1 mM sodium orthovanadate, 2.5 mM sodium pyrophosphate, 1 mM

β-glycerophosphate. The Urea Lysis Buffer should be prepared prior to each experiment (*see* **Note 1**). Aliquots of the Urea Lysis Buffer can be stored at −80 °C for up to 6 months.

7. DTT solution: to make 1.25 M stock, add 19.25 g/100 mL. Divide into 0.20 mL aliquots, store at −20 °C for up to 1 year. Thaw a fresh aliquot for each experiment.
8. Iodoacetamide solution: Dissolve 95 mg of iodoacetamide in water to a final volume of 5 mL. After weighing the powder, store in the dark and add water only immediately before use. The iodoacetamide solution should be prepared fresh prior to each experiment.
9. IAP buffer: 50 mM MOPS pH 7.2, 10 mM sodium phosphate, 50 mM NaCl.
10. 20 % trifluoroacetic acid (TFA): add 10 mL TFA to water to a total volume of 50 mL (CAUTION: TFA is a strong acid and is very toxic. Contact, inhalation, or ingestion may cause severe injury).
11. Solvent A (0.1 % TFA): add 5 mL of 20 % TFA to 995 mL water.
12. Solvent B (0.1 % TFA, 40 % acetonitrile): add 400 mL of acetonitrile (MeCN) and 5 mL of 20 % TFA to 500 mL of water; adjust final volume to 1 L.
13. Solvent C (0.1 % TFA, 40 % acetonitrile): add 0.1 mL TFA to 30 mL water, then add 60 mL acetonitrile; adjust the final volume to 100 mL by water.
14. Solvent D (0.1 % TFA, 50 % acetonitrile): add 0.1 mL TFA to 40 mL water, then add 50 mL acetonitrile; adjust the final volume to 100 mL by water.
15. Solvent E (0.1 % TFA): add 0.1 mL TFA to 50 mL water; adjust the final volume to 100 mL.

2.2 Peptide Preparation and IAP

1. TFA, Sequanal Grade.
2. Acetonitrile.
3. Sep-Pak® Classic C_{18} columns, 0.7 mL.
4. AcK beads. To prepare acK beads, wash protein A agarose in a microfuge tube three times with 1 mL of 1× PBS, add acK antibody at 4 mg/mL beads, and incubate 4 h at room temperature or overnight at 4 °C. After binding, wash the beads three times with 1 mL of 1× PBS and then two times with IAP buffer. Resuspend the beads as 1:1 slurry in IAP buffer and store at 4 °C for up to 1 week.

2.3 Liquid Chromatography (LC) Data Acquisition

1. Standard protein mixture (Waters Corporation, MassPrep Digestion Standard-Mix1, catalog# 186002865).

2.4 Stock Solutions for IP/IB Validation of MS Hits

1. acKIP buffer: 1 % Triton X-100, 0.5 % NP-40, 150 mM NaCl, 50 mM Tris (pH 7.4), 10 % glycerol, with complete protease inhibitor cocktail (Roche), 1 μM trichostatin A, and 10 mM nicotinamide [8].
2. Laemmli buffer (6×): 0.175 M Tris-base, pH 6.8, 30 % glycerol, 10 % (w/v) SDS, 9.3 % (w/v) DTT, and ~10 mg of bromophenol blue [9]. Vortex and warm to dissolve SDS. Aliquots should be frozen at −20 °C.
3. Running Buffer (10×): 1 % (w/v) SDS, 1.92 M glycine, 4 M Tris-base.
4. Transfer Buffer (10×): 1.92 M glycine, 4 M Tris-base. To make 1 L of 1× transfer buffer, add 700 mL of water to 100 mL of 10× transfer buffer, and then add 200 mL of methanol. Store 1× transfer buffer at 4 °C.
5. Tris buffered saline containing 0.1 % Tween-20 (TBST; 10×): dissolve 0.2 M Tris-base, 1.4 M NaCl, pH to 7.6, add 0.1 % (v/v) Tween-20.
6. Ponceau S Staining Solution: 0.1 % (w/v) Ponceau S, 5 % (v/v) acetic acid.

3 Methods

3.1 Preparation of Protein Lysate, Suspension Cells

1. Grow approximately $1–2 \times 10^8$ cells for each experimental condition (enough cells to produce approximately 10–20 mg of soluble protein). Alternatively, tissues may be used [1] (*see* **Note 2**).
2. Harvest cells by centrifugation at $130 \times g$ for 5 min at room temperature and then carefully remove supernatant. Wash cells with 20 mL of cold PBS, centrifuge and remove PBS wash. Add 10 mL Urea Lysis Buffer (at room temperature) to the cell pellet. Pipet the slurry up and down a few times (Do not cool lysate on ice as this may cause precipitation of the urea). At this point the harvested cells can be frozen and stored at −80 °C for several weeks.
3. Using a microtip, sonicate at 15 W output with 3 bursts of 15 s each. Cool on ice for 1 min between the bursts. Clear the lysate by centrifugation at $20{,}000 \times g$ for 15 min at 15 °C or room temperature and transfer the protein extract (supernatant) into a new tube (*see* **Note 3**).

3.2 Preparation of Cell Lysate, Adherent Cells

1. Grow $1–2 \times 10^8$ cells for each experimental condition (enough cells to produce approximately 10–20 mg of soluble protein). The cell number corresponds to approximately three to ten 150 mm culture dishes (depending on the cell type), grown to between 70 and 80 % confluence.

2. Take all 150 mm culture dishes for one sample, remove the medium from the first dish by decanting, and let stand in sharply tilted position for 30 s so that the rest of the medium flows to the bottom edge. Remove the remainder of the medium at the bottom edge with a P-1000 micropipettor. Wash each dish with 5 mL of cold PBS. Remove PBS as described above.
3. Add 10 mL of Urea Lysis Buffer (at room temperature) to the first dish, scrape the cells off in the buffer, let the dish stand in tilted position after scraping the buffer to the bottom edge of the tilted dish. Remove the medium from the second dish as above. Transfer the lysis buffer from the first dish to the second dish using a 10 mL pipette, then tilt the first dish with the lid on for 30 s and remove remaining buffer from the dish and collect. Scrape cells from the second dish and repeat the process until the cells from all the dishes have been scraped into the lysis buffer. Collect all lysate in a 50 mL conical tube (*see* **Note 4**). The yield will be approximately 9–12 mL lysate after harvesting all the culture plates. The cell lysate can be frozen and stored at −80 °C for several weeks.
4. Using a microtip, sonicate at 15 W output with 3 bursts of 15 s each. Cool on ice for 1 min between the bursts (*see* **Note 5**). Clear the lysate by centrifugation at 20,000 × *g*, for 15 min at 15 °C or room temperature and transfer the protein extract (supernatant) into a new tube.

3.3 Reduction and Alkylation of Proteins

1. Add 1/278 volume of 1.25 M DTT to the cleared cell supernatant (e.g., 36 μL of 1.25 M DTT for 10 mL of protein extract), mix well, place the tube into a 55 °C incubator for 30 min.
2. Cool the solution on ice briefly until it has reached room temperature (tube should feel neither warm nor ice-cold by hand).
3. Add 1/10 volume of iodoacetamide solution to the cleared cell supernatant, mix well, and incubate for 15 min at room temperature in the dark.

3.4 Dilution

1. Dilute lysate threefold with 20 mM HEPES pH 8.0 to a final concentration of 2 M urea, 20 mM HEPES, pH 8.0. For example, for 10 mL of lysate add 30 mL 20 mM HEPES pH 8.0.
2. After dilution and mixing, remove a small aliquot (e.g., 30 μL) for analysis on SDS-PAGE.

3.5 Trypsin Digestion

1. Add 1/100 volume of 1 mg/mL trypsin-TPCK (Promega) stock in 1 mM HCl and digest overnight at room temperature with mixing.
2. Analyze the lysate before and after digest by SDS-PAGE to check for complete digestion.
3. Continue through the Sep-Pak, IAP, and StageTip protocols prior to LC-MS analysis of enriched peptides.

3.6 Sep-Pak® C_{18} Purification of Lysate Peptides (See Note 6)

1. Add 1/20 volume of 20 % TFA to the digest for a final concentration of 1 % TFA. Check the pH by spotting a small amount of peptide sample on a pH strip (the pH should be under 3). After acidification, allow precipitate to form by letting stand for 15 min on ice (*see* **Note 7**).
2. Centrifuge the acidified peptide solution for 15 min at 1,780 × *g* at room temperature to remove the precipitate. Transfer peptide-containing supernatant into a new 50 mL conical tube without dislodging the precipitated material.
3. Connect a 10 cc reservoir (remove 10 cc plunger) to the SHORTER END of the Sep-Pak column (*see* **Note 8**).
4. Pre-wet the column with 5 mL 100 % MeCN (*see* **Note 9**).
5. Wash with sequentially with 1, 3, and 6 mL of solvent A (0.1 % TFA).
6. Load acidified and cleared digest (*see* **Notes 10** and **11**).
7. Wash with sequentially with 1, 5, and 6 mL of solvent A (0.1 % TFA).
8. Wash with 2 mL of 5 % MeCN, 0.1 % TFA.
9. Place columns above new 15 or 50 mL polypropylene tubes to collect eluate. Elute peptides with a sequential wash of 3 × 2 mL of solvent B (0.1 % TFA, 40 % acetonitrile).
10. Freeze the eluate on dry ice (or −80 °C freezer) for 2 h to overnight and lyophilize frozen peptide solution for a minimum of 2 days to assure all the TFA has been removed from the peptide sample (*see* **Note 12**).

3.7 Immunoaffinity Purification

1. Centrifuge the tube containing lyophilized peptide in order to collect all material to be dissolved. Add 1.4 mL IAP buffer. Resuspend pellets mechanically by pipetting repeatedly with a P-1000 pipette taking care not to introduce excessive bubbles into the solution. Transfer solution to a 1.7 mL Eppendorf tube (*see* **Notes 13** and **14**).
2. Clear solution by centrifugation for 5 min at 10,000 × *g* at 4 °C in a microcentrifuge (the pellet of insoluble matter may at times seem considerable, but this will not pose a problem since most of the peptide will be dissolved nonetheless). Cool on ice.
3. Transfer the peptide solution into the microfuge tube containing acK antibody beads (use 80 μL of a 1:1 slurry of beads:IAP buffer per IP). Pipet sample directly on top of the beads at the bottom of the tube to ensure immediate mixing. Avoid creating bubbles upon pipetting.
4. Incubate for 2 h on a rotator at 4 °C.
5. Centrifuge at 2,000 × *g* for 30 s and transfer the supernatant with a P-1000 micropipettor to a labeled Eppendorf tube to

save for future use. Flow-through material can be used for subsequent IAPs (*see* **Note 15**).

6. Add 1 mL of IAP buffer to the beads, mix by inverting tube five times, centrifuge for 30 s and remove supernatant with a P-1000 micropipettor (*see* **Note 16**).
7. Repeat **step 6** once for a total of *TWO* IAP buffer washes.
8. Add 1 mL chilled water to the beads, mix by inverting tube five times, centrifuge for 30 s and remove supernatant with a P-1000 micropipettor (*see* **Note 17**).
9. Repeat **step 8** three times for a total of *THREE* water washes. During the last water washes, the tube may need to be shaken while inverting in order to ensure efficient mixing (*see* **Note 18**).
10. Add 55 μL of 0.15 % TFA to the beads, tap the bottom of the tube several times (do not vortex) and let stand at room temperature for 10 min, mixing gently every 2–3 min (*see* **Note 19**).
11. Centrifuge 30 s at 2,000 × *g* and transfer supernatant to a new 1.7 mL Eppendorf tube.
12. Add 50 μL of 0.15 % TFA to the beads, and repeat the elution/centrifugation steps. Combine both eluents in the same 1.7 mL tube. Briefly centrifuge the eluent to pellet and remaining beads and carefully transfer eluent to a new 1.7 mL tube taking care not to transfer any beads.

3.8 Concentration and Purification of Peptides for LC-MS Analysis (see Notes 20–22)

1. Equilibrate the StageTip by passing 50 μL of Solvent D through followed by 50 μL of Solvent E twice.
2. Load sample by passing IP eluent through the StageTip. Load IAP eluent in two steps using 50 μL in each step.
3. Wash the StageTip by passing 55 μL of Solvent E through twice.
4. Elute peptides off the StageTip by passing 10 μL of Solvent C twice, pooling the resulting eluent.
5. Dry down the StageTip eluate in a vacuum concentrator (Speed-Vac) and redissolve the peptides in an appropriate solvent for LC-MS analysis such as 5 % acetonitrile, 0.1 % TFA.

3.9 LC Data Acquisition

1. Resuspend vacuum dried, immunoaffinity purified acetylated peptides in 0.125 % formic acid containing 100 fmol of a standard five protein mixture (e.g., MassPrep Digestion Standard-Mix1) and separate on a reverse phase column (75 μm inner diameter × 10 cm) packed into a PicoTop emitter (~8 μm diameter tip) with Magic C_{18} AQ (100 Å × 5 μm). Elute acetylated peptides using a 72 min gradient of acetonitrile (5–40 %) in 0.125 % formic acid delivered at 280 nL/min.

3.10 MS/MS Data Acquisition

1. Tandem MS/MS is collected in a data-dependent manner with an LTQ-Orbitrap Velos (or Elite) mass spectrometer running XCalibur software (version 2.0.7 SP1) using a top 20 MS/MS method, a dynamic repeat count of one, and a repeat duration of 30 s. Real time recalibration of mass error is performed using lock mass with a singly charged polysilaxane ion (m/z= 371.101237).

3.11 MS/MS Data Analysis

1. MS/MS spectra is searched using SEQUEST 3G and the SORCERER2 platform from Sage-N Research (version 4.0). Peptide assignments are generated using the following SEQUEST parameters for ESI-CID MS/MS spectra: peptide precursor mass tolerance of 10 ppm and fragment ion tolerance of 1.0 Da. A maximum of four amino acids are allowed per modification and up to four internal missed cleavage sites are permitted. Neutral loss of water and ammonia from corresponding b and y ions are considered in the correlation analysis for the matching peptide assignment and the proteolytic enzyme is specified as trypsin. SEQUEST searches are performed against a FASTA formatted, species-specific database. Cysteine carboxamidomethylation is specified as a static modification, oxidation of methionine residues is allowed as a variable modification and acetylation is allowed on lysine residues. A reverse decoy databases is included for all searches to estimate false positive rates and filtered using a 5 % false discovery rate in the Peptide Prophet module of SORCERER2. The final peptide assignment are further narrowed by applying a mass accuracy filter of ±2–5 ppm of the calculated m/z, depending on the distribution and symmetry of the XCorr versus ppm plot for a given search result (*see* Fig. 1a, b).
2. In a typical label-free LC-MS/MS study, replicate injections of each sample condition are collected to assess the analytical reproducibility of the LC-MS acquisition of the data set prior to conducting any further analysis. In addition, an external standard (e.g., Waters Corporation, MassPrep Digestion Standard-Mix 1, catalog# 186002865) is included along with the enriched acetylated peptide samples to serve as a performance check during the data acquisition phase of the study to help verify that the LC and MS systems are performing with acceptable specifications. Significant deviations in retention time reproducibility, signal intensity, mass resolution, and mass accuracy indicate that the LC-MS system should be inspected for system errors and/or recalibration. Typical performance metrics include retention time deviations of less than 2.5 % relative standard deviation (RSD), intensity measurement variation of less than 30 % RSD and mass measurement accuracy of within ±2–5 ppm (Fig. 2a, b).

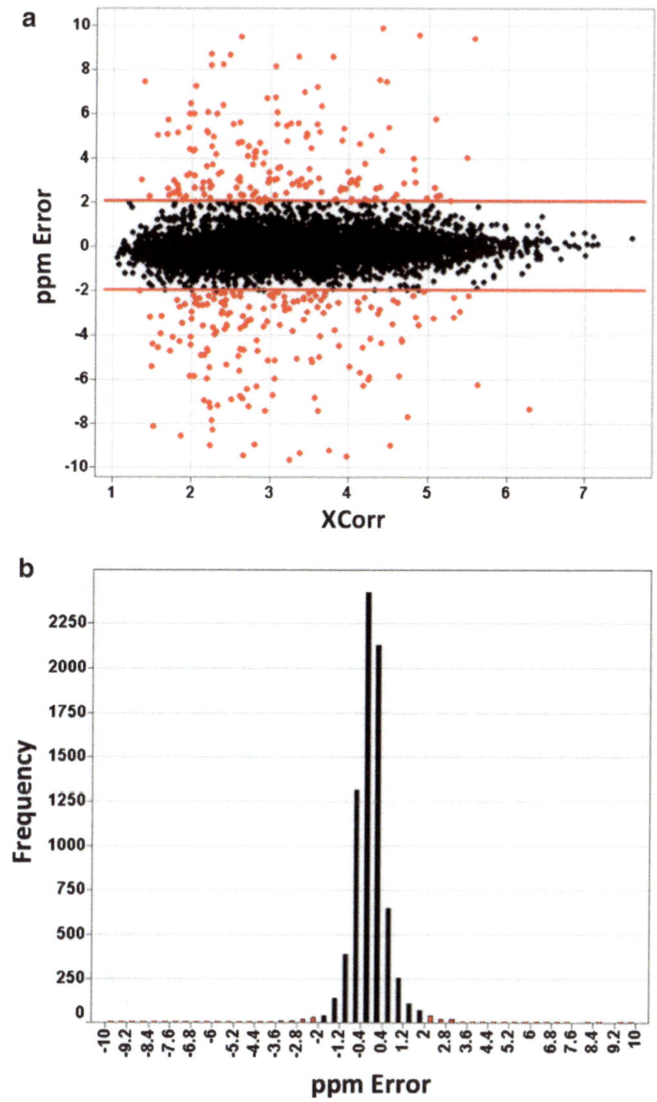

Fig. 1 Correlation of ppm Error and SEQUEST Cross Correlation Score, XCorr. (**a**) Scatter plot of ppm Error and XCorr value from SEQUEST showing 7886 peptide assignments (modified and unmodified, lysine acetylation) with 95 % of the peptide assignments within ±2 ppm error (in *black*) of the calculated monoisotopic mass. (**b**) Histogram plot showing the binned ppm Error for each peptide assignment

3. After confirming that the performance metrics for the LC-MS/MS data set are acceptable, the intensity measurements for all acetylated peptide assignments are evaluated between the analytical replicates. There are a variety of commercial and open source software platforms that can be used to extract the corresponding intensity measurements from the RAW LC-MS data files (Rosetta Elucidator System,

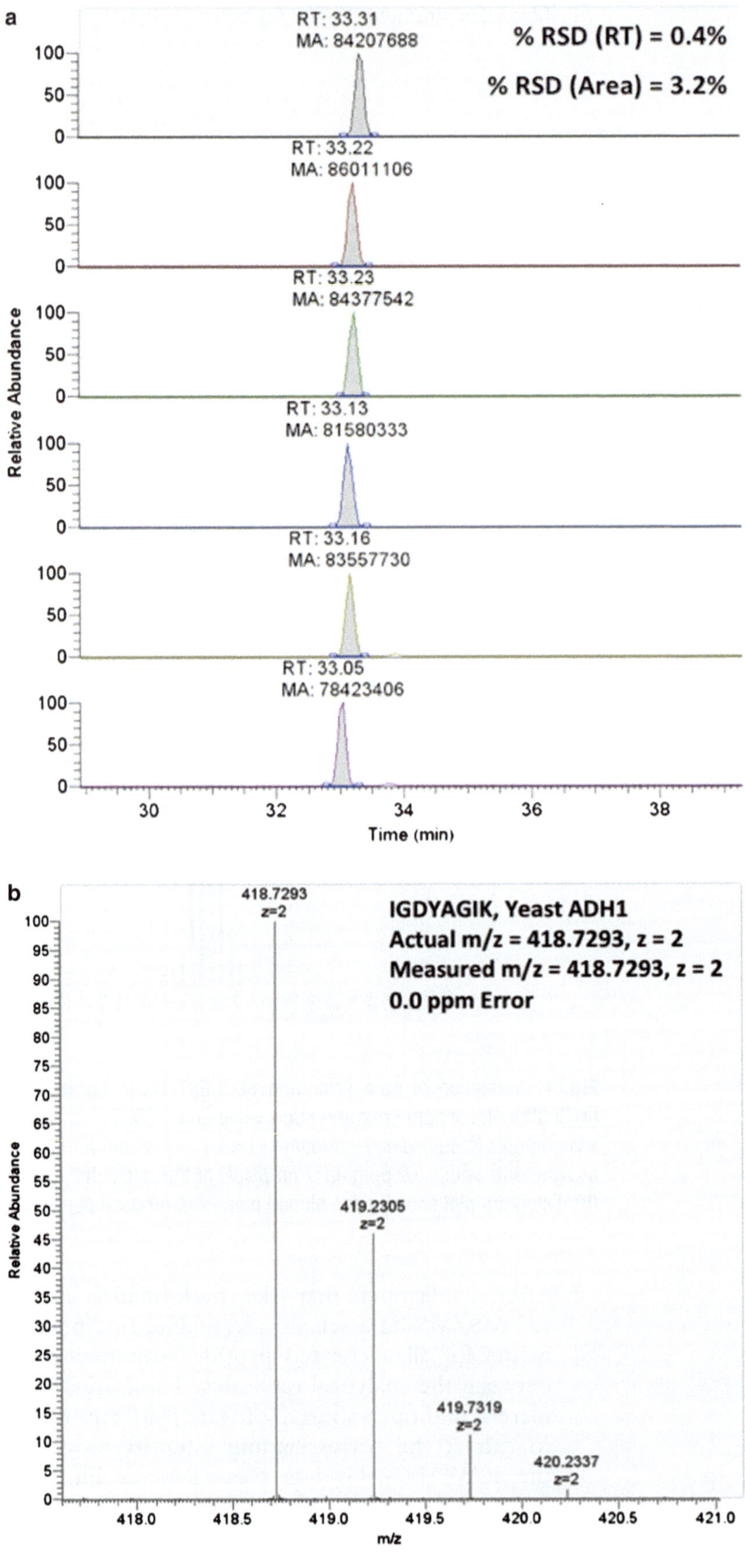

a
RT: 33.31
MA: 84207688
% RSD (RT) = 0.4%
% RSD (Area) = 3.2%
RT: 33.22
MA: 86011106
RT: 33.23
MA: 84377542
RT: 33.13
MA: 81580333
RT: 33.16
MA: 83557730
RT: 33.05
MA: 78423406
Relative Abundance
Time (min)
b
418.7293
z=2
IGDYAGIK, Yeast ADH1
Actual m/z = 418.7293, z = 2
Measured m/z = 418.7293, z = 2
0.0 ppm Error
419.2305
z=2
419.7319
z=2
420.2337
z=2
Relative Abundance
m/z

Progenesis, mzMine, ProteinLynx Global Server, GeneData Expressionist, LC-MSWarp, MaxQuant, ProteoIQ, PEAKS, Skyline, COMPASS, Scaffold, APEX, Census, and others). The main function of these software tools is to accurately extract peak height or intensity information for all identified peptides (or features) using the specificity afforded by the accurate mass measurements of the corresponding peptides along with their associated chromatographic retention times. The clustering algorithms associated with these software tools utilize the mass precision and accuracy of the mass spectrometer and the retention time reproducibility obtained from the chromatography to align identical peptides across all the samples in a particular study (Fig. 3a–d). Proper alignment is required to ensure that the correct intensity measurements are properly associated to the originating peptides (or features) in order to accurately determine the relative quantitation of acetylated peptides (proteins and their associated sites of modification) among the different conditions (Fig. 4).

4. Although the sample preparation protocol for AcetylScan incorporates a means to normalize against any sample bias among different conditions by processing identical amounts of soluble protein from each sample condition, there are additional steps one can employ to control for slight changes in protein amounts and injection variability that may have been introduced by the LC autosampler or systematic attenuation of the MS signal. In instances when it is expected that a majority of the proteins/sites of modification will not change upon the applied treatment condition, a median normalization technique can be applied between replicate injections and between control and treatment conditions. If the perturbation causes changes in many proteins/peptides, this normalization strategy may not be the best technique and will result in attenuation of the overall quantitative changes between control and treatment conditions and will underestimate the total number of proteins/sites affected in the study. Once all the peptide assignment are clustered across all replicate injections and experimental conditions, the relative ratio for each replicate injection can be compared independently to determine the correct offset correction for each sample normalization. In the example below (Fig. 5), the raw intensity ratio for all detected peptides (or features) is converted to a log2(Ratio) and the

Fig. 2 System performance metrics for LC-MS Data Acquisition. (**a**) Extracted ion chromatograms for a control peptide, IGDYAGIK (Yeast, ADH1, *m/z* 418.7293, $z=2$) among the duplicate injections of three experimental conditions. The relative standard deviation for the retention time and area measurements are 0.4 % and 3.2 %, respectively. (**b**) Mass spectrum of the yeast ADH1 control peptide, measured at 0.0 ppm mass accuracy

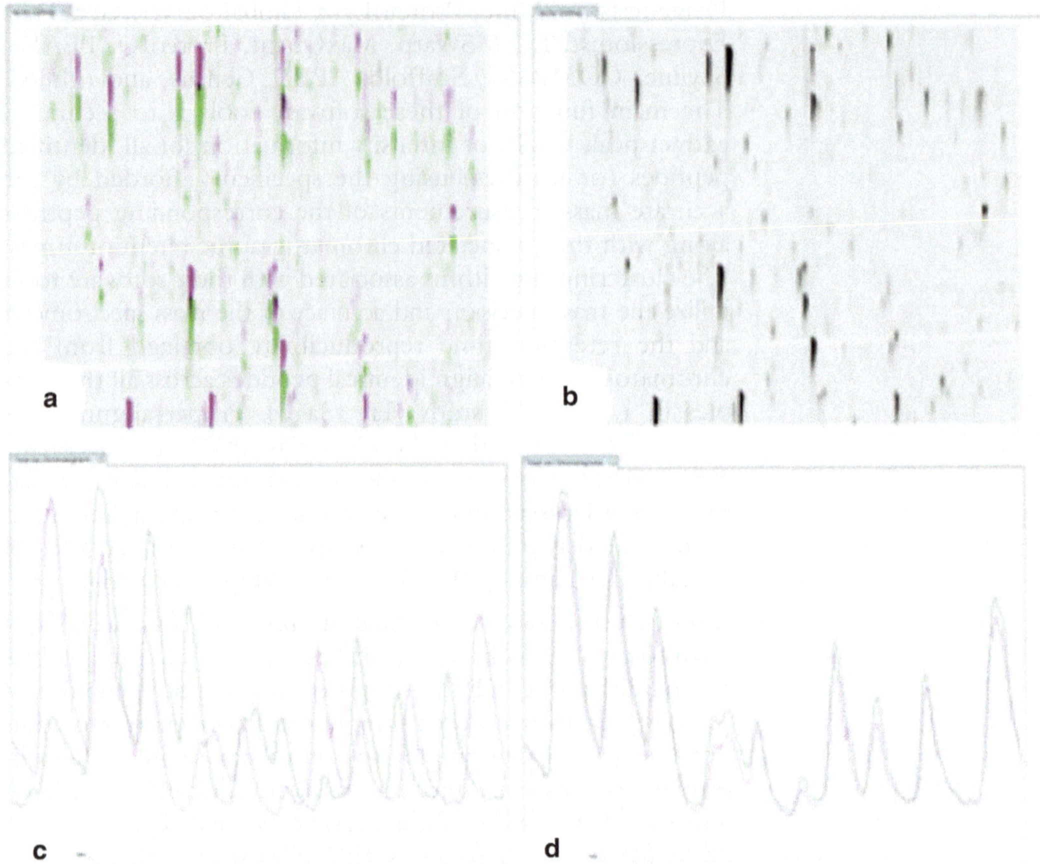

Fig. 3 Alignment of Label-Free LC-MS Data. An overlay of two LC-MS runs represented as *m/z* (*x*-axis) and retention time (*y*-axis) before (**a**) and after (**b**) alignment using Progenesis LC-MS software from Nonlinear Dynamics. An overlay of two LC-MS runs showing the total ion chromagram before (**c**) and after (**d**) alignment

corresponding median log2(Ratio) is determined for the entire data set between the two analytical replicates. An offset correction factor, equal to the antilog value of the median offset, is applied to one of the replicate injections to force the newly calculated median log2(Ratio) of the entire data set to be centered at zero. This is performed for each replicate pairs in the study. To normalize against sample bias, the average intensity value for each peptide is calculated within each sample condition using the same normalization routine between each control and treatment condition in a binary manner (Fig. 6).

5. After data normalization is complete, those modified peptide showing significant fold changes between treated and control conditions can be identified as potential biomarkers associated with the perturbation and potentially related to the mechanism of action (i.e., inhibitor treatment, overexpression of ligase or transferase, comparison of sensitive versus resistant cell lines or tissues).

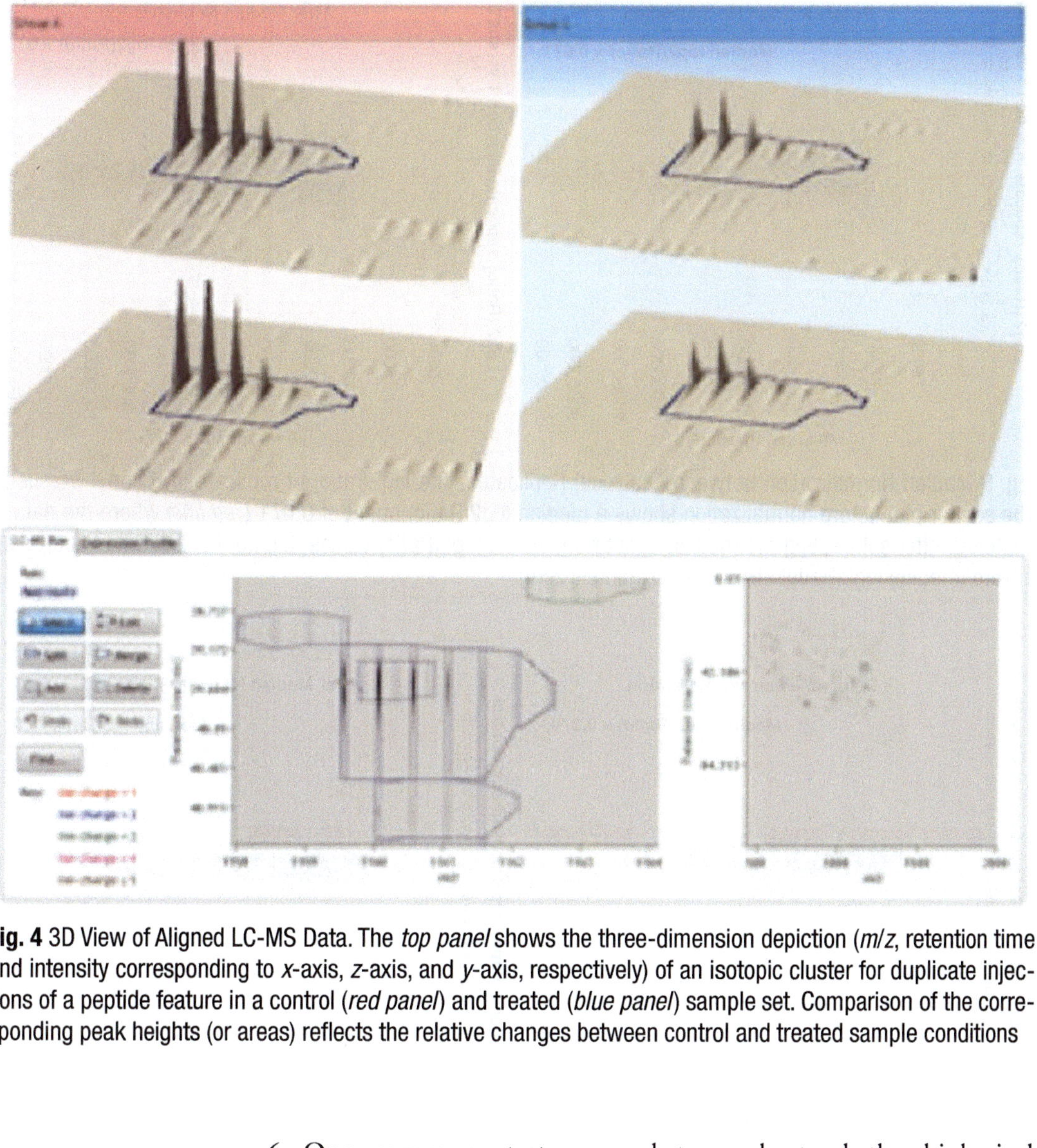

Fig. 4 3D View of Aligned LC-MS Data. The *top panel* shows the three-dimension depiction (*m/z*, retention time and intensity corresponding to *x*-axis, *z*-axis, and *y*-axis, respectively) of an isotopic cluster for duplicate injections of a peptide feature in a control (*red panel*) and treated (*blue panel*) sample set. Comparison of the corresponding peak heights (or areas) reflects the relative changes between control and treated sample conditions

6. One common strategy used to understand the biological significance of a given perturbation is to query the list proteins identified proteins from the set of enriched acetylated peptides using pathway analysis tools such as STRING, Cytoscape, Ingenuity or Reactome, to name a few examples. In Fig. 7, the proteins (or accession numbers) identified from the set of enriched acetylated peptides obtained from mouse liver samples to obtain the illustrated set of predicted protein interactions obtained from experimental and database evidence at EMBL. The same list of proteins was also submitted to Reactome's expression analysis data tool to provide a representative list of signaling pathways associated with those immunoaffinity enriched acetylated peptides (Table 1). PhosphoSitePlus is another commonly used bioinformatics resource that can

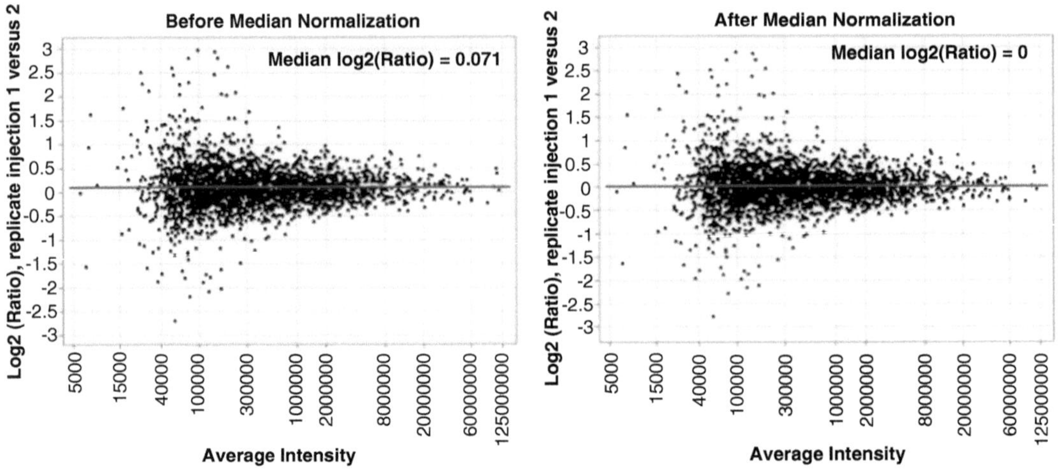

Fig. 5 Median Normalization between Analytical Replicates. The log2(ratio) of replicate injection 1 versus 2. The scatter plot before normalization shows a median log2(Ratio) offset of 0.071 (*red line*) where the data is centered. After subtracting the median offset from all the log2(Ratio) values, the normalized data is centered at zero (median log2(Ratio) after normalization is zero)

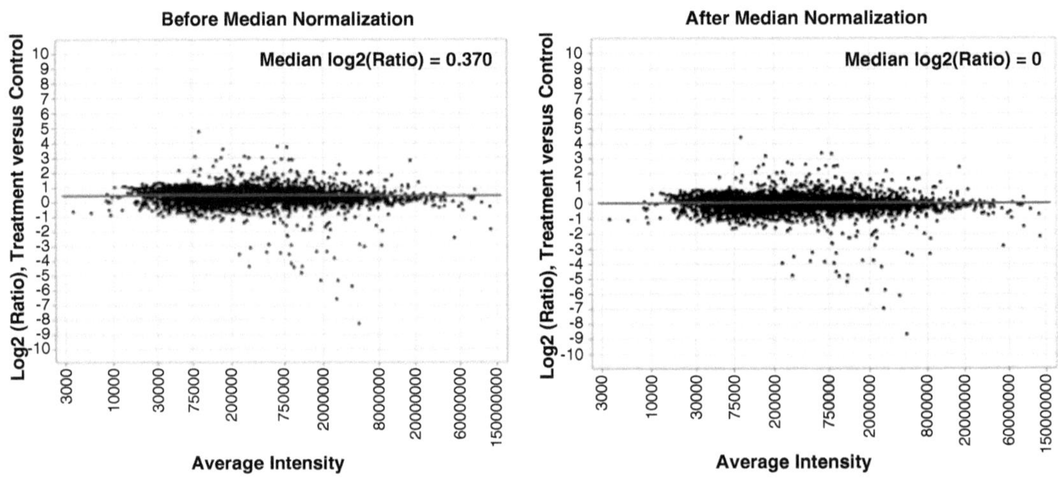

Fig. 6 Median Normalization between Sample Conditions. The log2(ratio) of treated versus control conditions. The scatter plot before normalization shows a median log2(Ratio) offset of 0.370 (*red line*) where the data is centered. After subtracting the median offset from all the log2(Ratio) values, the normalized data is centered at zero (median log2(Ratio) after normalization is zero)

provide comprehensive information related to the biological role of associated PTMs (http://www.phosphosite.org) along with examples from a variety of sample data sets related to the characterization of acetylated and other posttranslational proteins (http://www.phosphosite.org/staticDownloads.do).

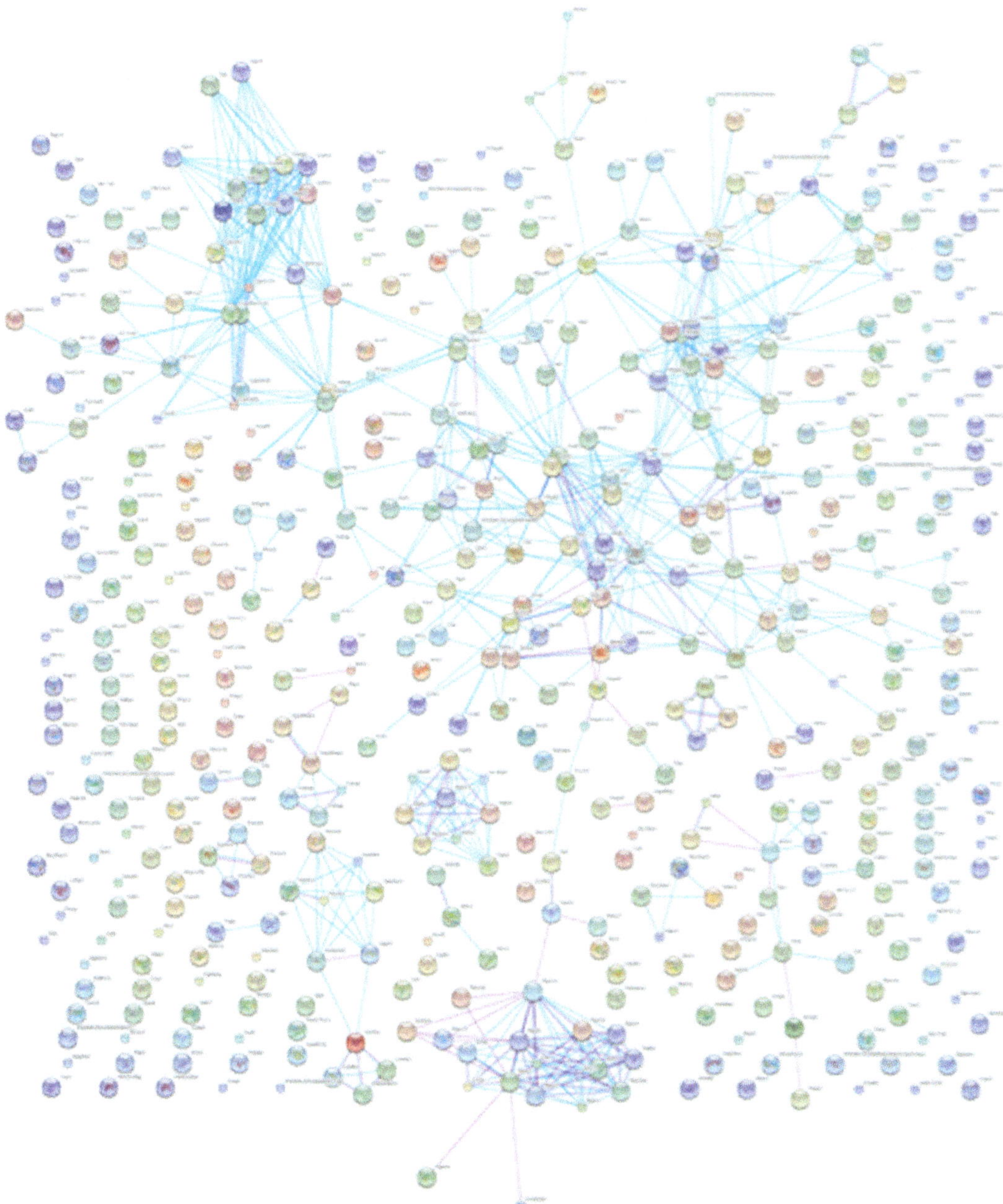

Fig. 7 STRING Confidence Plot of Acetylated Peptides to Proteins from enriched from mouse Liver. The proteins associated to acetylated lysine peptides enriched from mouse liver were submitted to STRING (version 9.0, http://string-db.org/) to illustrate known protein–protein interactions based on direct and indirect association from experimental and database evidence high confidence score of 0.700. Peptides were enriched using the acetylated lysine motif antibody from Cell Signaling Technology (http://www.cellsignal.com/services/ptmscan_antibodies.html)

Table 1
Signaling pathways associated to acetylated peptides from mouse liver

No	Pathway name	Total number of proteins	Matching proteins in data	% in data
1	Activation of chaperone genes by ATF6-alpha	5	3	60.00
2	Neurotransmitter uptake and metabolism in Glial Cells	2	1	50.00
3	Peroxisomal lipid metabolism	21	10	47.62
4	Alpha-linolenic (omega 3) and linoleic (omega 6) acid metabolism	12	5	41.67
5	Metabolism of amino acids and derivatives	206	74	35.92
6	Respiratory electron transport, ATP synthesis by chemiosmotic coupling, and heat production by uncoupling proteins	124	43	34.68
7	The citric acid (TCA) cycle and respiratory electron transport	165	57	34.55
8	O2/CO2 exchange in erythrocytes	13	4	30.77
9	Uptake of carbon dioxide and release of oxygen by erythrocytes	13	4	30.77
10	Uptake of oxygen and release of carbon dioxide by erythrocytes	13	4	30.77
11	Neurotransmitter clearance in the synaptic cleft	7	2	28.57
12	Bile acid and bile salt metabolism	44	12	27.27
13	Abnormal metabolism in phenylketonuria	4	1	25.00
14	Phase II conjugation	76	19	25.00
15	tRNA Aminoacylation	44	10	22.73
16	Eukaryotic translation elongation	137	30	21.90
17	Metabolism of porphyrins	14	3	21.43
18	Eukaryotic translation termination	134	27	20.15
19	Fatty acid, triacylglycerol, and ketone body metabolism	189	37	19.58
20	Metabolism of nitric oxide	21	4	19.05
21	SRP-dependent cotranslational protein targeting to membrane	165	31	19.79
22	Eukaryotic translation initiation	167	31	18.56
23	Metabolism of vitamins and cofactors	54	10	18.52
24	Metabolism of water-soluble vitamins and cofactors	54	10	18.52
25	Nonsense-mediated decay	162	30	18.52
26	Translation	207	38	18.36
27	Metabolism	1,671	284	17.00
28	Mitochondrial protein import	36	6	16.67
29	Biological oxidations	209	34	16.27
30	Meiotic synapsis	37	6	16.22
31	Metabolism of nucleotides	88	47	15.91
32	Cellular response to hypoxia	26	4	15.38
33	Cellular responses to stress	26	4	15.38
34	Regulation of hypoxia-inducible factor (HIF) by oxygen	26	4	15.38
35	Signaling by Hippo	20	3	15.00

The top 35 signaling pathways associated to acetylated lysine peptides obtained from immunoaffinity enrichment of mouse liver tissue. The list of signaling pathways were from Reactome's Expression Data tool, http://www.reactome.org/ReactomeGWT/entrypoint.html

3.12 Validation of MS/MS Results by IP/IB (See Note 23)

1. Resuspend cells/organelles of interest in acKIP buffer in 1.7 mL tubes, gently pipetting up and down (*see* **Notes 24** and **25**).
2. Rotate lysates for 20 min (*see* **Note 26**).
3. Spin at maximum speed in a refrigerated microcentrifuge at 4 °C.
4. Remove supernatant to a new prechilled 1.7 mL tube.
5. Measure protein concentration (*see* **Note 27**), keeping samples on ice.
6. Dilute samples in acKIP buffer to a uniform 1–2 mg in 1 mL acKIP buffer (*see* **Note 28**).
7. Remove 50 μL of extract and set aside at 4 °C for use later as a control.
8. Preclear extract for 2 h at 4 °C in a rotator with appropriate control beads (*see* **Note 29**).
9. Spin extracts at 2,300 × *g* for 5 min in a refrigerated microcentrifuge at 4 °C to pellet the beads. Transfer the supernatant to a new tube.
10. Add the acK antibody (or antibody conjugate) to tubes with extract; incubate overnight with rotation at 4 °C overnight (*see* **Note 30**). The optimal amount of antibody needed requires optimization for each different antibody and antigen, but as a rough guide, as a starting point, use 5 μg purified antibody/1 mg IP. For pre-conjugated acetyl-lysine beads (*see* **Note 30**), we use 25 μL slurry/1 mg IP (*see* **Note 31**).
11. The following day, if free antibody was used, add 25 μL Protein A- (for rabbit) or Protein G-conjugated agarose beads (e.g., EMD Millipore Corp. Protein A beads catalog #16-125; Roche Diagnostics Protein G beads catalog #1243233).
12. Incubate 1 h with rotation at 4 °C.
13. Spin the IP tubes at 2,300 × *g* for 5 min in a refrigerated microcentrifuge at 4 °C to pellet the beads. Remove the supernatant (*see* **Notes 32** and **33**).
14. Add 1 mL acKIP buffer and tap tubes gently to resuspend beads. Rotate 5 min at 4 °C.
15. Repeat **steps 13** and **14** for a total of three washes.
16. Following removal of the last wash, centrifuge the pellets again at 2,300 × *g* for 5 min in a refrigerated microcentrifuge at 4 °C. Then, using a 20 μL pipet tip, remove the last bit of liquid from the beads (*see* **Note 34**).
17. Add 80 μL 6× Laemmli, boil for 5 min (*see* **Notes 35** and **36**).
18. Fractionate samples by SDS-PAGE. Load 35 μL of each sample and each control, and 8 μL of protein standard ladder (e.g., Bio-Rad Precision Plus Kaleidoscope™ #1610375) on a 4–20 % polyacrylamide precast gel (*see* **Note 36**).

19. Run the gel at 200 V in 1× running buffer for roughly 50 min (until the dye just runs off of the bottom of the gel).
20. Presoak four pieces of Whatman paper and two transfer pads in 1× transfer buffer for the transfer of fractionated proteins from the gel to a PVDF membrane (e.g., Fisher/Pierce catalog #0088518). The presoaked Whatman paper should be cut to the size of the corresponding transfer pads.
21. Hydrate the PVDF membrane (which should be cut to the approximate size of your gel) by soaking it in methanol for 1 min and then rinse the membrane with 1× transfer buffer. Use forceps only to handle the membrane.
22. On the black (internal) portion of the transfer cassette, place one presoaked transfer pad followed by two pieces of presoaked Whatman paper.
23. Remove the polyacrylamide gel from the casting rig and position it on the Whatman paper in the cassette so that the gel is oriented face-down toward the black side of the cassette where the top of the gel is nearest the cassette clamp and the bottom of the gel is nearest the cassette hinge (*see* **Note 37**).
24. Using forceps, place the PVDF membrane on top of the gel, followed by two pieces of presoaked Whatman paper. Gently run a small roller over the Whatman paper to remove any bubbles that may have accumulated between the gel and the PVDF membrane and assure uniform transfer.
25. Place the final presoaked transfer pad on top of the Whatman paper and close the cassette. There should be a snug fit when closing the clamp of the cassette.
26. Insert the cassette into the blotting apparatus, aligning the black side of the cassette with the black side of the box; the other side will be either red or clear depending on the apparatus used. Fill to mark with 1× transfer buffer. Add a magnetic stir bar and an ice pack to the apparatus.
27. Run the transfer at 4 °C at 350 mA for 1.5 h on a stir plate. Alternatively, gels can be transferred at 150 mA overnight.
28. Remove the PVDF membrane with forceps and rinse in purified water by rocking in a small tray for 1 min. Repeat two more times.
29. Remove water from the tray and add just enough Ponceau S staining solution to cover the membrane and rock for 5 min.
30. Wash in purified water until transferred proteins are visualized and the background is removed. Over-washing can remove the Ponceau stain entirely.
31. Place the membrane in clear plastic and make a digital scanned image of the stained membrane for later reference.

32. Rock the membrane in 20 mL of 5 % milk in 1× TBST for 30 min to block nonspecific binding sites.
33. Discard blocking solution and incubate blot in primary antibody (in 5 % milk or 5 % BSA in TBST—check manufacturer's instructions for appropriate starting concentration) for 1 h at room temperature or overnight at 4 °C. Remove the primary antibody for later use (*see* **Note 38**).
34. Wash the membrane in 20 mL of 1× TBST at room temperature for 5 min. Repeat two more times.
35. Incubate the membrane with appropriate HRP-conjugated secondary antibody diluted in 20 mL of 5 % milk in 1× TBST for 1 h at room temperature. Check manufacturer's instructions for appropriate starting concentration of secondary antibody.
36. Wash membrane 3× for 10 min each in 20 mL 1× TBST.
37. Develop signal using appropriate substrate (e.g., Fisher #WBKLS0500).

4 Notes

1. Dissolving urea is an endothermic reaction. Preparing the Urea Lysis Buffer can be facilitated by placing a stir bar in the beaker and by using a warm (not hot) water bath on a stir plate. 9 M urea is used so that upon lysis, the final concentration is approximately 8 M. The urea lysis buffer should be used at room temperature. Placing the urea lysis buffer on ice will cause the urea to precipitate out of solution.
2. Cells should be washed with 1× PBS before lysis to remove any media containing protein contaminants. Elevated levels of media-related proteins will interfere with the total protein determination.
3. Centrifugation is performed at 15 °C or room temperature to prevent urea from precipitating out of solution. Centrifugation should be performed in an appropriate container rated for at least 20,000 ×*g*.
4. DO NOT place Urea Lysis Buffer or culture dishes on ice during harvesting. Harvest cells using Urea Lysis Buffer at room temperature. During lysis, the buffer becomes viscous due to DNA released from the cells.
5. As the viscous lysate material is sonicated, the DNA will be fragmented and the viscosity of the lysate will decrease. Ensure that the sonicator tip is submerged in the lysate. If the sonicator tip is not submerged properly, this may induce foaming and degradation of the sample (refer to instruction manual of the manufacturer of the sonication apparatus).

6. Sep-Pak® C_{18} purification entails reversed-phase (hydrophobic) solid-phase extraction. Peptides and lipids bind to the chromatographic material. Large molecules such as DNA, RNA, and most protein, as well as hydrophilic molecules such as many small metabolites are separated from peptides using this technique. Peptides are eluted from the column with 40 % acetonitrile (MeCN) and separated from lipids and proteins, which elute at approximately 60 % MeCN and above.
7. Before loading the peptides from the protein digest on the column, the digest must be acidified with TFA for efficient peptide binding. The acidification step helps remove fatty acids from the digested peptide mixture.
8. The purification of peptides is performed at room temperature on 0.7 mL Sep-Pak columns from Waters Corporation, WAT051910. About 20 mg of protease-digested peptides can be purified from one Sep-Pak column. Peptides should be purified immediately after the proteolytic digest.
9. Application of all solutions should be performed by gravity flow. Each time solution is applied to the column air bubbles form in the junction where the 10 cc reservoir meets the narrow inlet of the column. These must be removed with a gel-loader tip placed on a P-200 micropipettor, otherwise the solution will not flow through the column efficiently. Always check for appropriate flow.
10. In rare cases, if the flow rates decrease dramatically upon (or after loading of sample), the purification procedure can be accelerated by *gently* applying pressure to the column using the plunger that was originally inserted, after cleaning it with organic solvent. Again make sure to remove air bubbles from the narrow inlet of the column before doing so. Do not apply vacuum (as advised against by the manufacturer).
11. The lysate digest may have a much higher volume than the 10 cc reservoir will hold (up to 50–60 mL from adherent cells) and therefore the peptides must be applied in several fractions. If available a 60 cc syringe may be used in place of a 10 cc syringe to allow all sample to be loaded into the syringe at once.
12. The lyophilization should be performed in a standard lyophilization apparatus. *DO NOT USE A SPEED-VAC APPARATUS AT THIS STAGE OF THE PROTOCOL.* After the lyophilization step, the digested peptides are stable at −80 °C for several months (seal the tightly closed tube with parafilm for storage). The procedure can be interrupted before or after lyophilization. Once the lyophilized peptide is dissolved in IAP buffer (see next step), continue to the end of the procedure.
13. If the cells were directly harvested from culture medium without PBS washing, some of the Phenol Red pH indicator will remain (it co-extracts during the Sep-Pak® C_{18} purification of peptides)

and color the peptide solution yellow. This coloration has no effect on the IAP step.

14. After dissolving the peptide, check the pH of the peptide solution by spotting a small volume on pH indicator paper (the pH should be close to neutral, or no lower than 6). In the rare cases that the pH is more acidic (due to insufficient removal of TFA from the peptide under suboptimal conditions of lyophilization), titrate the peptide solution with a 1 M Tris-base solution that has not been adjusted for pH; 5–10 μL is usually sufficient to neutralize the solution.
15. In order to recover the beads quantitatively, do not spin the beads at lower *g*-forces than what is specified in this procedure. Avoid substantially higher *g*-forces as well, since this may cause the bead matrix to collapse. All centrifugation steps should be performed at the recommended speeds throughout the protocol.
16. All subsequent wash steps are at 0–4 °C. In all wash steps, the supernatant should be removed reasonably well. Avoid removing the last few microliters, except in the last step, since this may cause inadvertent carry over of the beads.
17. All steps from this point forward should be performed with solutions prepared with Burdick and Jackson or other HPLC grade water.
18. After the last wash step, remove supernatant with a P-1000 micropipettor as before, then centrifuge for 5 s to remove fluid from the tube walls, and carefully remove all remaining supernatant with a gel loading tip attached to a P-200 micropipettor.
19. In this step, the posttranslationally modified peptides of interest will be in the eluent.
20. We recommend concentrating peptides using this protocol by Rappsilber et al. [10].
21. There are many other routine methods for concentrating peptides using commercial products such as ZipTip® (http://www.millipore.com/catalogue/item/ZTC18S096) and StageTips (http://www.proxeon.com/productrange/sample_preparation_and_purification/stage_tips/index.html) that have been optimized for peptide desalting/concentration. Regardless of the particular method, we recommend that the method of choice be optimized for recovery and be amenable for peptide loading capacities of at least 10 μg.
22. Solvents A-E contain volatile organic solvent. Tubes containing small volumes of these solutions should be prepared immediately before use and should be kept capped as much as possible, because the organic components evaporate quickly.

23. In our experience, performing an IP reaction using an acK-directed antibody followed by IB for the protein of interest is a more robust approach for most proteins than the converse (i.e., performing an IP reaction for the candidate protein followed by acK IB). This approach offers the additional advantage that the resulting immunoblots can be cut into several sections and probed for different acetylated proteins, thus minimizing the number of IP reactions that need to be performed. However, there are numerous examples of both approaches working in the literature.
24. All subsequent steps until boiling in Laemmli should be conducted at 4 °C, ideally in the cold room and on ice whenever possible.
25. We have found that acK IPs are quite sensitive to the IP buffer used, and in our experience, do not work well in certain common IP buffers (e.g., RIPA). The ackIP buffer listed works well for immunopurification of acetylated proteins from cells, mitochondria, and tissue lysates
26. If necessary, sonication can be performed at this point. For mitochondria, we have found this step to be unnecessary.
27. Make sure that whatever protein quantification method is used is insensitive to the detergent in the lysis buffer (e.g., Bio-Rad DC Protein Assay Kit II, cat. # 500–0112). Also, it is critical that the blank and the standard reactions be performed in the same buffer as the samples.
28. AcK antibodies tend to be of low affinity; hence the need for relatively large amounts of starting material. We have been uniformly unsuccessful at immunopurifying acetylated proteins by acK IP using less than 1–2 mg of starting material.
29. This step removes proteins from the extract that bind nonspecifically to beads or immunoglobulin and produce artifactual results. For example, for subsequent IPs using pre-conjugated rabbit acK beads, we typically add 25 μL of rabbit IgG-agarose (Sigma A2909) for the preclearing step.
30. Different acK antibodies recognize different subsets of acetylation sites and proteins, so it is worth trying several different acK antibodies from different manufacturers at this step to optimize detection of acetylation on specific proteins of interest (e.g., Cell Signaling catalog #s 9814, 9681, 9441; Immunechem catalog # ICP0380; Millipore catalog #3879). One product that works extremely well is acK antibodies pre-conjugated to agarose beads (Immunechem catalog #ICP0388). This product has the additional advantage that it obviates the need for adding protein A/G-conjugated beads subsequently. As the pre-conjugated beads also minimize the amount of free immunoglobulin heavy and light chain in the

sample, they improve background in subsequent immunoblotting significantly.

31. To allow accurate pipetting of agarose beads, cut the tip off of a 200 μL pipet tip to widen the bore, and pipet the beads up and down several times prior to adding them to the sample.
32. If desired, a sample of the post-IP supernatant can be saved at this step and run alongside the IP reactions to check depletion of the antigen of interest by the IP. However, it is rare to observe such depletion, even with robust acK IPs.
33. Aspirate the supernatant using a 200 μL pipet tip attached to an aspirator hose. The narrow pipet tip minimizes bead losses, and the amount of vacuum can be lowered to achieve a slow rate of removal.
34. This last step is critical, since a significant amount of liquid remains after the last aspiration, which can dilute the sample.
35. To avoid the opening of tubes during boiling, lid-locks can be used.
36. It is possible to split each IP reaction onto two gels, loading 35 μL/well. If many different acetylated proteins are to be identified, it is desirable to load the IP reactions on gradient gels, enabling a better resolution across a wide range of molecular weights (e.g., Bio-Rad Criterion Gels, 4–20 % Tris–HCl, 12-well, 1.0 mm, precast polyacrylamide gels, catalog #345-0032).
37. When removing the polyacrylamide gel from the cassette, the Bio-Rad Criterion cell that is used for running the gel has a built-in cassette opening wedge in the lid, but we prefer to crack the welded joints with a metal spatula. The lower half of the gel is thicker and easier to lift for removal without tearing the gel. If tearing of the gel still occurs, it can also be floated off in 1× transfer buffer for gentle removal.
38. The primary antibody can be saved and reused. Add sodium azide to 0.05 % and store at 4 °C for future use. Alternatively, antibody solutions can be frozen.

Acknowledgements

The authors would like to thank members of the Lombard lab for helpful discussions, and the PTMScan Service Group at Cell Signaling Technology (Matt P. Stokes, Charles L. Farnsworth, and Hongbo Gu) for their assistance in reviewing the protocol. Work in the Lombard lab is supported by NIH grants R01GM101171 and R21CA177925 (Lombard), R01HL114858 (Lukacs), DP3DK094292 (Brosius and others), and a New Scholar in Aging award from the Ellison Medical Foundation (Lombard).

References

1. Schwer B, Eckersdorff M, Li Y, Silva JC, Fermin D, Kurtev MV, Giallourakis C, Comb MJ, Alt FW, Lombard DB (2009) Calorie restriction alters mitochondrial protein acetylation. Aging Cell 8(5):604–606. doi:ACE503 [pii] 10.1111/j.1474-9726.2009.00503.x
2. Norvell A, McMahon SB (2010) Cell biology. Rise of the rival. Science 327(5968):964–965. doi:327/5968/964 [pii] 10.1126/science.1187159
3. Baur JA, Ungvari Z, Minor RK, Le Couteur DG, de Cabo R (2012) Are sirtuins viable targets for improving healthspan and lifespan? Nat Rev Drug Discov 11(6):443–461. doi:10.1038/nrd3738 nrd3738 [pii]
4. Choudhary C, Kumar C, Gnad F, Nielsen ML, Rehman M, Walther TC, Olsen JV, Mann M (2009) Lysine acetylation targets protein complexes and co-regulates major cellular functions. Science 325(5942):834–840. doi:1175371 [pii] 10.1126/science.1175371
5. Zhao S, Xu W, Jiang W, Yu W, Lin Y, Zhang T, Yao J, Zhou L, Zeng Y, Li H, Li Y, Shi J, An W, Hancock SM, He F, Qin L, Chin J, Yang P, Chen X, Lei Q, Xiong Y, Guan KL (2010) Regulation of cellular metabolism by protein lysine acetylation. Science 327(5968):1000–1004. doi:327/5968/1000 [pii] 10.1126/science.1179689
6. Kendrick AA, Choudhury M, Rahman SM, McCurdy CE, Friederich M, Van Hove JL, Watson PA, Birdsey N, Bao J, Gius D, Sack MN, Jing E, Kahn CR, Friedman JE, Jonscher KR (2011) Fatty liver is associated with reduced SIRT3 activity and mitochondrial protein hyperacetylation. Biochem J 433(3):505–514. doi:BJ20100791 [pii] 10.1042/BJ20100791
7. Kim SC, Sprung R, Chen Y, Xu Y, Ball H, Pei J, Cheng T, Kho Y, Xiao H, Xiao L, Grishin NV, White M, Yang XJ, Zhao Y (2006) Substrate and functional diversity of lysine acetylation revealed by a proteomics survey. Mol Cell 23(4):607–618. doi:10.1016/j.molcel.2006.06.026
8. Ahn BH, Kim HS, Song S, Lee IH, Liu J, Vassilopoulos A, Deng CX, Finkel T (2008) A role for the mitochondrial deacetylase Sirt3 in regulating energy homeostasis. Proc Natl Acad Sci U S A 105(38):14447–14452. doi:0803790105 [pii] 10.1073/pnas.0803790105
9. Laemmli UK (1970) Cleavage of structural proteins during the assembly of the head of bacteriophage T4. Nature 227(5259):680–685
10. Rappsilber J, Ishihama Y, Mann M (2003) Stop and go extraction tips for matrix-assisted laser desorption/ionization, nanoelectrospray, and LC/MS sample pretreatment in proteomics. Anal Chem 75(3):663–670
11. Rush J, Moritz A, Lee KA, Guo A, Goss VL, Spek EJ, Zhang H, Zha XM, Polakiewicz RD, Comb MJ (2005) Immunoaffinity profiling of tyrosine phosphorylation in cancer cells. Nat. Biotechnol 23(1):94–101. doi:10.1038/nbt1046

Chapter 7

SILAC-Based Quantification of Sirt1-Responsive Lysine Acetylome

Yue Chen, Gozde Colak, and Yingming Zhao

Abstract

*S*table *I*sotope *L*abeling by *A*mino acids in *C*ell culture (SILAC) is one of the in vivo metabolic labeling methods widely used for dynamic analysis of protein modifications. Here, we describe a general approach to applying SILAC, in combination with affinity enrichment of acetyllysine peptides and mass spectrometry, to study the dynamic changes of the Lysine acetylome in response to Sirt1. The method should be applicable to quantify changes to other post translational modifications in diverse cellular systems.

Key words SILAC, Quantification, Sirt1, Immunoaffinity purification, Lysine acetylation, Nano HPLC mass spectrometry

1 Introduction

Lysine acetylation (Kac) is a reversible and highly conserved protein modification that occurs on the ε-group of the lysine side chain. Early discovery of this modification on histones and transcription factors established the fundamental role of Lys acetylation in chromatin-templated processes [1–3]. Because of its dynamic nature, unique nonpolar structure, and low cellular stoichiometry, proteome-wide analysis of lysine acetylation has been an analytical challenge. Recently, we have developed a strategy for analysis of Kac using affinity enrichment of acetylated peptides with pan anti-Kac antibodies [4]. In this strategy, proteins are first digested to generate peptides; therefore, modification sites that are buried in the native protein are exposed. Then, peptides are incubated with pan anti-Kac antibodies for immunoaffinity precipitation (IP). Finally, the peptide-bound antibodies are extracted by protein A agarose beads; the enriched Lys acetylated peptides are eluted and subjected to nanoLC-mass spectrometry analysis. Using this strategy, we and others have substantially expanded the scope of Lys acetylation and have revealed significant roles of Kac in cell physiology and energy metabolism [5–9].

Matthew D. Hirschey (ed.), *Sirtuins: Methods and Protocols*, Methods in Molecular Biology, vol. 1077,
DOI 10.1007/978-1-62703-637-5_7,

Regulation of in vivo lysine acetylation is achieved by two distinct groups of enzymes with opposing activities—acetyltransferases and de-acetylases. Sirt1 or its yeast homolog Sir2, is one of the class III de-acetylases that are highly conserved and NAD+ dependent. Numerous studies in model organisms including yeast, *C. elegans* and mouse have shown that Sirt1 is involved in diverse physiological functions such as aging and energy metabolism. Traditional approaches using biochemical and molecular biologic methods have been applied to discover novel Sirt1 substrates. These studies led to the discovery of p53, NF-κB, FOXO1, PGC1-α, and histones as Sirt1 substrates [10–16].

Despite the continually growing list of Sirt1 substrates, there is a lack of comprehensive understanding of Sirt1's impact on the Lys acetylome. Towards this goal, we developed a SILAC-based quantitative mass spectrometry approach to quantify changes in the Lys acetylome in response to knocking out Sirt1. Our studies using this approach resulted in the identification of more than 4,500 Lys acetylation sites, of which nearly 500 showed more than 100 % increase in acetylation abundance upon Sirt1 knockout, providing the first comprehensive map of Kac sites directly or indirectly regulated by Sirt1. In this chapter, we will describe our experimental strategy in detail, and point out important steps involved in sample analysis.

Briefly, our approach involves five major components: (1) SILAC cell culture of Sirt1 wild type (Sirt1 +/+) and Sirt1 knock out (Sirt1 −/−) mouse embryonic fibroblast (MEF) cells in heavy and light media; (2) Ion-exchange-based protein fractionation or isoelectric focusing-based peptide fractionation; (3) immunoaffinity precipitation of Lys acetylated peptides; (4) nanoLC-mass spectrometry analysis; (5) identification and quantification of Lys acetylation sites (Fig. 1). Although each step has to be carried out carefully, success of the whole experiment depends on two key factors. The first is the use of high quality antibody and an efficient procedure for enrichment of Lys acetylated peptides. Typically, pan anti-Kac antibodies have very low affinity towards acetylated lysine. To overcome this challenge, we perform immunoprecipitation of small peptides rather than of large proteins, which improves enrichment efficiency. In addition, we reason that longer incubation times will facilitate binding between the antibodies and the acetylated peptides. The second key factor is to carry out extensive protein/peptide fractionation. Some acetylated proteins such as histones are highly abundant in many types of cells. Acetylated histone peptides often interfere with the identification of low-abundance peptides because they can compete for the limited binding pockets on the antibodies or they might co-elute from the nanoLC column causing ion suppression. To overcome these problems, we offer two different strategies (Fig. 1). In Strategy I, we fractionate

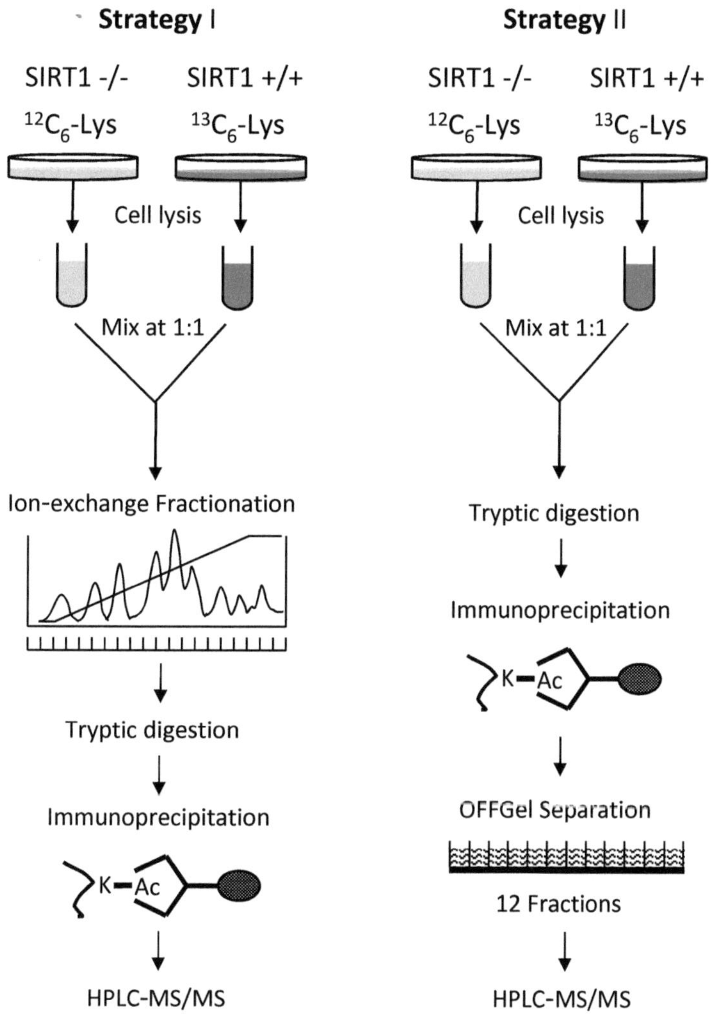

Fig. 1 Schematic workflow for SILAC-based identification and quantification of Lys acetylome in response to Sirt1. Strategy I (*left panel*) applies protein fractionation prior to immunoaffinity precipitation of Lys acetylated peptides. Strategy II (*right panel*) applies isoelectric focusing fractionation of peptides after immunoaffinity precipitation

proteins into 10–20 fractions using ion-exchange chromatography. Each fraction is tryptically digested and immunoprecipitated to enrich Kac peptides. In Strategy II, we first perform immunoaffinity precipitation to enrich Kac peptides, and then apply isoelectric focusing to fractionate peptides into 12 fractions for analysis. Both strategies offer significant improvements in the identification of low-abundance Lys acetylated peptides, and avoid interference from highly abundant acetylated proteins.

2 Materials

2.1 SILAC Cell Culture, Labeling Efficiency Check, Cell Lysis, and Tryptic Digestion

1. Dulbecco's Modified Eagle Medium (D-MEM) for SILAC: D-MEM SILAC cell culture medium without L-Lysine or L-Arginine (pH 7.0–7.4), supplemented with 2 mM L-GLUTAMINE, 4.5 g/L glucose, and dialyzed 10 % FBS.
2. $^{13}C_6$-Lysine and $^{13}C_6{}^{15}N_2$-Arginine: 99 % isotope purity.
3. Sirt1+/+ and Sirt1−/− MEF cells.
4. NETN cell lysis buffer: 40 mM Tris–HCl, pH 8.0, 200 mM NaCl, 2 mM EDTA, 1 % NP-40.
5. 37 °C water bath.
6. 150 mm cell culture plates.
7. Sterile falcon tubes and pipettes.
8. 70 % Ethanol (v/v).
9. 4 °C Refrigerator.
10. Orbital shaker.
11. Humidified 37 °C incubator containing 5 % carbon dioxide.
12. Laminar biosafety cabinet class II A/B3.
13. Phosphate-buffered saline (PBS), pH 7.4.
14. 4 °C Micro-centrifuge.
15. Benchtop centrifuge at room temperature.
16. Microfuge tubes (1.5 mL).
17. Bradford assay kit.
18. UV spectrometer.
19. pH paper (pH 1–14) or pH meter.
20. Ultrasonicator.
21. Trichloroacetic acid (TCA).
22. Acetone (store at −20 °C).
23. Digestion buffer: 50 mM Ammonium bicarbonate (pH 8.0).
24. 100 mM Dithothreitol (DTT) stock solution. Dissolve 15.4 mg of DTT in 1 mL digestion buffer.
25. 300 mM Iodoacetamide (IA) stock solution. Dissolve 55.5 mg of IA in 1 mL digestion buffer.
26. 300 mM L-Cysteine stock solution. Dissolve 36.3 mg of Cys in 1 mL digestion buffer.
27. Trypsin (sequencing grade, enzyme-to-substrate ratio ~1/50, w/w).
28. Trifluoroacetic acetic acid (TFA) stock solution (10 %, v/v).
29. C18 desalting tips (Millipore C18 Zip-tips or Waters Oasis HLB cartridges, 1 cc).

30. Desalting buffer A: HPLC grade acetonitrile.
31. Desalting buffer B: Acetonitrile in water (50 %, v/v).
32. Desalting buffer C: TFA in water (0.1 %, v/v).
33. SpeedVac vacuum centrifuge concentrator.

2.2 Protein Fractionation with Ion-Exchange Chromatography

1. HPLC buffer A: 10 mM 2-(*N*-morpholino)ethanesulfonic acid (MES) in water. Prepare 1 L HPLC buffer A by dissolving 1.95 g MES in 900 mL water. Adjust pH to 6.0 at 25 °C. Bring volume to 1 L with water.
2. HPLC buffer B: 10 mM MES, 800 mM NaCl, in water. Prepare 1 L HPLC buffer B by dissolving 1.95 g MES and 46.64 g NaCl in 900 mL water. Adjust pH to 6.0 at 25 °C. Bring volume to 1 L with water.
3. Vacuum-assisted filtration device with 0.22 μm membrane.
4. Water-bath sonicator.
5. Preparative HPLC system with UV detector and fraction collector.
6. PolyCATWAX ion-exchange column (200 mm × 9.4 mm, 5 μm, 1,000 Å, PolyLC).
7. Syringe filter system (0.22 μm).

2.3 Immunoaffinity Purification (IP)

1. Pan anti-acetyllysine antibody-conjugated agarose beads (PTM Biolabs).
2. Benchtop centrifuge.
3. NETN IP buffer: 50 mM Tris–HCl, pH 8.0, 0.5 % w/v NP-40, 100 mM NaCl, 1 mM EDTA.
4. ETN buffer: 50 mM Tris–HCl, pH 8.0, 100 mM NaCl, 1 mM EDTA.
5. Elution buffer: 0.1 %, v/v TFA, in water.

2.4 Peptide Fractionation with Isoelectric Focusing

1. Agilent 3100 OFFGel fractionator with accessories provided by the manufacturer (Agilent Technology).
2. 12-well immobilized pH gradient strip (pH 3–10, Agilent Technology).
3. OFFGel stock solution for peptide fractionation (1.25×, Agilent Technology).

2.5 Reverse-Phase NanoLC (RPLC)-Mass Spectrometry

1. Nano-flow HPLC such as Eksigent 1D-plus NanoLC.
2. RPLC solvent A: 0.1 % v/v formic acid, in water.
3. RPLC solvent B: 0.1 %, v/v formic acid, in acetonitrile.
4. Commercial fused-silica capillary C18 HPLC column (75 μm ID, 12 cm length) with nanospray needle tip or similar-sized in-house packed capillary C18 HPLC column (Luna C18, 5 μm, 100 Å pore size, Phenomenex Inc.) with laser-pulled tip.

5. Manual injector or auto-sampler.
6. Nanospray source.
7. High resolution mass spectrometer such as LTQ Orbitrap mass spectrometer.

2.6 Quantitative Data Analysis

1. Multi-core computer with at least 2 GB RAM, installed with 32-bit Windows operating system (Windows XP or above) and .NET framework 2.0.
2. Xcalibur software (Thermofisher, v2.1 or above).
3. Mascot database searching software (Matrix Science, v2.1 or above).
4. Raw data conversion tools (such as extract_msn.exe by Thermo Fisher).
5. Mouse protein sequence database in FASTA format.
6. Maxquant software for quantification (http://maxquant.org, v1.0.13.13).

3 Methods

3.1 SILAC Cell Culture

1. Prepare heavy and light D-MEM SILAC media following manufacturer's instructions (*see* **Notes 1–4**).
2. Filter both media with 0.22 μm filter, and warm the media in 37 °C water bath.
3. Remove cryovials of Sirt1−/− and Sirt1+/+ MEF cells from liquid nitrogen storage, and thaw rapidly in 37 °C water bath for less than 1 min. Wipe the outside of tubes with 70 % ethanol before placing inside a laminar flow biosafety cabinet.
4. Transfer thawed cells into individual sterile falcon tubes and add 5 mL light SILAC medium to Sirt1−/− MEF cells and heavy SILAC medium to Sirt1+/+ MEF cells. Centrifuge at 800 × *g* for 5 min to remove supernatant containing DMSO.
5. Label two 60 mm plates as "Sirt1+/+ Heavy" and "Sirt1−/− Light." Plate the Sirt1+/+ cells in 4 mL heavy SILAC medium and the Sirt1−/− cells in 4 mL light SILAC medium.
6. Culture cells in 37 °C humidified incubator (5 % carbon dioxide) for half a day before changing medium once to further remove DMSO. Store SILAC media bottles in a clean 4 °C refrigerator.
7. Continue culturing cells and check daily for confluence.
8. Split cells when the confluence reaches 90 %, and allow the cells to grow for at least 6 doublings in heavy medium. Grow at least 10 large dishes of cells in light medium and 10 large dishes of cells in heavy medium to 90 % confluence.

3.2 Check for SILAC Labeling Efficiency, Cell Lysis, and Tryptic Digestion

1. To check labeling efficiency of heavy amino-acid incorporation, take one plate of heavy-labeled cells and rinse the cells in ice-cold PBS buffer to remove culture medium.
2. Add PBS to the plates and carefully scrape to remove cells. Collect the cells in 15 mL falcon tubes.
3. Centrifuge at 1,000 × *g* to remove supernatant, wash the cells with ice-cold PBS three times, and resuspend the cell pellet in 2 volumes NETN buffer. Transfer the cells to a 1.5 mL microfuge tube.
4. Lyse the cells in NETN buffer for 30 min on ice.
5. Centrifuge at 16,000 × *g* for 10 min at 4 °C.
6. Save the supernatant and resuspend the pellet in NETN buffer.
7. Set the sonicator to 10 % power and sonicate the pellet for a 5 s pulse followed by a 15 s pause. Repeat sonication twice.
8. Centrifuge at 16,000 × *g*, 10 min, 4 °C.
9. Save the supernatant, and combine with supernatant from previous extraction; measure protein content using Bradford assay.
10. Precipitate protein by adding TCA dropwise to a final concentration of 10 % (v/v).
11. Continue precipitation by incubating proteins overnight on an orbital shaker at 4 °C.
12. Centrifuge the protein pellet at 16,000 × *g* for 10 min at 4 °C and discard the supernatant.
13. Resuspend protein pellets in ice-cold acetone and centrifuge at 16,000 × *g* for 10 min at 4 °C. Discard the supernatant.
14. Repeat **step 13** two more times.
15. Resuspend the protein pellet in 500 μL Digestion buffer.
16. Add trypsin at enzyme-to-substrate ratio of 1:50 (w/w).
17. Incubate overnight in 37 °C water bath.
18. Centrifuge and add DTT solution to a final concentration of 5 mM, and incubate in a 55 °C water bath for 30 min.
19. Centrifuge and add IA stock solution to a final concentration of 15 mM. Incubate at room temperature for 30 min in dark.
20. Centrifuge and add Cysteine stock solution to a final concentration of 15 mM and incubate at room temperature for 30 min.
21. Add a second aliquot of trypsin at enzyme-to-substrate ratio of 1:100 (w/w) and incubate at 37 °C water bath for 3 h.
22. Stop the reaction by adjusting to pH 3 with TFA stock solution.
23. Dry the tryptic digests in SpeedVac.

24. To desalt the peptides using Millipore Zip-tips, wash the tip with 10 μL desalting buffer A and 10 μL desalting buffer B. Equilibrate with 10 μL desalting buffer C three times. Dissolve the peptides in 10 μL desalting buffer C and load onto the tip by pipetting up and down. Wash the tip with 10 μL desalting buffer C three times and elute with 10 μL desalting buffer B.
25. Dry the eluted peptides in SpeedVac.
26. Perform LC-MS/MS analysis (*see* Subheading 3.6).
27. Perform protein quantification and determine labeling efficiency by checking manually (*see* **Note 5**) and calculating normalized protein quantification H/L ratios (*see* Subheading 3.7).
28. If the labeling efficiency is not satisfactory, continue to culture cells for one or two more doublings and check labeling efficiency again.
29. When the cells have reached satisfactory labeling efficiency, collect the cells and perform **steps 2–9**.
30. Combine equal amounts (5–10 mg) of light and heavy cell lysates and mix well.
31. To perform protein fractionation before immunoprecipitation (Fig. 1, Strategy I), follow Subheading 3.3. Samples can be stored at −80 °C for later use.
32. To perform peptide fractionation after immunoprecipitation (Fig. 1, Strategy II), digest the protein samples by following **steps 15–25**, and then follow Subheading 3.4 for immunoaffinity purification.

3.3 Protein Fractionation with Ion-Exchange Chromatography

Use HPLC grade or Milli-Q water for protein fractionation with ion-exchange chromatography.

1. Prepare fresh 1 L HPLC buffer A (*see* **Note 6**).
2. Prepare fresh 1 L HPLC buffer B.
3. Filter HPLC buffer A and B through 0.22 μm filter before use.
4. Degas the solvents by sonicating the buffers in water-bath sonicator for 20 min.
5. Thaw the frozen protein samples and add 3× volumes of buffer A. Mix well.
6. Centrifuge at 16,000 × *g*.
7. Carefully extract the supernatant and pass it through 0.22 μm filter. Keep the precipitates as one fraction and store at −20 °C.
8. Prepare glass vials for fractionation collector and set up the following HPLC application: Flow rate 4 mL/min; gradient: 0 min—0 % B, 13 min—70 % B, 16 min—100 % B, 25 min—100 % B (*see* **Note 7**).

9. Flush the HPLC column with 100 % buffer B for 3 bed volumes.
10. Equilibrate the column with 100 % buffer A for 5 bed volumes until the UV absorption at 280 nm stabilizes.
11. Load samples through manual injector. If the sample volume is larger than the loop size, load multiple times and wait until the UV absorption stabilizes again. In the meantime, collect all the flow-through solvents in a 250 mL glass container labeled as flow-through fraction. This fraction contains proteins that are not bound to the ion-exchange column.
12. Start the HPLC and collect all fractions based on equal time intervals.
13. After HPLC is complete, wash the pumps and column with water for 10 bed volumes (*see* **Note 8**).
14. Combine adjacent fractions to produce 10–15 large fractions.
15. Measure protein concentration using Bradford assay, and calculate the amount of protein in each fraction.
16. Perform TCA precipitation, tryptic digestion, and desalting according to **steps 10–25** in Subheading 3.2.
17. Perform immunoaffinity purification of Lys acetylated peptides from each fraction according to Subheading 3.4.

3.4 Immunoaffinity Purification (IP)

1. Dissolve digested peptides (1–5 mg) in 1 mL NETN buffer.
2. Centrifuge at 16,000 × *g* for 10 min and carefully remove supernatant.
3. Keep 50 μL (5 %) of the supernatant as "Normalization fraction" prior to IP, and follow **steps 24–27** in Subheading 3.2 for protein quantification analysis.
4. Add 20–50 μL anti-acetyllysine antibody-conjugated agarose beads and incubate at 4 °C for 6 h or overnight with gentle shaking on an orbital shaker.
5. Centrifuge at 1,000 × *g* and carefully remove the supernatant.
6. Wash the beads with 1 mL ice-cold NETN buffer three times.
7. Wash the beads with 1 mL ice-cold ETN buffer once.
8. Wash the beads with 1 mL ice-cold water twice.
9. Add 50 μL elution buffer to elute peptides.
10. Centrifuge at 1,000 × *g* 1 min, and carefully collect the supernatant.
11. Repeat elution two more times and pool collected supernatants.
12. Dry supernatant in SpeedVac.
13. Carry out nano HPLC-MS/MS analysis (*see* **steps 24–26** in Subheading 3.2) or follow Subheading 3.5 for peptide fractionation with isoelectric focusing.

3.5 Peptide Fractionation with Isoelectric Focusing

1. To desalt the peptides using Oasis HLB cartridge prior to peptide fractionation, condition the cartridge with 1 mL desalting buffer A and 1 mL desalting buffer B. Equilibrate with 1 mL desalting buffer C three times. Dissolve the peptides in 1 mL desalting buffer C and load onto the cartridge. Wash the cartridge with 1 mL desalting buffer C three times and elute with 1 mL desalting buffer B. Completely dry the eluted peptides in a SpeedVac (*see* **Note 9**).
2. Prepare peptide OFFGel stock solution and IPG strip rehydration solution according to manufacturer's instructions.
3. Dissolve the dried peptides in 0.36 mL water, and add 1.44 mL OFFGel fractionation stock solution to the peptide solution and mix well.
4. Assemble 12-well Agilent OFFGel pH gradient strip (pH 3–10) with frames and electrodes according to the manufacturer's instructions.
5. Add 150 μL peptide solution to each well, and check the solution levels to ensure no leaking.
6. Start the isoelectric focusing at constant current of 50 μA for 50 kV h (*see* **Note 10**).
7. Isoelectric focusing is completed when the blue light indicator starts blinking. Carefully transfer solution from each well to a separate tube.
8. Wash each well with 100 μL desalting buffer A twice and combine with previous fractions.
9. Desalt each fraction with Oasis HLB cartridge following **step 1**.
10. Completely dry the eluted peptides and follow steps in Subheading 3.6 for Reverse-Phase nano HPLC-mass spectrometry analysis.

3.6 Reverse-Phase Nano HPLC (RP-HPLC)-Mass Spectrometry

RP-HPLC/MS/MS analysis can be performed on an Eksigent nanoLC-1D plus pump (Eksigent Technologies) coupled online through Thermofisher's nanospray source to an LTQ Orbitrap Velos mass spectrometer (ThermoFisher). Fused-silica capillary C18 RPLC column can be purchased commercially or packed in-house.

1. Equilibrate the RPLC column with 5 % RPLC solvent B in solvent A (v/v) for 10 min at 1 μL/min.
2. Set up mass spectrometer instrument method (*see* **step 6** for parameters).
3. Dissolve peptides in 3 μL RPLC solvent A and spin down.
4. Load the sample onto the capillary RPLC column through manual injector or auto-sampler at a flow rate of 1 μL/min.
5. Elute the peptides with a linear gradient of 5–35 % RPLC solvent B for 90 min at a constant flow rate of 200 nL/min.

6. Eluted tryptic peptides are electrosprayed into LTQ Orbitrap Velos mass spectrometer operating in a data-dependent mode that acquires MS/MS spectra for the top 20 most intense ions. The spray voltage is about 1.8–2.0 kV and capillary temperature is about 200 °C. Full MS can be acquired using FTMS in the Orbitrap at 60,000 resolution at 400 m/z with a mass range of 300–1,800 m/z, and MS/MS can be acquired using collision-induced dissociation (CID) in the Velos at 35 % normalized collision energy. A lock-mass ion from ambient air (m/z 445.120024) can be used for internal calibration of all full-scan measurements with the Orbitrap detector [17]. Dynamic exclusion can be enabled with repeat count of 2, repeat duration of 8 s, exclusion duration of 60 s, maximum exclusion size of 500, and exclusion mass width including ±0.01 % relative to the reference mass.

3.7 Quantitative Data Analysis

MS/MS data and SILAC quantification can be analyzed by Mascot software [18] and MaxQuant [19] according to a previously described protocol [20].

1. Install Mascot database searching software according to the manufacturer's instructions.
2. Download and install mouse fasta database (http://www.Uniprot.org) in Mascot.
3. Install Xcalibur software (Thermofisher, version 2.1 or above) on a Windows-based PC with 32-bit operating system. Install raw data conversion tools on the computer according to manufacturer's instructions.
4. Download and extract MaxQuant software (version 1.0.13.13, http://maxquant.org) onto the same computer as the Xcalibur software. Copy the "enzymes," "mascot.dat," "unimod.xml," and "mod_file" from Mascot "\config" folder to Maxquant "\conf" folder. Install Mascot Daemon software on the computer.
5. Copy mouse fasta sequence to Maxquant folder. Generate decoy database using the "SequenceReverser.exe" tool in Maxquant folder and be sure to check "Include Contaminants." Install the decoy database in Mascot and update mascot.dat in Maxquant (*see* **Note 11**).
6. Copy all raw data into a folder on the same computer installed with Maxquant and Xcalibur. For Lys acetylation quantification, include all raw data for the analysis of enriched Lys acetylated peptides and "Normalization fraction."
7. Start Quant.exe in the MaxQuant software folder and load all raw files under "Raw files" tab.
8. Under "Parameters" tab, select "Orbitrap" as instrument, "Doublets" as SILAC type and check "Lys6" for U-$^{13}C_6$

Lysine labeling (check both "Lys6" and "Arg10" if double labeling with U-$^{13}C_6$ Lysine and U-$^{13}C_6{}^{15}N_2$ Arginine was performed).

9. Select Oxidation (M), Acetyl (Protein N-term), and Acetyl (K) as variable modifications and Carbamidomethyl (C) as fixed modification. Select the mouse decoy database as searching database, trypsin as enzyme, maximum number of missing cleavages as 3, maximum number of labeled amino acids as 4, MS/MS tolerance as 0.5 Da, and top 6 peaks per 100 Da window (*see* **Note 12**).
10. Under "Raw files" tab, select the number of cores on the computer as the number of threads and start the analysis.
11. The program will perform peak generation, precursor mass recalibration, extraction of SILAC pairs, and calculation of SILAC ratios.
12. Upon completion of Quant.exe analysis, a "combined" folder will be generated under the Raw files folder. For "Doublets" SILAC, three ".par" files will be generated—"allSpectra.iso.par," "allSpectra.sil0.par" and "allSpectra.sil1.par." Each ".par" file will have corresponding data files—"iso_0.msm," "sil0_0.msm," "sil1_0.msm."
13. Start Mascot Daemon software and start a new task under "Task editor" tab. For each new task, add one ".par" file and corresponding ".msm" files from the new "combined" folder. Click "Run" to start each task.
14. Upon completion of all database searches, copy the Mascot result files .dat to the Raw files folder on the computer installed with Maxquant.
15. Start "Identify.exe" program and load the Raw files, Mascot result files, and FASTA sequence files under "Input files" tab.
16. Under the "Parameters" tab, select 0.01 as false discovery rate (FDR) for protein, peptide, and site levels. Select 6 as minimum peptide length, "REV_" as reverse sequence labeling, and "CON_" as contaminant sequence labeling.
17. Enable "Use Razor and Unique peptide," use unmodified and peptides with Oxidation (M) and Acetyl (Protein N-term) for protein quantification, select 2 as minimum ratio count, and select "least modified peptides" for Site quantification. Check "Re-quantify," "Keep low-scoring versions of identified peptides," and "Filter labeled amino acids."
18. Select the number of threads according to the number cores of the computer and start the program.
19. Upon the completion of the program, a number of result files in .txt format will be generated in the combined folder. The file "Acetyl (K) Sites.txt" contains all the Lys acetylation site

identification and quantification information (*see* **Note 13**). The file "evidence.txt" contains all the peptide sequence identification and quantification, assigned protein groups, modification state, and spectra matching information. The file "proteinGroups.txt" contains all the protein identifications, quantifications, and available functional annotations (*see* **Notes 14** and **15**).

4 Notes

1. Excessive arginine may be converted to proline in some cell lines, such as Hela and HEK293A cells. Therefore, extra caution should be used when double labeling with both heavy lysine and arginine is applied. A titration of arginine concentration and conversion percentage may be necessary to determine appropriate labeling concentration for the cell lines of interest.
2. Impurities in the amino acids may affect cell growth or labeling efficiency. If possible, use high purity amino acids for SILAC cell culture.
3. Low quality dialyzed FBS may significantly affect labeling efficiency. Test with dialyzed FBS from different vendors.
4. Add penicillin-streptomycin to the culture media if it does not affect the cell growth or morphology. Due to the cost and scale of SILAC experiments, use extra caution to prevent bacterial or fungal contamination during cell culture.
5. To manually check the labeling efficiency, perform raw data conversion, and search with Mascot database searching software on the MS/MS data analysis of the heavy-labeled peptides. Select 10–20 high scoring peptides with strong precursor ions from a number of best identified proteins. Manually check the intensities of the heavy peptide ions and corresponding light ions. Be sure to use the strongest precursor intensities to calculate Light-to-Heavy intensity ratios. A light peptide ion with 5 % relative intensity to the heavy peptide ion indicates a 95.2 % $(100/(100+5)\times 100\ \%)$ labeling efficiency. For double labeling experiments, specifically select the peptides that contain proline to check [U-$^{13}C_6{}^{15}N_2$] arginine-to-[U-$^{13}C_5{}^{15}N_1$] proline conversion.
6. If possible, perform ion-exchange protein fractionation in a cold room (4 °C) to prevent protein degradation. If fractionation is to be performed at 4 °C, then when preparing HPLC buffer A and B at room temperature, titrate pH to 5.76.
7. Gradient and flow rate should be adjusted according to protein lysates and column types to achieve satisfactory performance. For the HPLC fractionation, a small scale test can be performed

by using lysates from the same cell line and a smaller column with the same packing material.

8. High salt buffer may be harmful to the stainless steel pump/ tubing and may cause corrosion. Wash the pumps and HPLC system with water sufficiently after each analysis.
9. Samples containing salts will significantly lower the isoelectric focusing efficiency and increase the time to complete the focusing procedure. Therefore, it is essential that the sample is desalted carefully prior to IEF fractionation.
10. After a continuous 24-h run, replace electrode pads, and if necessary, add additional mineral oil according to the manufacturer's instructions.
11. To install decoy database in Mascot, new parsing rules may need to be set up to accommodate "REV_" and "CON_" prefixes in fasta sequence entries. Detail instructions can be found in the Mascot software manual.
12. Mascot does not allow two types of variable modifications on the same residue, such as Acetylation (K) and Lysine-13C6 (K). To identify Lys acetylated peptides that are only identified in heavy form, add a new variable modification to Mascot and Maxquant analysis as AcetylLysine-13C6 (K) with a monoisotopic mass shift of 48.03066. Due to the nature of SILAC, peptides containing AcetylLysine-13C6 (K) modification should not contain light Lys. Therefore, it is important to check the "AcetylLysine-13C6 (K) Sites.txt" result file and remove all false positive peptides with light Lys.
13. For acetylated peptides identified with low Mascot scores (<30), manual evaluation of spectrum matching quality is advised. For example, the raw file and matching MS/MS spectrum scan number can be found in the "Best Identification Raw File," and "Best Identification Scan Number" columns in "Acetyl (K) Sites.txt."
14. Remove hits indicated by "+" signs in the "Reverse" column and in most cases, remove hits indicated by "+" signs in the "Contaminant" column unless proteins such as keratins or trypsin are the targets of studies. The original heavy-to-light (H/L) ratios and normalized H/L ratios can be found in "Ratio H/L" and "Ratio H/L Normalized" columns. The Lys acetylation site in each peptide is indicated in the "Position in peptide" column and its corresponding positions in the leading proteins are indicated in the "Positions" column. As acetylated Lys cannot be cleaved by trypsin, it is important to filter out all the false positive Lys acetylation identifications on the C-terminus of the tryptic peptides unless the C-terminus is the C-terminus of the corresponding proteins.

15. If protein fractionation before IP is not applicable, alternative strategies include peptide fractionation using strong-cation-exchange (SCX); if SILAC-based quantification is not applicable, alternative strategies include label-free quantification or isotope-tagging quantification such as ITRAQ (*I*sobaric *T*ags for *R*elative and *A*bsolute *Q*uantification) and TMT labeling methods.

Acknowledgements

Research reported in this publication was supported by the National Cancer Institute of the National Institutes of Health under Award Number CA126832 and RR020839 to Y. Zhao. The content is solely the responsibility of the authors and does not necessarily represent the official views of the National Institutes of Health. Y. Zhao is also supported by the Nancy and Leonard Florsheim Family Fund. Y. Chen is supported by NIH Fellowship (T32 HL094282).

References

1. Gershey EL, Vidali G, Allfrey VG (1968) Chemical studies of histone acetylation. The occurrence of epsilon-N-acetyllysine in the f2a1 histone. J Biol Chem 243(19):5018–5022
2. Gu W, Roeder RG (1997) Activation of p53 sequence-specific DNA binding by acetylation of the p53 C-terminal domain. Cell 90(4):595–606. doi:S0092-8674(00)80521-8 [pii]
3. Roth SY, Denu JM, Allis CD (2001) Histone acetyltransferases. Annu Rev Biochem 70:81–120. doi:70/1/81 [pii] 10.1146/annurev.biochem.70.1.81
4. Kim SC, Sprung R, Chen Y, Xu Y, Ball H, Pei J, Cheng T, Kho Y, Xiao H, Xiao L, Grishin NV, White M, Yang XJ, Zhao Y (2006) Substrate and functional diversity of lysine acetylation revealed by a proteomics survey. Mol Cell 23(4):607–618. doi:10.1016/j.molcel.2006.06.026
5. Choudhary C, Kumar C, Gnad F, Nielsen ML, Rehman M, Walther TC, Olsen JV, Mann M (2009) Lysine acetylation targets protein complexes and co-regulates major cellular functions. Science 325(5942):834–840. doi:1175371 [pii] 10.1126/science.1175371
6. Zhang J, Sprung R, Pei J, Tan X, Kim S, Zhu H, Liu CF, Grishin NV, Zhao Y (2009) Lysine acetylation is a highly abundant and evolutionarily conserved modification in Escherichia coli. Mol Cell Proteomics 8(2):215–225. doi:M800187-MCP200 [pii] 10.1074/mcp.M800187-MCP200
7. Wang Q, Zhang Y, Yang C, Xiong H, Lin Y, Yao J, Li H, Xie L, Zhao W, Yao Y, Ning ZB, Zeng R, Xiong Y, Guan KL, Zhao S, Zhao GP (2010) Acetylation of metabolic enzymes coordinates carbon source utilization and metabolic flux. Science 327(5968):1004–1007. doi:327/5968/1004 [pii] 10.1126/science.1179687
8. Zhao S, Xu W, Jiang W, Yu W, Lin Y, Zhang T, Yao J, Zhou L, Zeng Y, Li H, Li Y, Shi J, An W, Hancock SM, He F, Qin L, Chin J, Yang P, Chen X, Lei Q, Xiong Y, Guan KL (2010) Regulation of cellular metabolism by protein lysine acetylation. Science 327(5968):1000–1004. doi:327/5968/1000 [pii] 10.1126/science.1179689
9. Chen Y, Zhao W, Yang JS, Cheng Z, Luo H, Lu Z, Tan M, Gu W, Zhao Y (2012) Quantitative acetylome analysis reveals the roles of SIRT1 in regulating diverse substrates and cellular pathways. Mol Cell Proteomics 11(10):1048–1062. doi:M112.019547 [pii] 10.1074/mcp.M112.019547
10. Luo J, Nikolaev AY, Imai S, Chen D, Su F, Shiloh A, Guarente L, Gu W (2001) Negative control of p53 by Sir2alpha promotes cell survival under stress. Cell 107(2):137–148. doi:S0092-8674(01)00524-4 [pii]

11. Vaziri H, Dessain SK, Ng Eaton E, Imai SI, Frye RA, Pandita TK, Guarente L, Weinberg RA (2001) hSIR2(SIRT1) functions as an NAD-dependent p53 deacetylase. Cell 107(2):149–159. doi:S0092-8674(01)00527-X [pii]
12. Peng L, Yuan Z, Ling H, Fukasawa K, Robertson K, Olashaw N, Koomen J, Chen J, Lane WS, Seto E (2011) SIRT1 deacetylates the DNA methyltransferase 1 (DNMT1) protein and alters its activities. Mol Cell Biol 31(23):4720–4734. doi:MCB.06147-11 [pii] 10.1128/MCB.06147-11
13. Yeung F, Hoberg JE, Ramsey CS, Keller MD, Jones DR, Frye RA, Mayo MW (2004) Modulation of NF-kappaB-dependent transcription and cell survival by the SIRT1 deacetylase. EMBO J 23(12):2369–2380. doi:10.1038/sj.emboj.7600244 7600244 [pii]
14. Brunet A, Sweeney LB, Sturgill JF, Chua KF, Greer PL, Lin Y, Tran H, Ross SE, Mostoslavsky R, Cohen HY, Hu LS, Cheng HL, Jedrychowski MP, Gygi SP, Sinclair DA, Alt FW, Greenberg ME (2004) Stress-dependent regulation of FOXO transcription factors by the SIRT1 deacetylase. Science 303(5666):2011–2015. doi:10.1126/science.1094637 1094637 [pii]
15. Nemoto S, Fergusson MM, Finkel T (2005) SIRT1 functionally interacts with the metabolic regulator and transcriptional coactivator PGC-1{alpha}. J Biol Chem 280(16):16456–16460. doi:M501485200 [pii] 10.1074/jbc.M501485200
16. Imai S, Armstrong CM, Kaeberlein M, Guarente L (2000) Transcriptional silencing and longevity protein Sir2 is an NAD-dependent histone deacetylase. Nature 403(6771):795–800. doi:10.1038/35001622
17. Olsen JV, de Godoy LM, Li G, Macek B, Mortensen P, Pesch R, Makarov A, Lange O, Horning S, Mann M (2005) Parts per million mass accuracy on an Orbitrap mass spectrometer via lock mass injection into a C-trap. Mol Cell Proteomics 4(12):2010–2021. doi:T500030-MCP200 [pii] 10.1074/mcp.T500030-MCP200
18. Perkins DN, Pappin DJ, Creasy DM, Cottrell JS (1999) Probability-based protein identification by searching sequence databases using mass spectrometry data. Electrophoresis 20(18):3551–3567.doi:10.1002/(SICI)1522-2683(19991201)20:18<3551::AID-LPS3551>3.0.CO;2-2 [pii] 10.1002/(SICI)1522-2683(19991201)20:18<3551::AID-ELPS3551>3.0.CO;2-2
19. Cox J, Mann M (2008) MaxQuant enables high peptide identification rates, individualized p.p.b.-range mass accuracies and proteome-wide protein quantification. Nat Biotechnol 26(12):1367–1372. doi:nbt.1511 [pii] 10.1038/nbt.1511
20. Cox J, Matic I, Hilger M, Nagaraj N, Selbach M, Olsen JV, Mann M (2009) A practical guide to the MaxQuant computational platform for SILAC-based quantitative proteomics. Nat Protoc 4(5):698–705. doi:nprot.2009.36 [pii] 10.1038/nprot.2009.36

Chapter 8

Targeted Quantitation of Acetylated Lysine Peptides by Selected Reaction Monitoring Mass Spectrometry

Matthew J. Rardin, Jason M. Held, and Bradford W. Gibson

Abstract

Mass spectrometry (MS) allows for the large-scale identification of multiple peptide analytes in complex mixtures. However, the low abundance of acetylated peptides in the overall mixture requires an enrichment step. After enrichment, the resulting acetylated peptides of interest can be quantitated using selected reaction monitoring (SRM)-MS with stable isotope dilution. Here, we describe the enrichment of lysine acetylated peptides from typsin digested mouse liver mitochondria, and the targeted quantitation of a known lysine acetylation site in succinate dehydrogenase A using SRM-MS on a triple quadrupole instrument.

Key words Selected reaction monitoring, Targeted quantitation, Peptide immunoprecipitation, Acetylation, Sirtuin

1 Introduction

Large-scale affinity enrichment of posttranslational modifications (PTMs) from sirtuin knockout models, in combination with advancements in mass spectrometry, is rapidly identifying a large number of potential sirtuin substrates [1, 2]. However, quantitative data from discovery-based shotgun proteomic data sets can be limited by instrument sensitivity and sampling efficiency. Selected reaction monitoring mass spectrometry (SRM-MS) with stable isotope dilution (SID) has emerged as one of the most sensitive techniques for the precise quantitative analysis of predetermined peptide analytes from multiple samples [3–5]. SRM takes advantage of unique features available in a triple quadrupole mass spectrometer. As shown in Fig. 1, in SRM the first quadrupole is set to pass a peptide analyte of a known mass-to-charge ratio (m/z) into the second quadrupole (Q2). In Q2, the peptide precursor(s) (P) contained in the mass selection window, typically ±1 m/z, undergo collision-activated dissociation (CAD) generating a set of fragment

Matthew D. Hirschey (ed.), *Sirtuins: Methods and Protocols*, Methods in Molecular Biology, vol. 1077,
DOI 10.1007/978-1-62703-637-5_8, © Springer Science+Business Media, LLC 2013

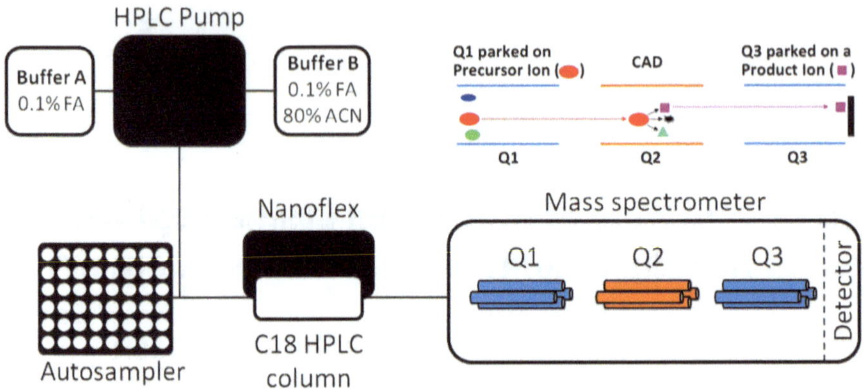

Fig. 1 Schematic of LC-SRM instrumentation. Samples are introduced into the LC-MS system via the autosampler. The HPLC pumps deliver a mixture of Buffer A (0.1 % formic acid) and Solvent B (0.1 % formic acid, 80 % acetonitrile) to the system. In a typical reversed-phase LC-MS analysis, the sample binds to the C18 HPLC column in buffer and peptides are gradually eluted by increasing the percentage of the organic-containing Solvent B. A triple quadrupole mass spectrometer is connected inline with the HPLC system. A triple quadrupole isolates a specific precursor ion with a known mass-to-charge (*m/z*) ratio in the first quadrupole (Q1). The precursor ion is fragmented in Q2 by collision-activated dissociation (CAD) to form product ions. A product ion of a specific *m/z* is isolated in Q3 and allowed to pass through the detector. Each Q1/Q3 *m/z* combination is called an SRM transition

ions ($F_{1,\,2,\,\text{etc.}}$). One or more of these fragments ions can then be selected by the third quadrupole to pass onto the detector thus generating a chromatographic trace for each $P>F_n$ transition. These traces are then used to determine the relative abundance of peptide analytes among samples. In practice, one often targets several peptides at a time in parallel acquisitions of SRM transitions, although this multiplexing approach comes at a cost of reduced sensitivity.

In SID, a synthetic peptide containing an isotopically labeled amino acid such as arginine ($^{13}C_6$,$^{15}N_4$-Arg) or lysine ($^{13}C_6$,15N2-Lys) is first used to optimize the $P>F$ transitions and can then be used to generate a standard curve for determining the limit of detection (LOD) and quantitation (LOQ). This standard is then subsequently spiked into the peptide mixture at a known concentration and later used for normalization before quantitation.

Here we demonstrate a method for the targeted quantitation of a previously identified lysine acetylated peptide ($_{172}$AFGGQSLK^{Ac}FGK$_{182}$) in wild-type (WT) and SIRT3 knockout (KO) animals on the mitochondrial protein succinate dehydrogenase A (SDHA) [2]. To control for biological variability we started with mitochondrial samples from five WT and five KO animals. The steps for proteolysis, solid phase extraction, peptide enrichment, and quantitation by LC-SRM using a heavy labeled synthetic peptide ($_{172}$AFGGQSLK^{Ac}FG**K**[+8] $_{182}$) are described. Using this technique we demonstrate a 6.0-fold increase (p value—2E−9) in acetylation at lysine-179 on SDHA in the SIRT3 KO animals.

2 Materials

2.1 Protein Digestion

1. Prepare crude or gradient purified mitochondria according to Rardin et al. [6] in the presence of deacetylase inhibitors: 10 mM nicotinamide and 0.5 μM trichostatin A.
2. 0.1 M Triethylammonium bicarbonate solution (TEAB), pH 8.5, in water.
3. 1 % (w/v) *n*-Dodecyl-β-D-maltoside in 0.1 M TEAB, store at −20 °C (*see* **Note 1**).
4. 10 M urea prepared fresh in 0.1 M TEAB.
5. Reducing reagent: 0.5 M Tris (2-carboxyethyl) phosphine Bond Breaker solution (Thermo Scientific, Rockford, IL, USA).
6. Alkylating reagent: 0.21 M iodoacetamide prepared fresh in 0.1 M TEAB.
7. Proteolysis: Sequencing grade modified trypsin.
8. Formic acid (FA) (*see* **Note 2**).
9. Low range pH strips.

2.2 Desalting

1. Solid phase extraction (SPE) cartridges: Oasis HLB 1 cc (30 mg) Extraction Cartridges (Waters, Milford, MA, USA).
2. Vacuum extraction manifold.
3. Acetonitrile (ACN).
4. Speedvac concentrator.

2.3 Immuno-precipitation

1. Immunoprecipitation buffer: 50 mM Tris–HCl, pH 8.0, 100 mM NaCl, 1 mM EDTA (NET) (*see* **Note 3**). Store at 4 °C.
2. Protein G agarose beads.
3. 1.5 mL polypropylene siliconized micro centrifuge tubes (*see* **Note 4**).
4. Anti-acetyl lysine antibody (*see* **Note 5**).
5. Peptide elution buffer: 40 % (v/v) ACN, 1 % (v/v) trifluoroacetic acid (TFA) in HPLC grade water.
6. Flat gel loading pipet tips.
7. ZipTipC18 pipette tips (Millipore, Billerica, MA, USA).

2.4 Mass Spectrometry

1. HPLC buffers: Buffer A (0.1 % formic acid) and Buffer B (0.1 % formic acid, 90 % acetonitrile).
2. Reversed-phase HPLC column: Eksigent Nano cHiPLC ChromXP C18 column, 75 μM inner diameter, 15 cm length, 3 μM particle size, designed for use with the Eksigent cHiPLC-Nanoflex System (*see* **Note 6**).

3. HPLC: Eksigent NanoLC-Ultra 2Dplus (*see* **Note 6**).
4. Triple quadrupole mass spectrometer: 5500QTRAP, with Analyst data acquisition software.
5. Skyline: http://proteome.gs.washington.edu/software/skyline

3 Methods

3.1 Denaturation, Reduction, Alkylation, and Digestion of Mitochondrial Protein

1. Centrifuge 1 mg of mitochondrial protein per sample for 10 min at 15,000 × *g* in a 1.5 mL microfuge tube. Aspirate off the supernatant and resuspend the pellet in 100 μL of 1 % maltoside. Vortex the tube as necessary until the lysate becomes clear. Add 100 μL of 10 M urea and mix by pipetting up and down 5–10 times.
2. Bring the volume up to ~990 μL using 0.1 M TEAB and add 9 μL of the TCEP reducing agent for a working concentration of 4.5 mM (*see* **Note 7**). Briefly vortex the sample to mix, then incubate the sample at 37 °C for 1 h.
3. Remove the sample from the incubator and briefly centrifuge to ensure the sample is at the bottom of the tube to prevent sample loss when opening the lid. Add 50 μL of the alkylating reagent for a final concentration of 10 mM. Briefly vortex the sample to mix, then incubate the sample in the dark at room temperature for 30 min.
4. Remove the sample from the dark and add 20 μg of trypsin resuspended in 50 μL of 0.1 M TEAB for a working trypsin to protein ratio of 1:50. Incubate the sample overnight at 37 °C.
5. Following proteolysis, remove the sample from the incubator and add concentrated FA in a stepwise fashion (2–4 μL at a time) (*see* **Note 8**) until the pH < 3 (*see* **Note 9**).

3.2 Desalting and Solid Phase Extraction (SPE)

1. Place SPE column into the vacuum manifold and add 3 × 400 μL of 0.1 % FA/80 % ACN in water to condition the column (*see* **Note 10**).
2. Turn on the vacuum pump and adjust the flow rate to a flow rate of 0.5 drops per second.
3. Add 4 × 400 μL of 0.1 % FA in water to equilibrate the column.
4. Add the acidified peptides to the column.
5. Add 4 × 400 μL of 0.1 % FA in water to wash the column. Turn off the vacuum pump when the wash solution is 0.5 cm from the top of the column. Remove the lid and place a 1.5 mL microfuge tube into the manifold to collect the sample during elution.
6. Add 3 × 400 μL of 0.1 % FA/80 % ACN in water and turn the vacuum pump on to elute peptides from the column into the

1.5 mL microfuge tube. Turn off the vacuum pump prior to removal of the column (*see* **Note 11**).

7. Place the sample into the Speedvac concentrator and evaporate the sample to near dryness.

3.3 Lysine Acetylated Peptide Enrichment

1. Peptides are resuspended in 500 μL of NET buffer.
2. Place 40 μL of protein G agarose slurry in a separate siliconized microfuge tube and wash the agarose beads 3 × 500 μL with NET buffer. Beads are then resuspended in 500 μL of NET buffer and the resuspended peptides from **step 1**.
3. Add 80 μg of anti-acetyl lysine antibody to the sample and incubate overnight at 4 °C with gentle rocking to mix the sample.
4. Samples are washed 3 × 1 mL with NET buffer for 5 min at 4 °C with gentle rocking (*see* **Note 12**).
5. Aspirate off the supernatant of the final wash of the NET buffer and add 200 μL of the peptide elution buffer.
6. Incubate the sample at 30 °C with gentle agitation for 5 min. Pellet the beads by centrifugation at 4,000 × *g* for 2 min, and remove 100 μL of the supernatant. Save the supernatant containing the eluted peptides in a separate microfuge tube. Add 100 μL of the peptide elution buffer to the beads and incubate the sample at 30 °C with gentle agitation for 5 min.
7. Repeat **step 6**.
8. Centrifuge the final elution at 3,000 × *g* for 2 min and collect the remaining supernatant of ~200 μL and combine with the previous elutions.
9. Speedvac the solution containing the eluted peptides to near dryness and resuspend in 30 μL of 0.1 % FA and 1 % ACN in HPLC grade water (*see* **Note 13**).
10. Desalt peptides two times using a C18 ZipTip (*see* **Note 14**).
11. Speedvac desalted peptides to dryness and resuspend samples in 20 μL of 0.1 % FA, 1 % ACN, in HPLC grade water (*see* **Note 15**).

3.4 Define SRM Transitions for the Peptide AFGGQSLK^{Ac}FGK Using Skyline

1. In the Transition settings menu, Prediction tab, choose the default collision energy equation for the ABI QTRAP 5500. The declustering potential used for all SRM transitions on the QTRAP 5500 is 100 and the collision cell exit potential is 40.
2. In the Transition settings menu, Filter tab, choose ions from m/z > precursor to last ion (*see* **Note 16**).
3. In the Transition settings menu, Library tab, pick four product ions (*see* **Note 17**).
4. For peptide AFGGQSLK^{Ac}FGK, chose SRM transitions corresponding to fragment ions (F) y9, y8, y6, and y4 (*see* **Note 18**).

5. A stable isotope-modified version of the AFGGQSLK^{Ac}FGK peptide synthesized with a $^{13}C_6{}^{15}N_2$-Lys at the C terminus is spiked into all samples. To make SRM transitions for this stable isotope-labeled peptide with an 8 Da mass shift, AFGGQ SLK^{Ac}FG**K**[+8], add a heavy isotope label modification in Skyline and include this peptide in the peptide list. Use the same collision energy, declustering potential, and transitions as the light (endogenous) AFGGQSLK^{Ac}FGK peptide.

3.5 Optimize and Validate LC-SRM Assay

1. A typical LC-MS instrument configuration for targeted assays is shown in Fig. 1 (*see* **Note 19**).
2. Use Skyline to optimize the collision energy and/or declustering potential for each transition using a synthetic peptide (*see* **Note 20**). The retention time of the peptide is used for retention time scheduling in the final assay.
3. The final SRM mass spectrometry parameters for AFGGQSLK^{Ac}FGK are:

Q1	Q3	Retention time	Peptide	Fragment ion	Declustering potential	Collision energy	Collision cell exit potential
591.3	963.5	31.76	AFGGQSLK^{Ac}FGK	y9	78.1	32.1	40
591.3	906.5	31.76	AFGGQSLK^{Ac}FGK	y8	88.1	30.1	40
591.3	721.4	31.76	AFGGQSLK^{Ac}FGK	y6	88.1	28.1	40
591.3	521.3	31.76	AFGGQSLK^{Ac}FGK	y4	98.1	36.1	40
595.3	971.5	31.76	AFGGQSLK^{Ac}FGK [+8]	y9	78.1	32.1	40
595.3	914.5	31.76	AFGGQSLK^{Ac}FGK [+8]	y8	88.1	30.1	40
595.3	729.4	31.76	AFGGQSLK^{Ac}FGK [+8]	y6	88.1	28.1	40
595.3	529.3	31.76	AFGGQSLK^{Ac}FGK [+8]	y4	98.1	36.1	40

4. Import the transition list from Skyline into Analyst for data acquisition on the 5500 QTRAP. The final assay uses retention time scheduling based on the peptide retention determined in the collision energy and declustering potential optimization analyses (*see* **Note 21**).
5. Perform a standard curve in triplicate by spiking in the stable isotope-labeled peptide to determine the linear range, limit of detection (LOD), and limit of quantitation (LOQ) (*see* **Note 22**). Use 250 ng of crude mitochondria as a sample matrix for each concentration point. Perform three independent calibration curves using with calibration points of 0, 4, 12, 40, 110, 330, 1,000, 3,000, and 25,000 amol per injection.
6. Integrate peaks with Skyline to determine the peak area of the stable isotope-labeled AFGGQSLK^{Ac}FG**K** peptide where **K** is $^{13}C_6$,$^{15}N_2$-Lys.

7. Calculate the LOD and LOQ with the following equations:

$$\mathrm{LOD} = x_b + 3s_b$$

$$\mathrm{LOQ} = x_b + 10s_b$$

where x_b is the mean concentration of the blank and s_b is the standard deviation of the blank.

3.6 Apply LC-SRM Assay to Quantify Targeted Acetylation Sites in Samples

1. Analyze each sample with 25 fmol heavy labeled AFGGQSLK^{Ac}FG**K** spiked in using the validated, retention time scheduled SRM assay.
2. Perform two replicate LC-SRM analyses of each sample.
3. Integrate peak areas in Skyline. Representative results are shown in Fig. 2b.

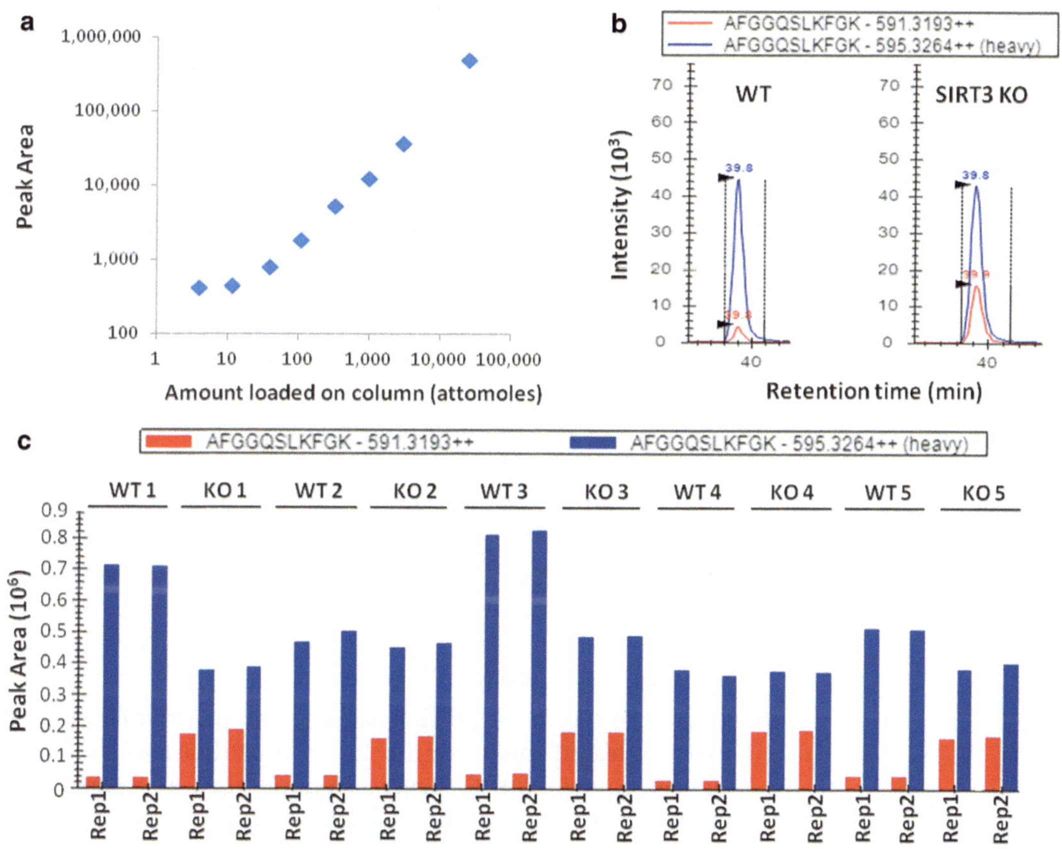

Fig. 2 LC-SRM analysis of peptide AFGGQSLK^{Ac}FGK from SDHA. (**a**) Concentration curve analysis of a synthetic, stable isotope-labeled AFGGQSLK^{Ac}FG**K** peptide spiked into a mitochondrial lysate is used to assess the linear range, limit of detection (LOD), and limit of quantitation (LOQ). (**b**) LC-SRM data of a wild-type (WT) and SIRT3 knockout mouse liver mitochondria. Demonstrating increased endogenous AFGGQSLK^{Ac}FGK peptide relative to the spiked-in synthetic, stable isotope-labeled AFGGQSLK^{Ac}FG**K** peptide. (**c**) Two LC-SRM technical replicates are performed per sample. In this study, five WT and five SIRT3 KO mouse livers were analyzed. The SIRT3 knockout mice have a 6.0-fold increase in levels of the acetylated AFGGQSLK^{Ac}FGK peptide

4. Use the peak area ratio of the light to heavy peptide to determine the molar amount of the AFGGQSLK^{Ac}FG**K** peptide in each sample (*see* **Note 23**).

4 Notes

1. Maltoside precipitates at 4 °C. Therefore it is necessary to warm it prior to use.
2. When working with concentrated formic acid (FA), work in a chemical fume hood or use a mask as it can be severely irritating to skin and mucous membranes.
3. Small amount of nonionic detergents such as NP-40 may be used during the immunoprecipitation step, but may cause ion suppression if not properly removed during the washing step. Therefore, we chose to leave out detergents from our sample preparation and did not observe a significant difference in peptide enrichment.
4. Siliconized tubes provide less friction than more commonly used microfuge tubes and allow for adequate mixing of the beads in the absence of detergent.
5. The amount of antibody needed is often greater then the normal aliquot sold by various vendors and may require contacting the vendor to inquire about availability.
6. This protocol for SRM quantitation of acetylated peptides uses a nanoflow HPLC (nanoliters per minute flow rate) since this approach maximizes assay sensitivity. The specific system described in this method uses an Eksigent Nanoflex as part of the LC system; however any C18 column and HPLC system compatible with flow rates from 250 to 500 nL/min can be used.
7. Concentrated denaturants, such as urea and maltoside, can inhibit trypsin activity during proteolysis. Therefore it is necessary to dilute out the sample to 1 M and 0.1 % (w/v) respectively, prior to the addition of trypsin.
8. The acidification of TEAB results in the production of CO_2 that is released following the addition of FA. This should be allowed to bleed off for several minutes at room temperature prior to moving on to the solid phase extraction step.
9. Low range pH strips provide the most efficient method for monitoring changes in pH while minimizing sample loss.
10. To maximize binding efficiency the column should not be allowed to dry out between changes in solutions once it has been wetted. Once the solution reaches ~0.5 cm above the column the next solution may be added.

11. To prevent sample loss following the elution of peptides from the column, turn off the vacuum pump prior to removal of the column. Allow pressure to equilibrate using the pressure release tab and *not* by removing the column above the eluate.
12. We find the use of a flat gel loading pipet tip minimizes the amount of agarose beads that are lost during removal of the supernatant.
13. If there is large carry over of beads from extraction of the supernatant, samples can be centrifuged for 5 min at 15,000 × g and the supernatant transferred to a new microfuge tube prior to desalting with the C18 ZipTips.
14. It is beneficial to perform two ZipTips to maximize the efficiency of peptide recovery. In addition, peptides are eluted from the ZipTip into the same elution buffer prior to concentrating.
15. It is essential to resuspend all samples with equal volumes of buffer to ensure the same amount of sample is loaded onto the mass spectrometer for quantitation.
16. Fragment ions with an m/z larger than the precursor are optimal for SRM transitions since they decrease interferences from singly charged precursor ions.
17. Three SRM transitions are typically used per peptide in the final assay. Therefore it is best to begin assay development with four or more transitions per peptide in case there are interferences from the matrix.
18. These fragment ions have the highest intensity in the MS/MS spectra of the peptide, although the highest abundance fragments do not always generate the $P>F$ transitions with the highest selectivity and lowest coefficient of variation.
19. This step is optional. The benefits of optimizing the collision energy and/or declustering potential for each transition varies by mass spectrometer. For the 5500 QTRAP, collision energy and declustering potential optimization improves sensitivity for very small and very large peptides. For collision energy optimization in Skyline, use a step size of 2 and for declustering potential use a step size of 10. Use five steps for each optimization step.
20. For this assay the autosampler is configured in "direct injection" mode, without a trap column, where the sample is transferred from the autosampler directly to the HPLC column inline with the mass spectrometer. A 1 μL sample loop is used for the autosampler. The LC flow rate is 300 nL/min. The LC gradient is 97 % Buffer A at 0–5 min, 85 % A at 8 min, 65 % A at 42 min, 10 % A at 45 min to 49 min, 97 % A at 50 min and equilibrated for at least 30 min prior to the next analysis.

21. Standard curves for peptides analysis typically begin at 1 amol and go as high at 1 pmol for nanoLC-SRM. The optimal experimental design is to spike in a stable isotope-labeled synthetic peptide into a matrix similar to the assay matrix in order to identify potential interferences. Complete each concentration curve in its entirety before acquiring additional curves. Perform at least three standard curves. There are several ways to calculate the LOD and LOQ [7], each with their own strengths and weaknesses.
22. Retention time scheduling methods acquire each SRM transition during a small window of time when the peptide of interest elutes. Since the peak width and the number of data points per peak that must be collected (7–9, typically) determine the maximum duty cycle of the assay, acquiring additional transitions decreases the dwell time per SRM transition. Therefore, retention time scheduling, rather than acquisition throughout the entire LC-Ms analysis, increases the number of transitions that can be measured during a single LC-MS analysis. Retention time scheduling is optional.
23. Absolute quantitation using heavy peptides have several caveats, including digestion efficiency and accurate quantitation of the synthetic peptides. In this assay, organelle enrichment and peptide immunopurification efficiency are additional variables. Therefore, the molar concentration in each sample is a "best estimate" and not a completely rigorous value.

Acknowledgments

This work was supported by NIH grant R24 DK085610 (B.W.G.) and the NCRR shared instrumentation grant S10 RR027953 (B.W.G.) for the 5500 QTRAP system.

References

1. Chen Y, Zhao W, Yang JS, Cheng Z, Luo H, Lu Z, Tan M, Gu W, Zhao Y (2012) Quantitative acetylome analysis reveals the roles of SIRT1 in regulating diverse substrates and cellular pathways. Mol Cell Proteomics 11(10):1048–1062. doi:10.1074/mcp.M112.019547, [pii] M112. 019547
2. Schilling B, Rardin MJ, MacLean BX, Zawadzka AM, Frewen BE, Cusack MP, Sorensen DJ, Bereman MS, Jing E, Wu CC, Verdin E, Kahn CR, Maccoss MJ, Gibson BW (2012) Platform-independent and label-free quantitation of proteomic data using MS1 extracted ion chromatograms in skyline: application to protein acetylation and phosphorylation. Mol Cell Proteomics 11(5):202–214. doi:10.1074/mcp.M112.017707, [pii] M112.017707
3. Addona TA, Abbatiello SE, Schilling B, Skates SJ, Mani DR, Bunk DM, Spiegelman CH, Zimmerman LJ, Ham AJ, Keshishian H, Hall SC, Allen S, Blackman RK, Borchers CH, Buck C, Cardasis HL, Cusack MP, Dodder NG, Gibson BW, Held JM, Hiltke T, Jackson A, Johansen EB, Kinsinger CR, Li J, Mesri M, Neubert TA, Niles RK, Pulsipher TC, Ransohoff D, Rodriguez H, Rudnick PA, Smith D, Tabb DL, Tegeler TJ, Variyath AM, Vega-Montoto LJ, Wahlander A, Waldemarson S, Wang M, Whiteaker JR, Zhao L, Anderson NL, Fisher SJ,

Liebler DC, Paulovich AG, Regnier FE, Tempst P, Carr SA (2009) Multi-site assessment of the precision and reproducibility of multiple reaction monitoring-based measurements of proteins in plasma. Nat Biotechnol 27(7):633–641. doi:10.1038/nbt.1546, [pii] nbt.1546

4. Keshishian H, Addona T, Burgess M, Kuhn E, Carr SA (2007) Quantitative, multiplexed assays for low abundance proteins in plasma by targeted mass spectrometry and stable isotope dilution. Mol Cell Proteomics 6(12):2212–2229. doi:10.1074/mcp.M700354-MCP200, [pii] M700354-MCP200
5. Gerber SA, Rush J, Stemman O, Kirschner MW, Gygi SP (2003) Absolute quantification of proteins and phosphoproteins from cell lysates by tandem MS. Proc Natl Acad Sci U S A 100(12):6940–6945. doi:10.1073/pnas.0832254100, [pii] 0832254100
6. Rardin MJ, Taylor GS, Dixon JE (2009) Distinguishing mitochondrial inner membrane orientation of dual specific phosphatase 18 and 21. Methods Enzymol 457:275–287. doi:10.1016/S0076-6879(09)05015-0, [pii] S0076-6879(09)05015-0
7. Shrivastava A, Gupta VB (2011) Methods for the determination of limit of detection and limit of quantitation of the analytical methods. Chron Young Sci 2(1):21–25. doi:10.4103/2229-5186.79345

Chapter 9

Identification of Deacetylase Substrates with the Biotin Switch Approach

J. Will Thompson, Alex Robeson, and Joshua L. Andersen

Abstract

The identification of lysine-acetylated proteins and deacetylase substrates has primarily relied on protein immune-affinity techniques with antibodies that recognize acetylated lysine residues (Kac antibodies). While these antibody-based techniques are continuously improving, they can be limited by the narrow and many times unknown epitope specificity of Kac antibodies. An alternative approach is the biotin switch capture of deacetylated proteins. Similar in part to other biotin switch methodologies, this technique relies on the blocking of native lysine residues and removal of the modification of interest in vitro, after which the newly deacetylated proteins can be captured and identified by mass spectrometry (MS). In this chapter, we cover the essential steps of the procedure, highlight key points in the assay to reduce false positive protein identification, and discuss the quantitative MS methods useful for identifying the captured deacetylase substrates. We also discuss potential strategies and future improvements to overcome current limitations of the assay.

Key words Acetylation, Deacetylase, Sirtuin, Proteomics, Mass spectrometry, Biotin switch

1 Introduction

Protein lysine acetylation is a reversible posttranslational modification (PTM) commonly considered in the context of histone modifications and transcriptional regulation. However, recent reports [1–4] describing the acetylation of numerous cytoplasmic and metabolic proteins have suggested that this PTM may be more abundant across the proteome than originally thought. Acetylation has been shown to modulate protein–protein interactions, DNA–protein interactions, subcellular localization of proteins, transcriptional activity, enzyme activity, and protein stability (reviewed in ref. 5). Yet, considering the number of identified acetylation events in the proteome, it continues to be a somewhat underexplored PTM; particularly in comparison with more commonly studied PTMs, such as phosphorylation and ubiquitylation. Nevertheless, the clinical significance of deacetylases (HDACs and Sirtuins), the enzymes

Matthew D. Hirschey (ed.), *Sirtuins: Methods and Protocols*, Methods in Molecular Biology, vol. 1077,
DOI 10.1007/978-1-62703-637-5_9, © Springer Science+Business Media, LLC 2013

that catalyze the removal of acetyl groups from lysine residues, is evidenced by the numerous ongoing clinical cancer trials examining HDAC inhibitors, and the role of Sirtuins in metabolism and various human diseases, including cancer and neurodegenerative disorders [6].

Current unbiased approaches to identify deacetylase substrates are primarily peptide-based immuno-affinity techniques [1, 7, 8] and, although powerful, these experiments can be hindered by the narrow epitope range of commercially available Kac antibodies. Other approaches to identify specific deacetylase substrates are more targeted, and commonly rely on detecting physical interaction between the enzyme and its substrate, or through a process of elimination (e.g., targeting deacetylases with RNAi) in the analysis of a given acetylated protein—thus published substrates likely represent a small fraction of those actually targeted by these enzymes. In this regard, unbiased proteomics and chemical biology-based approaches may complement the current standard approaches for identifying deacetylase substrates and enhance our understanding of this PTM.

The acetyl-biotin switch technique (Acetyl-BST) was developed to address some of these limitations. Acetyl-BST is based upon the combination of two approaches which were developed independently; namely the biotin switch technique of Jaffrey and the enzymatic removal of PTMs such as deglycosylation via PNGase F for mapping of protein glycosylation sites on a proteomic scale [9–11]. Jaffrey and Snyder (2001) first described the use of biotin switch as a method for identifying S-nitrosylated proteins in cell lysates [9]. In their method, S-nitrosothiols are "switched" for a biotinylated thiol in a three-step process. First, thiols on free (reduced) cysteines are blocked with a methylthiolating agent under denaturing conditions. Second, the S-nitrosylated thiols are reduced to thiols with ascorbate treatment and finally the newly freed cysteine residue is derivatized with biotin using a sulfhydryl-specific reagent. The biotin-tagged proteins are then purified on streptavidin resin and analyzed by gel or mass spectrometry (MS). This biotin switch methodology was extended to mapping S-acylation sites in a technique termed the Acyl-Biotin Exchange (ABE), by substituting the thioester-specific reagent hydroxylamine for ascorbate to cleave the targeted modification [12].

These biotin switch protocols utilize chemical cleavage of the targeted PTM, and we reasoned that enzymatic cleavage of specific modifications of interest might also work for this purpose. As an example, enzymatic cleavage using PNGase F after glycoprotein capture has been previously used to quantify sites of N-linked glycosylation in plasma and other matrices [10, 11, 13, 14], showing that enzymatic release of modifications can be efficient and reproducible. Enzymatic reactions have the additional advantage of being inherently specific to the substrate of interest. We therefore

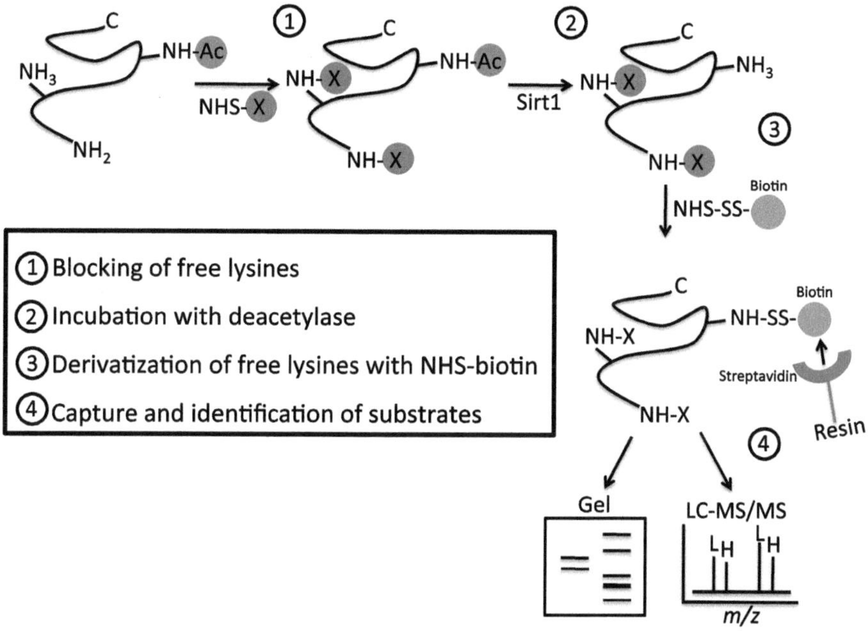

Fig. 1 Summary of the acetyl-biotin switch method

designed the Acetyl-BST as a combination of two techniques whereby we could block free (underivatized) lysine residues, cleave the target acetylation site(s) by enzymatic deacetylation, then, using amine-directed reagents, biotinylate these newly freed lysine residues, and capture using streptavidin.

The Acetyl-BST is performed in cell extract or lysate at neutral pH, at which *N*-hydroxysuccinimidyl ester (NHS) forms a stable amide bond specifically with the ε-amine of lysine residues. As shown in Fig. 1, the assay involves three main steps: In short, cell lysates are treated with *N*-succinimidyl-*N*-methyl carbamate (*N*-SMMC) to stably block PTM-free (non-derivatized) lysines, producing a mixture of proteins with blocked lysines or lysines previously acetylated by endogenous acetylases. The lysate is then incubated with recombinant, exogenous deacetylase and the lysines freed by deacetylase activity are derivatized with NHS-SS-biotin and captured on streptavidin resin. The putative deacetylase substrates are then eluted with a reducing agent and analyzed by SDS-PAGE or trypsinized for characterization by LC–MS/MS, which will determine the identity of the protein and allow identification of the specific lysine residue targeted by the deacetylase (Fig. 1).

The biotin switch methodology presented here provides a proteome-scale approach to identifying substrates of specific deacetylases, as well as the lysine residues they target. Such chemical biology/proteomics-based strategies offer the advantage of being unbiased and are capable of quickly generating a proteome-wide set of data. These approaches are particularly useful in cases where more common methods (e.g., molecular and immuno-affinity-based strategies) are unavailable or still early in development (e.g., methods to assess S-nitrosylation and protein acylation).

As with any developing methodology, this approach has weaknesses, including the capture of nonspecific background proteins, as well as the loss of potential deacetylase substrates at the biotin/streptavidin capture step. We propose that certain steps may be made to overcome these issues. The original biotin switch approach of Jaffrey and Snyder [9], for identifying S-nitrosylated proteins in cell lysates, has been improved substituting the biotin derivatization with a resin-assisted capture (RAC) step, in which the ascorbate-treated cysteines are purified on a thiol-reactive resin, thus allowing the biotin/streptavidin step to be bypassed [15]. This change in methodology allows for better coverage of high-MW proteins, and the ability to perform on-resin digestion and therefore enrich specifically for *peptide* substrates as opposed to the entire protein, thus giving deeper overall proteome coverage because of reduced sample complexity. Similarly, the development of a RAC analog to the acetyl-biotin switch would likely be a significant improvement for the methodology. Additionally, to address inefficiency at the blocking step, it will be important to thoroughly evaluate different blocking reagents (various amine-reactive NHS reagents are commercially available) as some may work better in certain systems due to differences in their amine-targeted chemical moieties, solubility/activity in lysate, or propensity to cause protein precipitation. Finally, we expect that the core Acetyl-BST methodology could be extended to a variety of other PTMs for which robust enrichment methodologies do not currently exist, such as other lysine acyl modifications or sumoylation.

2 Materials

2.1 Stock Solutions

1. Egg lysis buffer: 250 mM sucrose, 2.5 mM MgCl, 1 mM DTT, 50 mM KCl, 10 mM HEPES pH 7.7, phosphate-buffered saline (PBS) pH 7.4.
2. 2 % L-cysteine monohydrate (diluted in water) adjusted to a pH of 7.8 with NaOH.
3. Stock solutions of cyclohexamide (100 mg/ml), cytochalasin B (10 mg/ml), and aprotinin/leupeptin (1 mg/ml).
4. *N*-succinimidyl *N*-methyl carbamate (*N*-SMMC) stock: *N*-SMMC is solubilized in DMSO to a stock concentration of 400 mM. Make immediately prior to use.
5. Tris–HCl pH 8. Zebaspin desalting columns (Pierce, Part 89893).
6. His-tagged recombinant deacetylase (His-Sirt1 can be obtained commercially from Enzo Life Sciences).

7. 50 mM NAD^+: solubilize in water to a stock concentration of 50 mM.
8. Ni-NTA agarose resin.
9. Sulfo-NHS-SS-biotin is solubilized in DMSO to a stock concentration of 400 mM. Make immediately prior to use.
10. Tris–HCl pH 8. Zebaspin desalting columns.
11. Streptavidin agarose resin.
12. PBS wash buffer.
13. PBS wash buffer with 0.1 % SDS.
14. Standard SDS-PAGE electrophoresis sample buffer (LDS).
15. Dithiothreitol (DTT) is diluted in water to a stock concentration of 1 M.
16. 50 mM ammonium bicarbonate (AmBic).
17. 200 mM iodoacetamide.

2.2 LC–MS/MS Solutions

1. 1 % Rapigest SF.
2. Trypsin.
3. Sample acidification solution: 1 % v/v TFA and 2 % v/v MeCN.
4. ADH1_YEAST digest (MassPrep, Waters Corporation).
5. LC–MS/MS buffer: 0.1/2/98 v/v/v TFA/MeCN/water.

3 Methods

3.1 Xenopus Egg Extract

1. To generate the extract, eggs were crushed by centrifugation in egg lysis buffer and the cytoplasmic fraction was extracted according to standard Xenopus egg extract protocol [16] as summarized here (*see* **Notes 1–2**).
2. Xenopus eggs are collected in a 500 ml glass cylinder, excess water is removed, and eggs are dejellied for 3–5 min in 2 % cysteine pH 7.8.
3. Eggs are rinsed in egg lysis buffer five times (150 ml per wash) by gently pouring in buffer, swirling cylinder, and disposing of excess buffer (after final wash, the eggs should be in nearly 100 % egg lysis buffer).
4. Eggs are transferred to a 15 ml falcon tube and packed by centrifugation at 1,000 × *g* for 15 s in a clinical centrifuge. As much buffer as possible is then aspirated, leaving behind the packed eggs and trace amounts of egg lysis buffer (this step should simply pack eggs, not crush them).
5. Based on an estimation of the number of milliliters of packed eggs in the tube, cytochalasin B, aprotinin/leupeptin, and

cyclohexamide are added to a final concentration of 10, 1, and 100 μg/ml, respectively.

6. Eggs are then spin-crushed by centrifuging at 12,000 × *g* for 10 min.
7. The cytoplasmic layer (just below the top lipid layer) is then extracted from the tube via side puncture with a 20-G needle and 1 ml syringe.
8. Egg cytoplasm is then transferred to a 1.5 ml test tube and kept on ice until needed for the experiment.

3.2 Biotin Switch: Blocking

1. *N*-SMMC is prepared fresh by dissolving in DMSO to a concentration of 400 mM.
2. Prior to addition of *N*-SMMC to the samples, the lysate should be diluted to approximately 1 mg/ml protein concentration to limit protein precipitation (egg lysates were diluted in PBS pH 7.4).
3. *N*-SMMC is added to the lysate to a final concentration of 5 mM and incubated for 1 h at room temperature with gentle rocking (*see* **Notes 3–6**).
4. In order to quench the *N*-SMMC reagent, Tris–HCl at pH 8 is diluted in the *N*-SMMC-treated extract to a final concentration of 100 mM, followed by gentle rocking for 30 min at room temperature.
5. In order to remove excess Tris-HCl, the entire *N*-SMMC-treated lysate (2 ml) is run through a Zebaspin desalting column, which is pre-equilibrated with PBS and used according to the manufacturer's protocol (*see* **Note 7**).
6. The sample flow through is collected and the protein concentration is again determined to ensure that no protein was lost on the column.
7. At this point, a small amount of protein (~5 μg) can be reserved to confirm free lysine derivatization by MS.
8. The rest of the extract should be divided into two equal parts (one for enzyme addition and the other for a mock control).

3.3 Biotin Switch: Deacetylation

1. His-tagged deacetylase is added to the deacetylase extract (*see* **Note 8**).
2. For Sirt1, the extract is supplemented with 0.5 mM NAD^+.
3. NAD^+ is also added as a control to the "mock" extract (optimally, enzymatically inactive deacetylase could be used as a control—in our case, an elution from an untransformed bacterial preparation was used).
4. The final concentration of deacetylase must be determined empirically and will vary due to the relative activity of a given deacetylase preparation.

5. Incubate the enzyme 3–4 h at 30 °C (or whatever temperature best suits the enzyme of interest) with gentle rocking.
6. After deacetylase incubation, the His-tagged enzyme is removed by nickel resin. The amount of resin needed can be assessed by titrating the nickel resin in a pilot pull down experiment (20 μl under the conditions described).

3.4 Biotin Switch: Conjugation

1. Sulfo-NHS-SS-biotin is added to the mock and deacetylase-treated extracts at a final concentration of 1 mM, followed by incubation at room temperature for 30 min with gentle rocking (*see* **Note 9**).
2. As in Subheading 3.2, **step 4**, the NHS reagent (sulfo-NHS-SS-biotin, in this step) is quenched by adding Tris–HCl pH 8 to the extract at a final concentration of 100 mM (30 min incubation at room temperature with rocking).
3. The excess biotin is then removed with a Zebaspin desalting column according to manufacturer's instructions.

3.5 Biotin Switch: Immunoprecipitation

1. A small volume (~15–30 μl depending on the capacity of the resin) of streptavidin resin is added to the mock- and deacetylase-treated extract, followed by incubation at room temperature for 1 h on a tube rotator/rocker.
2. The streptavidin resin is retrieved by centrifugation and washed twice with cold PBS supplemented with 0.1 % SDS, followed by three washes with cold PBS (1 ml per wash and centrifugation at 10,000 rcf and aspiration of supernatant between washes).
3. At this step, a small portion of the resin slurry can be suspended in LDS loading buffer for visualization on SDS-PAGE.
4. The remaining resin is reserved for LC–MS/MS analysis.
5. The pelleted resin is washed twice with 500 μl 50 mM AmBic.
6. As with the PBS washing steps, one should be especially careful at this step not to aspirate the resin when removing the supernatant.
7. After the final wash, 50 μl of AmBic is added to yield a total volume of approximately 70 μl.

3.6 Mass Spec: Preparation

1. 7 μl of 100 mM DTT in AmBic is added to free the substrates from the NHS-SS-biotin (disrupting the disulfide link).
2. The DTT-treated samples are incubated with vigorous shaking at 900 rpm and 37 °C on a Thermomixer R (Eppendorf) for 45 min.
3. Samples are cooled to room temperature and treated with 7 μl of 200 mM iodoacetamide to alkylate all free sulfhydryls.

4. After pelleting the resin by centrifugation, 60 μl of the supernatant is transferred to another tube.
5. The remaining pellet is washed again with 50 μl AmBic followed by centrifugation.
6. After which, the supernatant (~50 μl) is combined with the other 60 μl of supernatant.
7. The combined supernatant (~110 μl) is supplemented with 10 μl of 1 % Rapigest SF and then digested overnight at 37 °C by adding 0.3 μg trypsin.
8. After digestion, the sample is acidified to 1 % v/v TFA and 2 % v/v MeCN, and then heated to 60 °C for 2 h to inactivate the surfactant.
9. Each sample for LC–MS/MS analysis is then typically spiked with a surrogate standard protein digest for use as a normalization control in the LC–MS assay.
10. Our standard approach is to use approximately 50 fmol of ADH1_YEAST digest per sample injection or approximately 100 fmol per sample.
11. In final preparation for analysis, the sample is then lyophilized and resuspended in 20 μl of a suitable LC–MS/MS buffer (e.g., 0.1/2/98 v/v/v TFA/MeCN/water).

3.7 Mass Spec: Analysis

1. Samples generated from the Acetyl-BST approach can be analyzed by most commonly utilized LC–MS/MS instrumentation and database searching algorithms.
2. We have demonstrated the approach using label-free quantitation previously; however, there are distinct advantages and disadvantages of each quantitative approach, which should be carefully considered within each analytical laboratory's workflow (*see* **Notes 10–17**).
3. Several strategies could be utilized to validate putative substrates. Our method of choice was to overexpress their epitope-tagged cDNAs in 293 T cells, followed by immunoprecipitation and immunoblotting with Kac antibodies (*see* **Note 18**).

4 Notes

1. The Acetyl-BST approach was designed to work in cell lysates. Since egg extracts are essentially purified concentrated cytosol, other cell lysis (or tissue homogenization) procedures that preserve protein integrity should work similarly. For example, a gentle hypotonic lysis protocol for tissue culture cells would produce a lysate similar (in terms of intact proteins and organelles) to egg extract.

2. Since the goal of the assay is to generate an array of candidate substrates of a given deacetylase, the first step is to consider a lysis procedure that will solubilize the subcellular compartment where the deacetylase in question resides (i.e., mitochondrial lysates should be used for mitochondrial deacetylases). Initial Acetyl-BST experiments were done in Xenopus egg extract. The Xenopus egg extract system offers the advantage of easily producing milligram quantities of purified cytosol (only trace amounts of buffer contamination are left in the final extract). For tissue culture cell application, lysis buffers will need to be evaluated for their compatibility with the Acetyl-BST reagents described below. Two key lysis buffer considerations are (1) whether the lysis buffer in question contains free amines, which react with and inactivate NHS reagents and (2) the pH of the buffer, since NHS reagents hydrolyze more readily at high pH.
3. Sufficient blocking of free lysine residues is critical for reducing false positive substrate identification, since any free lysines remaining will likely increase the background biotinylation and protein pull down, independent of the deacetylation reaction. Even with adequate blocking, proper controls must be used to ensure that biotinylated proteins (at the final step) are not misclassified as deacetylase substrates. Since lysine residues carry a primary ε amine, we chose to utilize a primary amine-targeted strategy that would block PTM-free lysines. Lysine blocking here prevents the reactivity of PTM-free lysines in downstream derivatization steps. Endogenous modification of the ε-amine of lysine by PTMs such as acetylation or ubiquitinylation also renders these residues inactive towards primary amine-directed reagents.
4. The fresh preparation and quick usage of *N*-SMMC is critical because it readily hydrolyzes and becomes inactive in aqueous solution. Also, the rate of hydrolysis increases with increasing pH, so this should be a consideration when choosing a buffer for the deacetylase reaction (pH 7.5 is recommended to limit speed of hydrolysis).
5. As shown in Fig. 2, incubating extract with increasing concentrations of *N*-SMMC, followed by protein capture on sulfo-NHS-SS-biotin, results in a reduction in background pull down proteins at 5 mM *N*-SMMC. The conditions given above for the egg extract (5 mM *N*-SMMC for 1 h at room temperature) were determined using a pilot experiment similar to that shown in Fig. 2, and this type of pilot screening of blocking conditions is recommended for each unique lystate and blocking reagent.

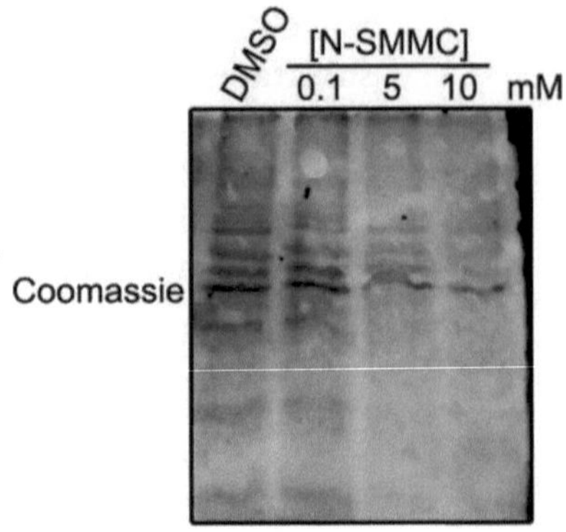

Fig. 2 Blocking of extract with *N*-SMMC reduces background binding of proteins to NHS-SS-biotin. 5 ml of extract (1 mg protein/ml) was treated with 5 mM *N*-SMMC, followed by *N*-SMMC inactivation by 100 mM Tris-HCl, Zebaspin desalting, and pull down on NHS-SS-biotin. Biotin-tagged proteins were captured on a streptavidin agarose resin and resolved by SDS-PAGE (12 % gel) and coomassie stained. The decrease in coomassie protein staining at 5 and 10 mM *N*-SMMC suggests that *N*-SMMC blocking is effective

6. Importantly, the concentration of *N*-SMMC will likely vary from lysate to lysate, and high concentrations (i.e., 10 mM) can produce visible protein precipitation and should be avoided. The concentration and incubation time should be determined empirically prior to performing the Acetyl-BST. Also, a variety of NHS and sulfo-NHS reagents are commercially available that could be used in place of *N*-SMMC, and may show more efficient blocking or less protein precipitation in a particular system of interest.
7. *N*-SMMC must be quenched prior to the enzyme incubation step in order to avoid unwanted amine labeling, which could result in negative effects on enzyme activity or potentially even blocking of lysine residues which were formerly acetylated and "freed" by the enzyme treatment.
8. Since the specificity of the pull down depends on adequate *N*-SMMC blocking of the lysate and any unnecessary proteins added along with the deacetylase (as well as the deacetylase itself) will not be blocked, it is critical that the enzyme preparation be as pure as possible (assessed by coomassie gel staining with infrared scanning or other methods). Optimally, only a single band for the deacetylase of interest should be visible on a gel. To achieve this level of purity, it will likely be necessary to wash the deacetylase resin (prior to elution) with a mild detergent buffer (such as 0.25 % NP40). A suitable affinity tag (e.g., 6xhis) on the deacetylase is also necessary since the enzyme must be removed by an affinity resin after this step.

For these reasons, we recommend the use of highly purified commercial deacetylase if it is available.

9. The addition of a sulfonate group to the NHS ring creates an NHS ester that is more soluble in water, which is important for this application. As mentioned above for *N*-SMMC, sulfo-NHS esters hydrolyze over time in water, rendering them useless in the assay. Thus, it is critical that the sulfo-NHS-SS-biotin be made just prior to the assay and used quickly. Also, the pH of the buffer here is a critical point, since NHS esters hydrolyze more readily at higher pH (pH 7.5 is recommended). In addition, while the disulfide linkage between NHS and biotin makes for easy dissociation of the pull down substrates from the streptavidin resin, no reducing agent can be used prior to that step due to the risk of prematurely disrupting the disulfide bond, which would result in a loss of any putative substrates.
10. Samples generated from the Acetyl-BST approach can be analyzed by most commonly utilized LC–MS/MS instrumentation and database searching algorithms, although we heavily encourage the use of high-resolution accurate mass instrumentation to lower false discovery rate for peptide identification, which becomes especially important for site-localization within the formerly acetylated peptides. Based on our experience with the Acetyl-BST approach, the protein yield from a relatively large preparation of several milligrams of starting material will be just a few micrograms of protein (or less). Therefore, we recommend analysis of approximately ¼ to ½ of the material isolated from the preparation as a starting point when using standard nanoscale LC coupled to high-resolution tandem mass spectrometers such as QToF or Orbitrap instruments [17–20]. As an example, our published work for Acetyl-BST utilized commercial nanoscale chromatography (nanoAcquity, Waters Corporation) with 75 μm × 25 mm column packed with 1.7 μm BEH C18 particles (Waters) and a high-resolution accurate mass QToF MS system (Synapt HDMS, Waters) [21]. If pilot experiments indicate significant protein yield, then a peptide-level micro-BCA assay (Pierce) can be used to determine peptide content and normalize appropriately, although typically protein yield may not be sufficient for this measurement.
11. Our standard protocol (*N*-SMMC blocking of total protein, deacetylase treatment, and NHS-SS-biotin derivatization of acetyl sites) requires that the database search parameters include a number of amino acid modifications, since the entire protein is being isolated for mass spectrometry and not just the modified peptide of interest. First, cysteine residues should be essentially 100 % alkylated because of the robust nature of DTT reduction and alkylation with iodoacetamide; therefore, we recommend searching carbamidomethyl Cys (+57.0215) as

a fixed modification. Next, lysine residues will have the possibility of existing in at least three states, and we recommend allowing for each of these as a variable search modification. Lysine residues may remain free ("unblocked") or be blocked with methyl carbamate (+57.0215, Unimod Acc #4). Additionally, the NHS-SS-biotin is cleaved by dithiothreitol; then the free sulfhydryl is alkylated, generating a CAM thiopropanoyl modification of Lys (+145.0197, Unimod Acc #293). CAMthiopropanoyl-modified Lys residues are the putative sites of deacetylase activity, but should be verified by quantitative comparison between treated sample and untreated control, as discussed below. Finally, when performing database searching we recommend using trypsin enzyme specificity but allowing at least three missed cleavages, since the modified lysine residues will not be recognized by trypsin but will still be called as "missed" in the database search scoring algorithms.

12. Proteomic identification of bona fide deacetylase substrates will be most rigorous when a comparison can be made to a negative control. The corresponding untreated sample (no exogenous deacetylase) can serve as an appropriate control, and, in many cases, this control will contain low levels of protein/peptide of interest versus the treated sample [9, 15]. Confidence in the identification of acetylated proteins can be enhanced by verification using other approaches such as immunoprecipitation followed by Western blotting [21]. Similarly, validation of acetyl site assignment can be accomplished by immunoprecipitation of the candidate protein followed by digestion and LC–MS/MS analysis to identify the acetyl-modified peptides. In addition, recent global acetylome measurements can serve as a useful resource to confirm that substrates identified by Acetyl-BST are indeed acetylated [1, 22, 23].

13. Acetyl-BST is amenable to most popular quantitative LC–MS/MS approaches including label-free (spectral counting and peak intensity quantitation), stable isotope labeling of amino acids in cell culture (SILAC), isotopic and isobaric labeling of resin-bound S-nitrosothiol (SNO)-site peptides (iTRAQ/TMT), as well as label-free quantification. We will give brief recommendations on how one would perform the approach with SILAC and iTRAQ labeling, but will focus on label-free quantification based on peak intensity (area-under-the-curve, AUC) measurements.

14. SILAC is a preferred MS-based peptide and protein quantitation strategy for many mass spectrometry laboratories and has been utilized for quantification of S-nitrosothiols using the biotin switch technique [24]. This method requires conditioning of cells with media containing light or heavy (i.e., $^{13}C^{15}N$) Arg and Lys and, for this reason, is most applicable to transformed

cell lines. Nonetheless, the stable endogenous isotope labeling of protein provided by SILAC should allow for relative quantification of deacetylase substrates between conditions or between test (+HDAC) and control (−HDAC). Because exogenous treatment with deacetylase must be performed on these two samples separately, the majority of the assay prior to the biotinylation step must be performed independently on test and control samples. This limits one of the traditional attractive qualities of SILAC, which is to combine samples early in the protocol to limit the effects of sample handling protocol on relative quantitation.

15. We have formerly utilized isobaric tagging (iTRAQ/TMT) in the RAC and relative quantitation of S-nitrosothiols (SNO-RAC) and S-acylations (acyl-RAC) but have not done so for the Acetyl-BST [15, 25]. The rationale for this lies in the fact that iTRAQ labeling in RAC samples allows for on-resin labeling after digestion (also on-resin) and subsequent combination of the resins prior to elution with reductant. This can lower the variability in recovery between samples and in theory reduce the quantitative variability between conditions. However, in the Acetyl-BST the capture and elution is performed at the protein-level, and labeling would have to be performed after digestion and essentially immediately prior to LC–MS/MS analysis. Therefore, we only recommend isobaric tagging approaches for Acetyl-BST for laboratories which prefer this quantitative approach over SILAC or label-free quantitative approaches.
16. Unlike the SILAC or isobaric tagging methods, label-free quantitative proteomic approaches do not require additional tagging, can be used over a wide of biological samples which are not amenable to metabolic labeling, and have no inherent sample size limitations. An important distinction is that each sample is assayed independently, which requires more care with respect to protocol and analytical variability. While spectral counting-based label-free quantitation can be utilized with Acetyl-RAC, we recommend using it primarily for presence/absence type determinations because of the relatively poor analytical reproducibility especially at low spectral counts [26]. AUC label-free quantitation is a robust alternative which typically relies on alignment of peptides across replicate analyses using their accurate mass and retention time, and can be accomplished using a number of commercial or free software packages [27–33].
17. As with other quantitative approaches for Acetyl-BST, the main contributor to analytical variability is upstream of the LC–MS/MS analysis in the sample handling steps. One critical step, as mentioned above, is to avoid aspiration of the resin

during the final resin washing steps. A surrogate standard should also be added equally to all samples at a low but measurable level, to be utilized for intensity normalization across LC–MS analyses; this can be especially important for the negative control sample, as there may be very little other signal in the sample and the standard is needed to ensure accurate injection and data collection. As described above, MassPrep alcohol dehydrogenase (or similar exogenous protein digest) can be added at 20–25 fmol per injection volume (typically 2–4 μl). Finally, when analyzing a number of conditions it is best to randomize sample injections to avoid temporal bias and to perform technical replicate injections of each sample in addition to biological replicates to obtain statistical confidence.

18. As with many unbiased screening approaches, LC–MS/MS analysis of Acetyl-BST-captured substrates has the potential for generating an overwhelming amount of data, some of which may be false positive. This puts the onus on substrate validation. The most practical method of validation is likely the overexpression of a given candidate substrate in tissue culture cells treated with and without a deacetylase inhibitor or deacetylase RNAi, followed by affinity purification and immunoblotting with Kac antibodies. This approach works well for many overexpressed proteins, but has the significant caveat of not working for every protein due to the narrow epitope range of many Kac antibodies. This is a common complaint by many in the acetylation field. Alternatively, the above experiment could be performed in cells incubated with ^{14}C-labeled acetyl-CoA and the affinity-purified overexpressed protein could be evaluated for acetylation by autoradiography.

Acknowledgements

We thank Drs. Arthur Moseley, Sally Kornbluth, and Erik Soderblom for their scientific input. This work was supported by an Alexander and Margaret Stewart Trust grant to J.L.A.

References

1. Choudhary C, Kumar C, Gnad F, Nielsen ML, Rehman M, Walther TC, Olsen JV, Mann M (2009) Lysine acetylation targets protein complexes and co-regulates major cellular functions. Science 325(5942):834–840. doi:10.1126/science.1175371, [pii] 1175371
2. Kim SC, Sprung R, Chen Y, Xu Y, Ball H, Pei J, Cheng T, Kho Y, Xiao H, Xiao L, Grishin NV, White M, Yang XJ, Zhao Y (2006) Substrate and functional diversity of lysine acetylation revealed by a proteomics survey. Mol Cell 23(4):607–618. doi:10.1016/j.molcel.2006.06.026
3. Wang Q, Zhang Y, Yang C, Xiong H, Lin Y, Yao J, Li H, Xie L, Zhao W, Yao Y, Ning ZB, Zeng R, Xiong Y, Guan KL, Zhao S, Zhao GP

(2010) Acetylation of metabolic enzymes coordinates carbon source utilization and metabolic flux. Science 327(5968):1004–1007. doi:10.1126/science.1179687, [pii] 327/5968/1004
4. Zhao S, Xu W, Jiang W, Yu W, Lin Y, Zhang T, Yao J, Zhou L, Zeng Y, Li H, Li Y, Shi J, An W, Hancock SM, He F, Qin L, Chin J, Yang P, Chen X, Lei Q, Xiong Y, Guan KL (2010) Regulation of cellular metabolism by protein lysine acetylation. Science 327(5968):1000–1004. doi:10.1126/science.1179689, [pii] 327/5968/1000
5. Yang XJ, Seto E (2008) Lysine acetylation: codified crosstalk with other posttranslational modifications. Mol Cell 31(4):449–461. doi:10.1016/j.molcel.2008.07.002
6. Guarente L (2011) Franklin H. Epstein Lecture: Sirtuins, aging, and medicine. N Engl J Med 364(23):2235–2244. doi:10.1056/NEJMra1100831
7. Cohen T, Yao TP (2004) AcK-knowledge reversible acetylation. Sci STKE 2004(245):pe42. doi:10.1126/stke.2452004pe42, [pii] stke.2452004pe42
8. Lombard DB, Alt FW, Cheng HL, Bunkenborg J, Streeper RS, Mostoslavsky R, Kim J, Yancopoulos G, Valenzuela D, Murphy A, Yang Y, Chen Y, Hirschey MD, Bronson RT, Haigis M, Guarente LP, Farese RV Jr, Weissman S, Verdin E, Schwer B (2007) Mammalian Sir2 homolog SIRT3 regulates global mitochondrial lysine acetylation. Mol Cell Biol 27(24):8807–8814. doi:10.1128/MCB.01636-07, [pii] MCB.01636-07
9. Jaffrey SR, Snyder SH (2001) The biotin switch method for the detection of S-nitrosylated proteins. Sci STKE 2001(86):pl1. doi:10.1126/stke.2001.86.pl1, [pii] 2001/86/pl1
10. Tian Y, Zhou Y, Elliott S, Aebersold R, Zhang H (2007) Solid-phase extraction of N-linked glycopeptides. Nat Protoc 2(2):334–339. doi:10.1038/nprot.2007.42, [pii] nprot.2007.42
11. Zhang H, Li XJ, Martin DB, Aebersold R (2003) Identification and quantification of N-linked glycoproteins using hydrazide chemistry, stable isotope labeling and mass spectrometry. Nat Biotechnol 21(6):660–666. doi:10.1038/nbt827, [pii] nbt827
12. Roth AF, Wan J, Bailey AO, Sun B, Kuchar JA, Green WN, Phinney BS, Yates JR III, Davis NG (2006) Global analysis of protein palmitoylation in yeast. Cell 125(5):1003–1013. doi:10.1016/j.cell.2006.03.042, [pii] S0092-8674(06)00515-0
13. Zhang H, Liu AY, Loriaux P, Wollscheid B, Zhou Y, Watts JD, Aebersold R (2007) Mass spectrometric detection of tissue proteins in plasma. Mol Cell Proteomics 6(1):64–71. doi:10.1074/mcp.M600160-MCP200, [pii] M600160-MCP200
14. Zhou Y, Aebersold R, Zhang H (2007) Isolation of N-linked glycopeptides from plasma. Anal Chem 79(15):5826–5837. doi:10.1021/ac0623181
15. Forrester MT, Thompson JW, Foster MW, Nogueira L, Moseley MA, Stamler JS (2009) Proteomic analysis of S-nitrosylation and denitrosylation by resin-assisted capture. Nat Biotechnol 27(6):557–559. doi:10.1038/nbt.1545, [pii] nbt.1545
16. Nutt LK, Margolis SS, Jensen M, Herman CE, Dunphy WG, Rathmell JC, Kornbluth S (2005) Metabolic regulation of oocyte cell death through the CaMKII-mediated phosphorylation of caspase-2. Cell 123(1):89–103
17. Patel K, Lucas JE, Thompson JW, Dubois LG, Tillmann HL, Thompson AJ, Uzarski D, Califf RM, Moseley MA, Ginsburg GS, McHutchison JG, McCarthy JJ (2011) High predictive accuracy of an unbiased proteomic profile for sustained virologic response in chronic hepatitis C patients. Hepatology 53(6):1809–1818. doi:10.1002/hep.24284
18. Roulhac PL, Ward JM, Thompson JW, Soderblom EJ, Silva M, Moseley MA III, Jarvis ED (2011) Microproteomics: quantitative proteomic profiling of small numbers of laser-captured cells. Cold Spring Harb Protoc 2011(2):pdb prot5573
19. Soderblom EJ, Philipp M, Thompson JW, Caron MG, Moseley MA (2011) Quantitative label-free phosphoproteomics strategy for multifaceted experimental designs. Anal Chem 83(10):3758–3764. doi:10.1021/ac200213b
20. Reidel B, Thompson JW, Farsiu S, Moseley MA, Skiba NP, Arshavsky VY (2011) Proteomic profiling of a layered tissue reveals unique glycolytic specializations of photoreceptor cells. Mol Cell Proteomics 10(3):M110.002469. doi:10.1074/mcp.M110.002469, [pii] M110.002469
21. Andersen JL, Thompson JW, Lindblom KR, Johnson ES, Yang CS, Lilley LR, Freel CD, Moseley MA, Kornbluth S (2011) A biotin switch-based proteomics approach identifies 14-3-3zeta as a target of Sirt1 in the metabolic regulation of caspase-2. Mol Cell 43(5):834–842. doi:10.1016/j.molcel.2011.07.028, [pii] S1097-2765(11)00629-0
22. Chen Y, Zhao W, Yang JS, Cheng Z, Luo H, Lu Z, Tan M, Gu W, Zhao Y (2012) Quantitative acetylome analysis reveals the roles of SIRT1 in regulating diverse substrates and cellular pathways. Mol Cell Proteomics 11(10):1048–1062. doi:10.1074/mcp.M112.019547, [pii] M112.019547

23. Norris KL, Lee JY, Yao TP (2009) Acetylation goes global: the emergence of acetylation biology. Sci Signal 2(97):pe76. doi:10.1126/scisignal.297pe76, [pii] scisignal.297pe76
24. Benhar M, Thompson JW, Moseley MA, Stamler JS (2010) Identification of S-nitrosylated targets of thioredoxin using a quantitative proteomic approach. Biochemistry 49(32):6963–6969. doi:10.1021/bi100619k
25. Forrester MT, Hess DT, Thompson JW, Hultman R, Moseley MA, Stamler JS, Casey PJ (2011) Site-specific analysis of protein S-acylation by resin-assisted capture. J Lipid Res 52(2):393–398. doi:10.1194/jlr.D011106, [pii] jlr.D011106
26. Collier TS, Sarkar P, Franck WL, Rao BM, Dean RA, Muddiman DC (2010) Direct comparison of stable isotope labeling by amino acids in cell culture and spectral counting for quantitative proteomics. Anal Chem 82(20):8696–8702. doi:10.1021/ac101978b
27. Karbowski M, Lee YJ, Gaume B, Jeong SY, Frank S, Nechushtan A, Santel A, Fuller M, Smith CL, Youle RJ (2002) Spatial and temporal association of Bax with mitochondrial fission sites, Drp1, and Mfn2 during apoptosis. J Cell Biol 159(6):931–938. doi:10.1083/jcb.200209124
28. Strittmatter EF, Ferguson PL, Tang K, Smith RD (2003) Proteome analyses using accurate mass and elution time peptide tags with capillary LC time-of-flight mass spectrometry. J Am Soc Mass Spectrom 14(9):980–991
29. Jaffe JD, Mani DR, Leptos KC, Church GM, Gillette MA, Carr SA (2006) PEPPeR, a platform for experimental proteomic pattern recognition. Mol Cell Proteomics 5(10):1927–1941
30. Neubert H, Bonnert TP, Rumpel K, Hunt BT, Henle ES, James IT (2008) Label-free detection of differential protein expression by LC/MALDI mass spectrometry. J Proteome Res 7(6):2270–2279
31. Neilson KA, Ali NA, Muralidharan S, Mirzaei M, Mariani M, Assadourian G, Lee A, van Sluyter SC, Haynes PA (2011) Less label, more free: approaches in label-free quantitative mass spectrometry. Proteomics 11(4):535–553. doi:10.1002/pmic.201000553
32. Schilling B, Rardin MJ, MacLean BX, Zawadzka AM, Frewen BE, Cusack MP, Sorensen DJ, Bereman MS, Jing E, Wu CC, Verdin E, Kahn CR, Maccoss MJ, Gibson BW (2012) Platform-independent and label-free quantitation of proteomic data using MS1 extracted ion chromatograms in skyline: application to protein acetylation and phosphorylation. Mol Cell Proteomics 11(5):202–214. doi:10.1074/mcp.M112.017707, [pii] M112.017707
33. MacLean B, Tomazela DM, Shulman N, Chambers M, Finney GL, Frewen B, Kern R, Tabb DL, Liebler DC, MacCoss MJ (2010) Skyline: an open source document editor for creating and analyzing targeted proteomics experiments. Bioinformatics 26(7):966–968. doi:10.1093/bioinformatics/btq054, [pii] btq054

Chapter 10

Assaying Chromatin Sirtuins

Lei Zhong, Barbara Martinez-Pastor, Dafne M. Silberman, Carlos Sebastian, and Raul Mostoslavsky

Abstract

Most of the sirtuins' nuclear substrates identified so far are histones or other chromatin-associated proteins and, thus, it is of special relevance the development of good biochemical techniques to analyze the biology of these proteins in the context of chromatin. Here, we describe several of the chromatin-based techniques to identify sirtuins' substrates, including a chromatin immunoprecipitation (ChIP) protocol, an acid-extraction protocol, and a nucleosomal immunoprecipitation protocol to analyze putative sirtuin chromatin interactors.

Key words Sirtuins, Chromatin, Chromatin immunoprecipitation (ChIP), Acid-extraction, Nucleosomal immunoprecipitation

1 Introduction

Four of the seven mammalian sirtuins have nuclear localization. SIRT1, SIRT6, and SIRT7 are mostly nuclear, whereas SIRT2, a cytoplasmic sirtuin, can be found also in the nucleus during mitosis. Most of the sirtuins' nuclear substrates indentified so far are histones or other chromatin-associated proteins and, thus, it is of special relevance the development of good biochemical techniques to analyze the biology of these proteins in the context of chromatin. Here, we describe several of the chromatin-based techniques to identify sirtuins' substrates that we have been routinely using in our laboratory. Those include a chromatin immunoprecipitation (ChIP) protocol to analyze sirtuin-bound DNA regions as well as histone modifications that might be affected by sirtuins, an acid-extraction protocol to purify chromatin from cells (a method that allows us to analyze changes in protein chromatin localization and histone modifications in bulk chromatin), and a nucleosomal immunoprecipitation protocol to analyze putative sirtuin chromatin interactors. Finally, we detail a cell-based approach to analyze sirtuins' activity qualitatively and quantitatively.

Lei Zhong and Barbara Martinez-Pastor have contributed equally to this work

Matthew D. Hirschey (ed.), *Sirtuins: Methods and Protocols*, Methods in Molecular Biology, vol. 1077, DOI 10.1007/978-1-62703-637-5_10,

2 Materials

2.1 ChIP

1. Formaldehyde: 36.5 %.
2. 1× PBS.
3. 2.5 M glycine: For 1 L, dissolve 75 g glycine in 1 L of distilled water. Filter to sterilize. Store at room temperature.
4. Lysis buffer: 150 mM NaCl, 1 % v/v Nonidet P-40, 0.5 % w/v sodium deoxycholate, 0.1 % w/v SDS, 50 mM Tris–Cl pH 8.0, and 5 mM EDTA.
5. Protein A agarose beads.
6. Salmon sperm DNA.
7. Bovine serum albumin (BSA): For 10 mg/mL stock solution, dissolve 100 mg in 8 mL of distilled water. Adjust the volume to 10 mL with distilled water. Aliquot and store at −20 °C.
8. SIRT6 antibody (*see* **Note 1**).
9. High salt buffer: 100 mM Tris–Cl pH 8.5, 500 mM LiCl, 1 % Nonidet P-40, and 1 % w/v sodium deoxycholate. For 500 mL, add 50 mL of 1 M Tris–Cl pH 8.5, 50 mL of 5 M LiCl (for 5 M LiCl, dissolve 21.2 g LiCl in 90 mL distilled water, adjust the volume to 100 mL, and store at 4 °C), 5 mL of 100 % Nonidet P-40, and 5 g of sodium deoxycholate. Adjust the volume to 500 mL with distilled water. Store at 4 °C.
10. Low salt buffer: 150 mM NaCl, 0.5 % w/v sodium deoxycholate, 0.1 % w/v SDS, 1 % Nonidet P-40, 1 mM EDTA pH 8.0, and Tris–Cl pH 8.0. For 500 mL, add 15 mL 5 M NaCl (final concentration: 150 mM), 2.5 g sodium deoxycholate, 2.5 mL 20 % SDS, 5 mL 100 % Nonidet P-40, 1 mL 0.5 M EDTA pH 8.0, and 25 mL 1 M Tris–Cl pH 8.0. Adjust the volume to 500 mL with distilled water. Store at 4 °C.
11. Tris–EDTA (TE, pH 8.0): 100 mM Tris–Cl pH 8.0 and 10 mM EDTA pH 8.0. For 500 mL, mix 50 mL 1 M Tris–Cl pH 8.0 and 10 mL 0.5 M EDTA pH 8.0. Adjust the volume to 500 mL with distilled water.
12. Talianidis Elution Buffer: 70 mM Tris–Cl pH 8.0, 1 mM EDTA pH 8.0, and 1.5 % w/v SDS. For 50 mL, mix 3.5 mL 1 M Tris–HCl pH 8.0, 100 μL 0.5 M EDTA pH 8.0, and 3.75 mL 20 % w/v SDS. Adjust the volume to 50 mL with distilled water. Store at room temperature.
13. DNA purification Kit.
14. Real-time PCR kit.

2.2 Chromatin Extraction

1. Lysis buffer: 10 mM HEPES pH 7.4, 10 mM KCl, and 0.05 % NP-40. For 50 mL, mix 500 mL of 1 M HEPES pH 7.4, 500 mL of 1 M KCl, and 25 mL of NP-40 (*see* **Note 2**). Make up to 50 mL with water and store at 4 °C. Prior to use, add protease, phosphatase (1 mM sodium orthovanadate), and deacetylase (5 mM TSA) inhibitors.
2. Low salt buffer: 10 mM Tris–HCl pH 7.4 and 0.2 mM $MgCl_2$. For 50 mL, mix 500 mL of 1 M Tris–HCl pH 7.4 and 100 mL of 1 M $MgCl_2$ and make up to 50 mL (*see* **Note 3**). Store at 4 °C. Prior to use, add protease, phosphatase (1 mM sodium orthovanadate), and deacetylase (5 mM TSA) inhibitors.
3. 0.2 N HCl: For 10 mL, dilute 165.3 μL of 12.1 N HCl in 10 mL of water.
4. 1 M Tris–HCl pH 8: For 10 mL, dissolve 1.21 g of Tris in 5 mL of water and adjust the pH to 8. Make up to 10 mL with distilled water.

2.3 Nucleosome Immunoprecipitation

1. Lysis buffer: 10 mM HEPES pH 7.4, 10 mM KCl, 0.05 % NP-40, 0.5 mM DTT, and 1 mM PMSF. For 50 mL, mix 500 mL of 1 M HEPES pH 7.4, 500 mL of 1 M KCl, 25 mL of NP-40, 25 μL of 1 M DTT, and 100 μL of 0.5 M PMSF (*see* **Notes 2** and **4**). Make up to 50 mL with water and store at 4 °C.
2. Wash buffer: 20 mM HEPES pH 7.4, 20 mM KCl, 0.5 mM EDTA, 0.5 mM DTT, and 0.5 mM PMSF. For 50 mL, mix 1 mL of 1 M HEPES pH 7.4, 1 mL of KCl, 50 μL of 0.5 M EDTA, 25 μL of 1 M DTT, and 50 μL of 0.5 M PMSF (*see* **Note 4**). Make up to 50 mL with water and store at 4 °C.
3. Buffer A: 20 mM HEPES pH 7.4, 20 mM KCl, 0.4 mM EDTA, and 0.4 mM DTT. For 50 mL, mix 1 mL of 1 M HEPES pH 7.4, 1 mL of KCl, 40 μL of 0.5 M EDTA, and 20 μL of 1 M DTT.
4. 5 M NaCl: For 50 mL, dissolve 14.62 g of NaCl in 50 mL of water.
5. 2 M $CaCl_2$: For 50 mL, dissolve 11.1 g of $CaCl_2$ in 50 mL of water.
6. Micrococcal nuclease.
7. 0.5 M EGTA: For 50 mL, dissolve 9.5 g of EGTA in 50 mL of water.
8. 10 % NP-40: For 50 mL, dilute 5 mL of NP-40 in 50 mL of water.
9. Agarose beads.

10. Laemmli buffer: 200 mM Tris–Cl pH 6.8, 8 % SDS, 40 % glycerol, 4 % β-mercaptoethanol, and 0.08 % bromphenol blue. For 20 mL, add 4 mL of 1 M Tris–Cl, 8 mL 20 % SDS, 8 mL glycerol, 0.8 mL β-mercaptoethanol, and 16 mg of bromphenol blue.

2.4 In Vivo SIRT6 Deacetylation Activity Assay by Immunofluorescence

1. Gelatin-coated 96-well black/clear plates suitable for fluorescence detection and imaging: Prepare a 0.2 % gelatin solution by weighting 1 g of gelatin (Sigma) in 500 mL of distilled water. Autoclave to allow the solution of the gelatin. Solution can be stored at room temperature. Use enough volume to cover the bottom of each well and let stand for 30 min at room temperature. Remove the gelatin before plating the cells.
2. Histone Deacetylase (HDACs) inhibitors: Suberoylanilide hydroxamic acid (SAHA, Cayman Chemical). For maximum solubility in aqueous buffers, dissolve first in dimethyl sulfoxide (DMSO) and then dilute in the aqueous buffer of choice. For a 1 M stock solution, add 189 μL of DMSO to a 50 mg vial and store at −20 °C. Trichostatin A (TSA): To make 5 mM TSA stock solution, dissolve 5 mg in 3.3 mL DMSO.
3. Fixing solution: Paraformaldehyde (PFA) 4 % in PBS. Make sure to use gloves when working with PFA. Use PFA 16 % solution, EM grade (Electron Microscopy Sciences) and make a 1:4 dilution to get a working solution (e.g., for 20 mL, take 5 mL and add 15 mL PBS). Store at room temperature.
4. Permeabilization solution: 0.2 % Triton X-100 (v/v) in PBS. For 100 mL of PBS, add 0.2 mL of Triton X-100. Store at room temperature.
5. Blocking solution: 1 % BSA (w/v) and 0.1 % Tween 20 (v/v) in PBS. Prepare freshly by weighting 0.5 g of BSA and diluting in 45 mL of PBS. Add 50 mL of Tween 20 and adjust to a final volume of 50 mL.
6. Antibodies: Prepare the working solutions diluting the antibody in blocking soluμtion. Make sure to keep secondary antibody in the dark.
7. Wash buffer: PBS and 0.1 % Tween 20 (v/v). For 100 mL of PBS use 0.1 mL of Tween 20. Store at room temperature.
8. Nuclei counterstaining: DAPI 100 ng/mL in PBS. Use DAPI (Sigma D9542) to prepare a stock solution (5 mg/mL) by weighting 10 mg of DAPI and dissolving in 2 mL of dimethylformamide (DMF). Aliquot and store at −20 °C. Make a working solution (100 ng/mL) by adding 4 microL of the stock solution in 50 mL of PBS. Store at 4 °C (use a brown bottle or wrap with aluminum foil to protect from light).
9. Fluorescent microscope with a ×20 or ×40 (dry) objective with suitable excitation and emission filters for DAPI and two

additional fluorophores. For automated acquisition of images, a fluorescence high-throughput microscope with the same characteristics and automated stage and focus must be used. Captured images in a variety of formats (.bmp, .jpg, .jpeg, .tif, .tiff, .zvi, among others) can be read by the image analysis software (*CellProfiler*).

10. *CellProfiler* software: The software can be downloaded for free at www.cellprofiler.org.
11. Example images and their corresponding *CellProfiler* pipelines can be downloaded from the Mostoslavsky laboratory webpage http://mostoslavskylab.mgh.harvard.edu.

3 Methods

3.1 Prepare Buffers

1. To prepare 500 mL of lysis buffer, mix 15 mL of 5 M NaCl. (For 1 L of 5 M NaCl, dissolve 292 g of NaCl in 800 mL of distilled water. Adjust the volume to 1 L with distilled water.)
2. Add 5 mL of 100 % Nonidet P-40.
3. Add 2.5 g sodium deoxycholate.
4. Add 2.5 mL of 20 % SDS. (For 1 L of 20 % SDS, dissolve 200 g of SDS in 900 mL of distilled water. Heat to 68 °C with constant stir. Adjust the volume to 1 L with distilled water.)
5. Add 5 mL of 1 M Tris–Cl pH 8.0. (For 1 L of 1 M Tris–Cl pH 8.0, dissolve 121.1 g of Tris base in 800 mL distilled water. Adjust the pH to 8.0 with concentrated HCl. Adjust the volume to 1 L with distilled water.)
6. Add 5 mL of 0.5 M EDTA. (For 1 L of 0.5 M EDTA, add 186.1 g of EDTA·2H_2O to 800 mL distilled water. Stir constantly. Adjust the pH to 8.0 with NaOH. Adjust the volume to 1 L with distilled water.)
7. Adjust the volume to 500 mL with distilled water. Store at 4 °C.
8. Right before use, add: Protease Inhibitors Cocktail (Roche Applied Science), Phosphatase Inhibitors: 50 mM NaF (for 5 mL of 500 mM NaF, dissolve 0.1 g in 5 mL distilled water, aliquot, and store at −20 °C), 0.2 mM sodium orthovanadate (for 5 mL of 200 mM sodium orthovanadate, dissolve 0.18 g in 5 mL distilled water, aliquot, and store at −20 °C), HDAC inhibitors: 5 μM TSA (for 5 mM TSA, dissolve 5 mg in 3.3 mL DMSO), 5 mM sodium butyrate (for 5 M, dissolve 2.75 g in 5 mL of distilled water, aliquot, and store at −20 °C), 0.5 mM PMSF (for 100 mM PMSF, dissolve 87.09 mg in 5 mL ethanol, aliquot, and store at −20 °C).

3.2 ChIP

This Sirt6 ChIP protocol is adapted with modifications from Joaquin Espinosa's laboratory [1].

1. Grow cells in 150 mm plates until cells are 50–70 % confluent.
2. Cross-linking: Remove media from the plates completely. Add 19 mL of 1 % formaldehyde in 1× PBS. Incubate on a shaking platform at room temperature for 15 min (*see* **Note 5**).
3. Quench formaldehyde with 125 mM glycine (1 mL of 2.5 M glycine per 19 mL of 1 % formaldehyde in 1× PBS) on a shaking platform at room temperature for 5 min.
4. Remove the formaldehyde solution completely from the plates.
5. Wash the plates twice with ice-cold 1× PBS. After the second wash, place the plates vertically on ice for 1 min to collect the leftover PBS at the bottom of the plates. Remove PBS by aspiration.
6. Put the plates on ice or at 4 °C. Add lysis buffer (1 mL per plate) to the first plate (*see* **Note 6**). Scrape the plate and transfer the lysate to the next plate, until all the plates are collected (*see* **Note 7**).
7. Sonicate the lysate to shear chromatin (*see* **Note 8**).
8. Centrifuge at 13,500 × *g* for 20 min at 4 °C to clear the lysate. Transfer the supernatant to a fresh tube. The lysate can be snap-frozen and stored at −80 °C (*see* **Note 7**).
9. Measure protein concentration and dilute the lysate to 2 mg/mL with complete lysis buffer. Aliquot ~1.1 mL/tube and snap freeze the lysate if it is not used immediately. Store aliquots at −80 °C.
10. Wash two sets of protein A agarose beads (50 μL per IP sample) three times with lysis buffer to remove ethanol. A wash consists of resuspending the beads with 1 mL lysis buffer, centrifuging at 1,500 × *g* for 1 min at 4 °C, and removing the supernatant. One set of protein A agarose beads will be used to preclear the lysate. The other set will be blocked and used in the IP.
11. Preclear lysate with 50 μL of 50 % slurry protein A agarose beads previously washed in lysis buffer, not blocked with salmon sperm DNA nor BSA. Rotate at 4 °C for 2 h. Centrifuge 1 min at 4,000 rpm at 4 °C. Transfer supernatant to a fresh tube.
12. During **step 11**, block the other set of washed agarose beads with 1 mg/mL BSA and 0.3 mg/mL salmon sperm DNA. Rotate for 2 h at 4 °C. Wash three times with lysis buffer.
13. Transfer 10 μL of precleared lysate to a new tube and store at −20 °C for later use. This is the Input sample.
14. Transfer 1 mL of precleared lysate to a new tube and add Sirt6 antibody or equivalent amount of nonspecific serum or IgGs.

The amount of Sirt6 antibody added should be determined empirically (*see* **Note 1**). Add 50 μL of blocked protein A agarose beads from **step 12**. Rotate at 4 °C overnight.

15. Wash two times with lysis buffer, four times with high salt buffer (*see* **Note 9**), two times with lysis buffer, and two times with TE.
16. Aspirate supernatant from the last wash down to the 100 μL mark in the microcentrifuge tube (take special care to be consistent between samples).
17. Add 200 μL of Talianidis Elution Buffer.
18. Elute immunocomplexes by incubating 10 min at 65 °C and vortex samples occasionally.
19. Transfer 240 μL to a fresh tube and add 10 μL of 5 M NaCl (200 mM final).
20. Take out the Input samples from −20 °C (from **step 13**). Add 10 μL of 5 M NaCl and 230 μL of Talianidis Elution Buffer.
21. Reverse-crosslink all the samples (including Inputs) by incubating them in a 65 °C water bath for 5 h. Spin occasionally to bring down liquid condensation formed on the lids of the tubes.
22. Purify DNA with commercially available DNA purification kit of your choice. Purified DNA can be stored at −20 °C.
23. Set up real-time PCR with the primer pairs designed to amplify genomic region of interest. Follow the PCR protocol of the PCR kit of your choice. For data analysis, follow standard procedure (e.g., http://www.invitrogen.com/site/us/en/home/Products-and-Services/Applications/epigenetics-noncoding-rna-research/Chromatin-Remodeling/Chromatin-Immunoprecipitation-ChIP/chip-analysis.html).

3.3 Chromatin Extraction

This protocol is adapted with modifications from Junjie Chen's laboratory [2].

1. Collect cells and wash them in 1× PBS.
2. Resuspend cell pellet in 2–5 volumes of lysis buffer.
3. Incubate 20 min on ice.
4. Centrifuge at 18,000 × *g* at 4°C, 10 min. The supernatant contains the cytoplasmatic proteins and the pellet contains the nuclei (Fig. 1, lanes 1 and 2).
5. Wash the nuclei once with lysis buffer and centrifuge at 18,000 × *g* at 4°C, 10 min. If nucleoplasmatic proteins are desired go to **step 6**, otherwise go to **step 8**.
6. OPTIONAL: Resuspend the nuclei with 2–5 volumes of low salt buffer + 1 % Triton X-100 and incubate 15 min on ice.

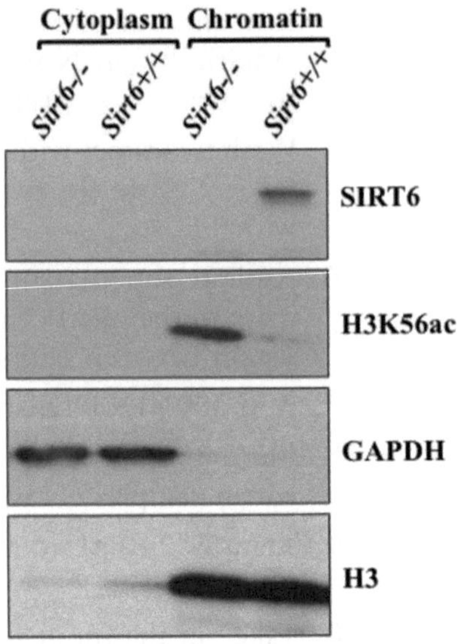

Fig. 1 Chromatin purification from *Sirt6* WT and KO ES cells. The supernatant (containing the cytoplasmatic proteins, 50 μg) and chromatin (20 μg) fractions were used to analyze by western blot the localization of SIRT6 (mostly chromatin-bound). Note that acetylation levels of H3K56 (a main substrate for SIRT6) are increased in *Sirt6* KO ES cells. GAPDH and H3 were used to confirm the purity of the fractions

7. Centrifuge at 18,000 × *g* at 4°C, 10 min. The supernatant contains the nucleoplasmatic proteins and the pellet contains the chromatin.
8. Resuspend chromatin with 2–5 volumes of HCl 0.2 N and incubate 20 min on ice.
9. Centrifuge at 18,000 × *g* at 4°C, 10 min.
10. Keep and neutralize SN (contains the acid-soluble proteins) with the same volume of 1 M Tris–HCl pH 8. This would be the chromatin fraction (Fig. 1, lanes 3 and 4).

3.4 Nucleosome Immunoprecipitation

This protocol is adapted with modifications from Don W. Cleveland's laboratory [3].

1. Collect cells (15–30 × 10^6 per IP) and wash them with 1 mL of 1× PBS.
2. Resuspend cell pellet in 5 volumes of lysis buffer.
3. Incubate 20 min on ice.
4. Pellet nuclei at 300 × *g* at 4 °C, 10 min.
5. Wash the nuclei once in 1 mL of wash buffer and centrifuge at 1,000 × *g* at 4 °C, 5 min.
6. Wash the nuclei once in 1 mL of wash buffer containing 300 mM NaCl and centrifuge at 1,000 × *g* at 4 °C, 5 min.

7. Resuspend nuclei in 1 mL of wash buffer containing 300 mM NaCl and 3 mM $CaCl_2$. Digest chromatin for 1 h at room temperature using 140 U/mL of micrococcal nuclease (*see* **Note 10**).
8. Stop reaction by adding 10 μL of 0.5 M EGTA (5 mM final) and 5 μL of 10 % NP-40 (0.05 % final) and centrifuge at 10,000 × *g* for 15 min at 4 °C. Keep the supernatant, which contains the soluble chromatin (mainly mononucleosomes) (*see* **Note 11**).
9. Incubate chromatin extract with the antibody overnight at 4 °C in an orbital rotator.
10. Add 40 μL of agarose beads (50 % slurry) and incubate 4 h at 4 °C in an orbital rotator.
11. Pellet the beads at 3,000 × *g* for 2 min at 4 °C.
12. Wash the beads sequentially with 1 mL of buffer A, 1 mL of buffer A with 300 mM KCl, and 1 mL of buffer A with 300 mM KCl, 1 mM DTT, and 0.1 % Tween 20. Centrifuge at 3,000 × *g* for 2 min between washes to pellet the beads.
13. Elute the immunocomplexes by boiling the beads with 50 μL of Laemmli buffer for 10 min.

3.5 In Vivo SIRT6 Deacetylation Activity Assay by Immunofluorescence

As observed for many proteins, recombinant purified SIRT6 exhibits low activity in vitro, likely due to either posttranslational modifications that are missing or else cofactors needed for its activity. In this regard, overexpressing SIRT6 in mammalian cells is a highly useful approach to assess its deacetylase activity in vivo. Indeed, commercially available antibodies specific for the acetylated form of the protein of interest make immunofluorescence a powerful tool that can be used to directly examine potential inhibiting or activating compounds. By using a FLAG-tagged SIRT6 expression system, double staining can be performed to determine both levels of expression of SIRT6 (anti-FLAG antibody) and levels of acetylation of a target protein (anti-acetyl-protein antibody). In this assay, fluorescence intensity of the acetyl-histone mark is inversely correlated with SIRT6 expression levels. We have developed a method where automated-microscopy-based analysis allows for a cell-based quantitative analysis of SIRT6 activity, which could be adapted for other HDACs as well (Fig. 2).

All procedures should be carried at room temperature unless otherwise specified.

1. Use FLAG-tagged SIRT6 fusion protein subcloned in the pCMV vector to transfect HeLa cells (5 μg of DNA in 10 cm plates) (*see* **Note 12**).
2. Plate FLAG-tagged SIRT6 expressing HeLa cells (100 μL, 2×10^4 cells/well) in 0.2 % gelatin-coated 96-well plates and allow to adhere overnight.

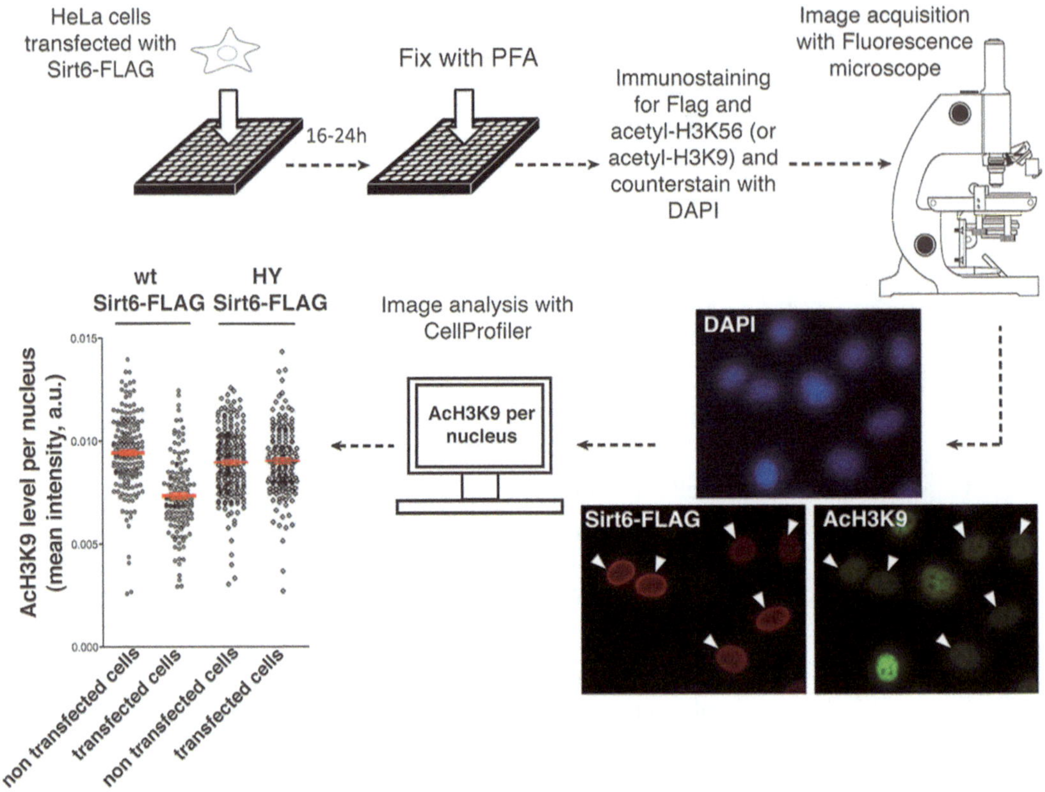

Fig. 2 SIRT6 overexpression inversely correlates with the levels of H3K9 acetylation. HeLa cells were transfected with a SIRT6-FLAG expressing vector and double immuno-stained with anti-FLAG and anti-AcH3K9 antibodies. The scheme depicts a summary of the analysis preformed after acquiring the images with a fluorescent microscope. *Arrows* indicate SIRT6-FLAG overexpressing cells. Images were analyzed using the *CellProfiler* software. Quantitation shows lower levels of H3K9 acetylation in SIRT6-FLAG overexpressing cells. This difference is abolished when a catalytic mutant of SIRT6 is used (HY-SIRT6-FLAG)

3. In order to increase basal levels of histone acetylation, preincubate transfected attached cells with HDAC inhibitors (SAHA, 1 μM or TSA, 200 nM) for 12–24 h. This pretreatment can be used for determining the potential effect of different compounds. For SAHA, prepare a fresh 10 μM solution for every experiment by making two consecutive dilutions 1:100 and 1:1000 in tissue culture medium from the 1 M stock and use 10 μL/well (1 μM). For TSA, make a 1:1,000 dilution (5 μM) of the 5 mM stock solution and add 4 μL/well.
4. Remove tissue culture media and wash the cells once with 100 μL of PBS/well. Fix with 100 μL/well of 4 % PFA (v/v) in PBS for 10 min.
5. Permeabilize with 100 μL/well of 0.2 % Triton X-100 (v/v) in PBS for 10 min.
6. Block with 100 μL/well of 1 % BSA (w/v) and 0.1 % Tween 20 (v/v) in PBS for 30 min with gentle shaking.

7. Stain by adding 50 μL/well of anti-Flag antibody and an antibody specific for the acetylated form of the protein of interest in blocking buffer (in the case of SIRT6, anti-H3K9Ac or anti-H3K56Ac antibody). Incubate for 1 h at room temperature (or overnight at 4 °C) with gentle shaking (*see* **Note 13**).
8. Wash three times with 100 μL/well of PBS and 0.1 % Tween 20 (v/v) for 5 min.
9. Incubate with fluorophore-conjugated secondary antibodies diluted in blocking buffer for 1 h at room temperature protected from light.
10. Wash two times with 100 μL/well of PBS and 0.1 % Tween 20 (v/v) for 5 min.
11. Counterstain nuclei with 100 μL/well of DAPI 100 ng/mL in PBS for 2 min at room temperature.
12. Wash with 100 μL/well of PBS and keep the cells in PBS at 4 °C until visualization with a fluorescence microscope.
13. Images from each well must be acquired using a fluorescent microscope with a ×20 or ×40 (air) objective using the suitable excitation and emission filters for the fluorophores that were used in the staining of Sirt6-Flag and the acetylated-protein of interest. It is recommended to acquire images from at least nine different fields per well when using a ×40 objective. Alternatively, images can be acquired with a high throughput microscope with automated stage and focus.
14. Cell image analysis is then performed using *CellProfiler* analysis software [4]. First, download and install *CellProfiler* analysis software from www.cellprofiler.org.
15. Download the file called *Cellular Measurement of Sirt6 activity* from Mostoslavsky lab webpage (http://mostoslavskylab.mgh.harvard.edu). After downloading the file, make sure that it is decompressed. The file contains example images and a pipeline made of different analysis modules placed in order.
16. Run the pipeline *Measurement of Sirt6 activity* on the example images you downloaded. To do this, follow the instructions *loading an existing pipeline* clicking in Help/Using CellProfiler/How to build a pipeline.
17. Adjust the main pipeline for your images using test images.
18. First, create a test input folder in your computer where you copy several test images into it and a test output folder. In the CellProfiler, set the default input folder and the default output folder to be your test input folder and test output folder, respectively.
19. Some adjustments might be done in the module LoadImages. This module loads an image set, which is a group of related images (in our case, the multiple channels at the same image location) and gives each image a name for further processing.

Our image set comprises three images: DAPI, Sirt6-Flag, and acetyl-protein. We use text matching to define the difference between images in a set: in the example images, all the images stained for nuclei have the text DAPI in the name, Sirt6-Flag images have the text Alexa555, and acetyl-protein images have Alexa488. Some adjustments might be done to load your images by matching files in the folder against the unique text pattern for each stain.

20. Additional modules will require some adjustments; the parameters that can be adjusted to run your images appear on the right after clicking on a module. In general, the modules that might require some adjustment are IdentifyPrimaryObjects and ClassifyObjects (*see* **Notes 14** and **15**).
21. Run the pipeline on your images.
22. Perform data acquisition (*see* **Note 16**).
23. Perform downstream data analysis (*see* **Note 17**).

4 Notes

1. The quality of the Sirt6 antibody is critical to the success of the ChIP. A desirable Sirt6 antibody would give one single clear band on a whole cell lysate western blot. It should also work well in an immunoprecipitation experiment. If such Sirt6 antibody is not available, proper controls are necessary to differentiate real Sirt6 binding signals from nonspecific binding. Always include an IgG control when planning the ChIP experiment. Another important control is to use Sirt6-deficient cells or Sirt6 knockdown cells (Fig. 3).

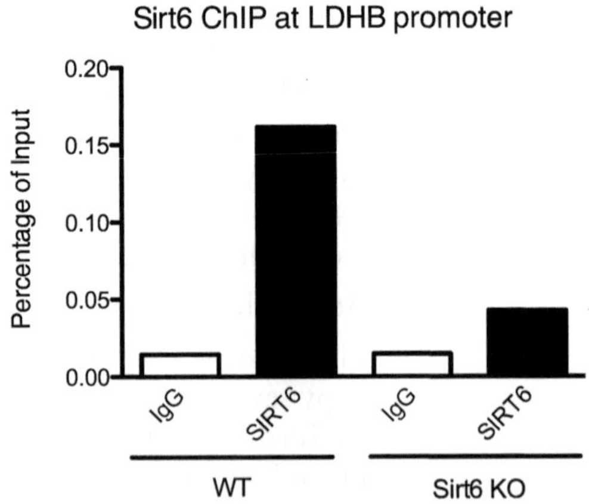

Fig. 3 An example of Sirt6 ChIP at LDHB promoter in mES cells. Controls include Sirt6-deficient cells (Sirt6 KO) and IgGs for both WT and Sirt6 KO cells. Note that even in Sirt6 KO cells, Sirt6 ChIP signal is fourfold higher than its IgG counterpart, possibly due to unspecific binding of Sirt6 antibody. This example highlights the necessity of a Sirt6 KO control in ChIP experiments

2. To prepare 1 L of 1 M HEPES pH 7.4, dissolve 238.3 g of HEPES in 800 mL of water, adjust pH at 7.4, and make up to 1 L with water. To prepare 1 L of 1 M KCl, dissolve 74.55 g of KCl and make up to 1 L with water.
3. Prepare 1 L of 1 M Tris–HCl pH 7.4 by dissolving 121.12 g of Tris base in water, adjust pH at 7.4, and make up to 1 L. To prepare 1 L of 1 M $MgCl_2$, weigh 95.21 g of $MgCl_2$ and dissolve with water to make 1 L of solution.
4. To prepare 1 M solution of DTT, weigh and dissolve in 10 mL of water 1.54 g of DTT. Make aliquots of 0.5 mL and store at −20 °C. To prepare 10 mL of 0.5 M PMSF, dissolve 0.87 g of PMSF in methanol and make up to 10 mL. Make aliquots of 0.5 mL and store at −20 °C. To prepare 1 L of 0.5 M EDTA, dissolve 146.12 g of EDTA and make up to 1 L with water.
5. The cross-linking time and the formaldehyde concentration are the key factors that affect cross-linking efficiency. Longer cross-linking time or higher formaldehyde concentration is more efficient, but may cause detection of nonspecific binding. On the other hand, shorter cross-linking time or lower formaldehyde concentration may increase chromatin shearing efficiency, but this may decrease the yield of precipitated chromatin for certain proteins, especially those that do not directly bind to DNA.
6. The protease/HDAC/phosphatase inhibitors should be added right before the harvest.
7. The lysate can be snap-frozen by dropping the tubes into liquid nitrogen and stored at −80 °C until ready to proceed. To thaw, put tubes in room temperature water bath and then put on ice.
8. The sonication conditions should be determined empirically for each cell type or sonicator model. To check the sonication efficiency, take an aliquot of 50 μL from the sonicated chromatin, add 3 μL of 5 M NaCl, and put on a 95 °C heatblock for 5 min. Run the mixture on a 1.5 % agarose gel. The fragment should be between 300 and 1,000 bp.
9. If the Sirt6 antibody is not strong, substitute high salt buffer with low salt buffer.
10. The time of digestion and the units of micrococcal nuclease used might be different among several enzymes. Testing and optimizing the right conditions to get all DNA in the mononucleosomal fraction is strongly recommended.
11. To confirm that most of the DNA is the mononucleosomal fraction, take an aliquot and purify the DNA as follows: add 1 volume of phenol/chloroform, mix well, and centrifuge at 14,000 rpm for 3 min. Take aqueous phase (upper phase) and precipitate DNA with 1 volume of isopropanol. Centrifuge at

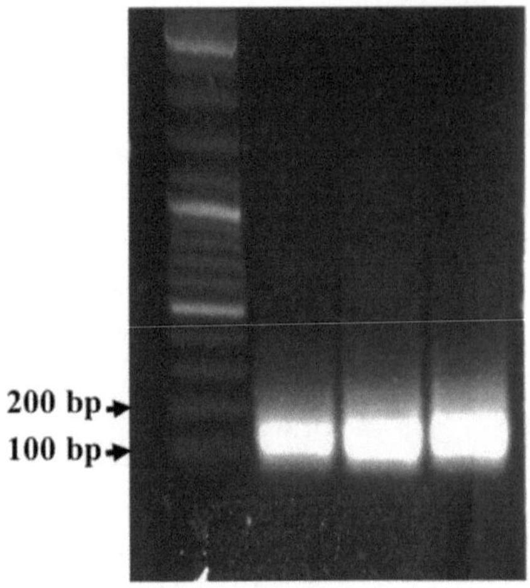

Fig. 4 Mononucleosome purification. Nuclei from 15 × 10^6 293T cells (*lanes 1–3*) were digested in 1 mL of digestion buffer (see main text) and an aliquot (100 μL) was taken to purify the DNA as described in Subheading 3. Note that almost all DNA is contained in the mononucleosomal fraction

14,000 rpm for 5 min and discard the supernatant. Wash the pellet with 70 % ethanol, centrifuge at 14,000 rpm for 3 min, and air-dry the pellet. Resuspend DNA in water and run the samples in a 1.2 % agarose gel. The DNA present in mononucleosomes is about 147 bp in size (Fig. 4).

12. The amount of DNA can be modified according to the size of the plates used (e.g., 2.5 mg of DNA for 6 cm plates). HEK 293T cells can be used but they easily detach, so especial care should be taken during washes. Volumes described in this protocol can be adapted for 384-well plates.
13. Make sure that different hosts were used to generate the primary antibodies.
14. In general, the IdentifyPrimaryObjects module often needs adjustment. IdentifyPrimaryObjects: the *typical diameter of objects* (will depend on the cell type and the magnification of the objective used for the acquisition), the *thresholding method*, *threshold correction*, the *lower and upper bounds on the threshold*, the *size of smoothing filter*, and *suppress local maxima*.
15. In general, the ClassifyObjects module often needs adjustment. ClassifyObjects: *Enter the custom threshold separating the values between bins.* The pipeline allows you to classify nuclei according to Sirt6-Flag staining into positive or negative for Sirt6-Flag. After measuring the intensity of nuclear Sirt6-Flag staining, the values are shown in the corresponding image (in the module DisplayDataOnImage) in order to visually determine a cutoff. The empirically determined cutoff value must be introduced in the setting *Enter the custom threshold separating the values between bins.*

16. The data files are exported to the designed output folder as .csv files that can be opened using Excel. For example, the DefaultOUT_Image.csv file contains per-image data. Each row represents one image and the parameters measured in each image are displayed in columns: total number of nuclei per image, number of Sirt6-Flag positive nuclei per image, number of Sirt6-Flag negative nuclei per image, and Sirt6-Flag positive and negative nuclei as percentage of total nuclei in the image. Also, DefaultOUT_Sirt6PosNuclei.csv and DefaultOUT_Sirt6NegNuclei.csv files contain per-cell data. Each row represents one nucleus and the parameters measured in each nucleus are displayed in columns. The columns *Intensity_MedianIntensity_OrigGreen* and *Intensity_MeanIntensity_OrigGreen* display the median and mean intensity of the acetyl-protein staining, respectively, and correlate with the acetylation levels of the target protein per nucleus. The *ImageNumber* column relates each nucleus to the image it belongs to.
17. Data analysis and plotting can be performed in programs like Excel or Prism. This cell-based image analysis allows us to detect changes in the whole population (if the median or mean intensity of acetylated-protein in Sirt6-Flag positive cells is used) but also allows us to detect small changes in a percentage of cells (if the data per individual cell is used). For robustness, it is strongly recommended a minimum of 200 cells/condition acquired, if the transfection efficiency was at least 30 % (Fig. 2).

Acknowledgements

Work in the Mostoslavsky lab is supported in part by NIH grants GM093072-01 and DK088190-01A1. R.M. is an MGH Research Scholar and a Howard Goodman Scholar Awardee. L.Z. is the recipient of the MGH-ECOR Tosteson Postdoctoral Fellowship Award. B.M.-P. is the recipient of a postdoctoral fellowship from the Spanish Ministry of Education. C.S. is the recipient of a Beatriu de Pinos Postdoctoral Fellowship (Generalitat de Catalunya).

References

1. Donner AJ, Ebmeier CC, Taatjes DJ, Espinosa JM (2010) CDK8 is a positive regulator of transcriptional elongation within the serum response network. Nat Struct Mol Biol 17:194–201. doi:10.1038/nsmb.1752
2. Huang J, Huen MS, Kim H, Leung CC, Glover JN, Yu X, Chen J (2009) RAD18 transmits DNA damage signalling to elicit homologous recombination repair. Nat Cell Biol 11:592–603. doi:10.1038/ncb1865
3. Foltz DR, Jansen LE, Black BE, Bailey AO, Yates JR III, Cleveland DW (2006) The human CENP-A centromeric nucleosome-associated complex. Nat Cell Biol 8:458–469
4. Carpenter AE, Jones TR, Lamprecht MR, Clarke C, Kang IH, Friman O, Guertin DA, Chang JH, Lindquist RA, Moffat J, Golland P, Sabatini DM (2006) CellProfiler: image analysis software for identifying and quantifying cell phenotypes. Genome Biol 7:R100

Part III

Methods to Measure Sirtuin Activity

Chapter 11

Measurement of Sirtuin Enzyme Activity Using a Substrate-Agnostic Fluorometric Nicotinamide Assay

Basil P. Hubbard and David A. Sinclair

Abstract

The sirtuins are NAD^+-dependent, multifunctional lysine deacylases that play key roles in cellular homeostasis. They are increasingly being found to target a variety of substrates including acetyl-, butyryl-, malonyl-, and succinyl-lysines. Early assays for measuring sirtuin activity in vitro were criticized for their use of fluorophores on the peptide substrates used, which may alter the results obtained and not be representative of the in vivo situation. We describe a new protocol for the measurement of sirtuin activity by detecting the production of nicotinamide (NAM). The assay is amenable to any substrate and any modification removed by sirtuins. The assay may also be used to measure glycohydrolase (e.g., CD38) and ADP-ribosyltransferase activity (e.g., mARTs and PARPs).

Key words Sirtuin assay, SIRT1, HDAC, Deacetylation, Aging, Desuccinylation, Depropionylation, Debutyrylation, Demalonylation, Glycohydrolase assay, PARP assay, Nicotinamide assay, yPnc1, Ortho-phthalaldehyde (OPT), Fluorometric detection of ammonia

1 Introduction

The sirtuin family of NAD^+-dependent histone deacetylases has been implicated in numerous cellular processes in yeast, worms, flies, and mammals involved in aging and organismal health [1]. For example, yeast Sir2 modulates budding yeast replicative life span by suppressing recombination within the rDNA array [2–4], while mammalian SIRT1 regulates DNA damage and apoptosis [5, 6], metabolism, and mitochondrial biogenesis [7–9]. Sirtuins catalyze the removal of acetyl groups from lysine residues via a multistep reaction using β-NAD as a cofactor [10–14]. The products of the reaction are the deacetylated protein substrate, nicotinamide (NAM), and 2′ *O*-acetyl-adenosine diphosphate-ribose (OAADPR) [10–14]. A schematic outlining the mechanism of deacetylation may be found in Fig. 1 [12]. First, the acetyl oxygen of the substrate initiates a nucleophilic addition on the 1′-carbon of nicotinamide ribose to form a C1′-*O*-alkylamidate intermediate [10–14].

Matthew D. Hirschey (ed.), *Sirtuins: Methods and Protocols*, Methods in Molecular Biology, vol. 1077, DOI 10.1007/978-1-62703-637-5_11, © Springer Science+Business Media, LLC 2013

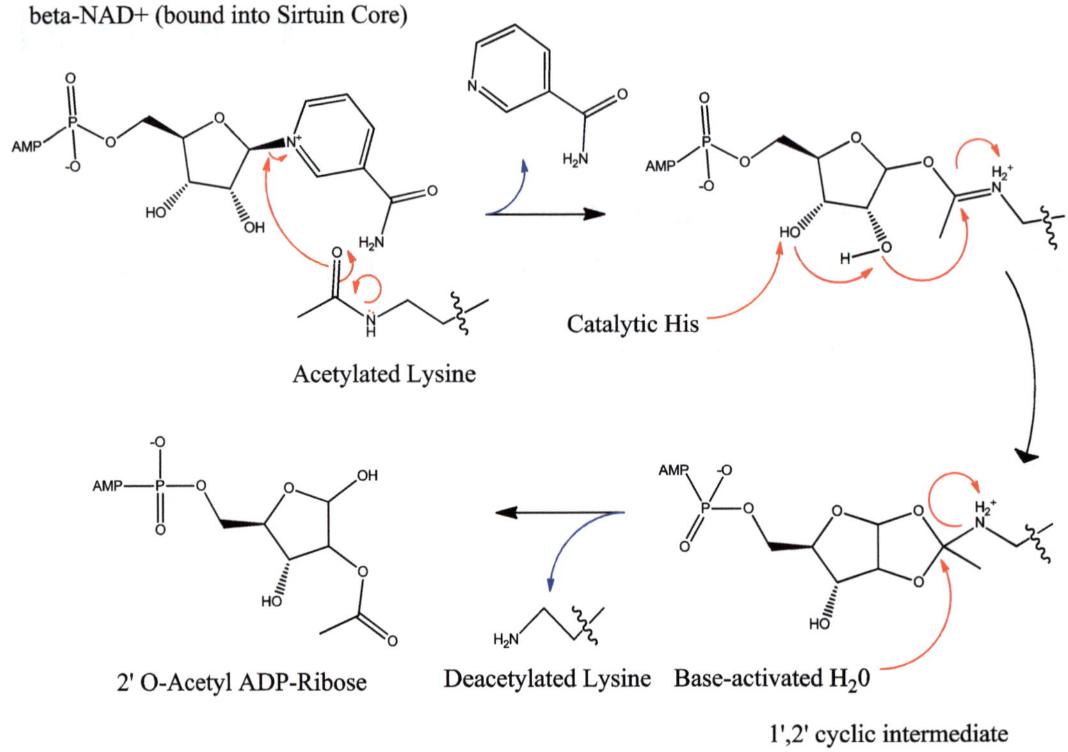

Fig. 1 Schematic outlining the biochemical mechanism of sirtuin deacetylation

This step is accompanied by release of free NAM [10–14]. Next, the 2′ hydroxyl group on the ribose ring is activated by a conserved catalytic histidine residue (His-135 in Hst2) and attacks the C1′-*O*-alkylamidate intermediate yielding a 1′2′-cyclic intermediate [11, 15]. This intermediate is then cleaved by a base-activated water molecule, resulting in the release of the deacetylated substrate and 2′-OAADPr [11, 16, 17]. Subsequently, nonenzymatic conversion of 2′OAADPr into 3′OAADPr may take place [11, 16]. In addition to catalyzing lysine deacetylation, specific sirtuins have been shown to possess depropionylase, debutyrylase [18, 19], desuccinylase, demalonylase [20], and ADP-ribosyltransferase activities [21, 22].

Traditional techniques used to measure sirtuin deacetylase activity include isotopic assays such as the charcoal binding assay, which employs a ^{3}H-labeled acetyl-lysine substrate to measure acetate release from hydrolyzed OAADPr [17], and [^{14}C] NAM release assays [23]. More commonly, two-step commercial assays such as the Fluor de Lys assay™ (BIOMOL/Enzo), which measures deacetylation of an acetyl-lysine peptide conjugated to aminomethylcoumarin (AMC), have been used [24]. This assay takes advantage of the property that while free AMC fluoresces, its peptide conjugate does not. Thus, following the deacetylation step, trypsin is subsequently added to cleave and release free AMC molecules from deacetylated peptides [24]. Importantly, this assay has been the subject of intense controversy due to its use of

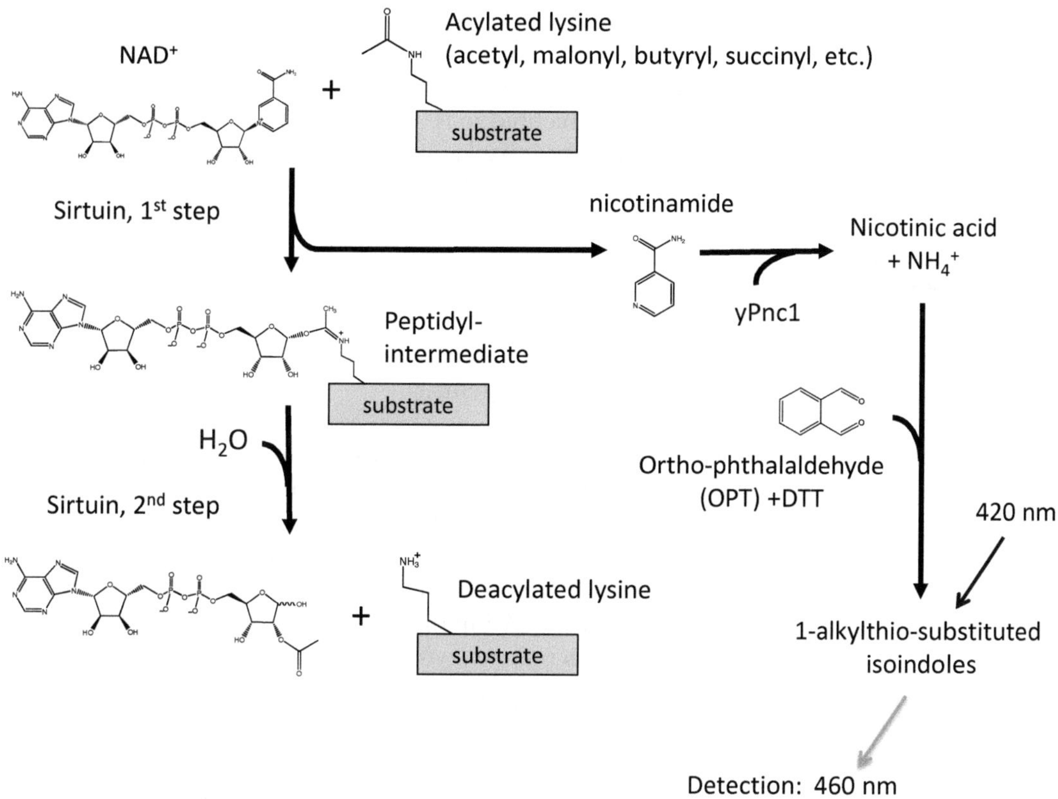

Fig. 2 Diagram illustrating the main steps in the PNC1-OPT assay

non-physiological peptides, and its potential to yield artifacts in drug screening assays [23, 25]. Assays employing electrophoresis [26], fluorometric detection of β-NAD [27], and mass spectrometry [28, 29] have also been reported.

Recently, two non-isotopic enzyme-coupled NAM assays have been described, PNC1-GDH [29, 30] and PNC1-OPT [29]. These assays are relatively convenient to perform, and overcome many of the shortcomings of classical deacetylase detection methods in that they are safe, cost-effective, versatile (any substrate may be used), and do not require the use of any specialized equipment. In both assays, the nicotinamidase enzyme from budding yeast, yPnc1 [31], is added to the reaction to convert the NAM produced by the sirtuin deacetylation reaction into ammonia. Ammonia may be either quantified spectrophotometrically via coupled reaction with glutamate dehydrogenase (GDH) [30, 32] or detected fluorometrically in a second step via chemical reaction with ortho-phthalaldehyde (OPT) [33] to produce 1-alkylthio-substituted isoindoles [29, 33]. While the PNC1-GDH assay has the advantage of being continuous rather than fixed-point, the PNC1-OPT assay offers a higher degree of sensitivity and is better suited for drug discovery since it requires the use of only two enzymes rather than three. Here we outline a detailed protocol for using the PNC1-OPT assay (schematic in Fig. 2) to measure SIRT1 enzyme activity.

We have previously used this assay to measure SIRT1 enzyme kinetics with several different natural amino acid peptides, and to study the effects of several SIRT1 inhibitors and activators in vitro [29]. Importantly, this technique is also applicable to the assay of different sirtuin isoforms, as well as other NAD-cleaving enzymes such as glycohydrolases (e.g., CD38) [34] and PARPs [35].

2 Materials

All solutions should be prepared using distilled or ultrapure water with analytical grade reagents.

2.1 Enzymes and Substrates

1. Recombinant SIRT1 and yPnc1 enzymes should be expressed and purified to >85 % using previously described protocols [29, 36, 37], aliquoted, and stored at −20 °C until use. Typically the final storage buffer consists of a 1:1 dilution of glycerol with elution buffer (250 mM imidazole, 40 mM Tris pH 8.0, 3 mM BME in distilled water). Repeated freeze–thaw cycles may decrease enzymatic activity.
2. Custom-acetylated peptide substrates may be purchased from commercial vendors or synthesized in house (>85 % purity), and should typically be dissolved in distilled water. In general, peptides should be between 5 and 15 amino acids with the acetylated lysine near the middle of the sequence, and should be flanked by an N-terminal acetyl group and a C-terminal NH_2 group. For example, an appropriate peptide substrate corresponding to the well-known SIRT1 substrate H3K9ac would be: Ac-TARK(ac)STG-NH_2 [29]. Sequence considerations should be taken into account in order to minimize substrate background fluorescence (*see* **Note 1**).

2.2 Buffers and Solutions

1. Reaction Buffer: Phosphate-buffered saline (PBS): 1 mM KH_2PO_4, 155 mM NaCl, 3 mM $Na_2HPO_4 \cdot 7H_2O$, pH 7.4 supplemented with 1 mM dithiothreitol (DTT) (add fresh from a frozen stock before each experiment). Depending on the enzyme being studied, alternative buffers may be used (*see* **Note 2**).
2. OPT Developer Reagent: A solution of 10 mM OPT and 10 mM DTT in 30 % ethanol/70 % PBS (pH 7.4) (*see* **Note 3**). Dissolve OPT in pure ethanol first (vortex on the highest setting) and then add this to a PBS/DTT solution. Cover with aluminum foil in order to minimize exposure to light and store at −20 °C until use. This solution will freeze and must be warmed at 37–42 °C prior to use; if precipitation of DTT is observed, vortex the mixture on the highest setting until no particulates remain.
3. β-NAD aliquots: 100 mM β-NAD dissolved in distilled water, aliquoted, and frozen at −20 °C. Thaw and dilute tenfold in

distilled water prior to assay. Use one aliquot per experiment and discard any remainder; avoid freeze–thaw cycles.

4. Stock solution of NAM for standard curve preparation (e.g., 100 mM). Aliquot and store at −20 °C.

2.3 Other Assay Equipment/Supplies

1. Aluminum foil.
2. 37 °C incubator and 37 or 42 °C water bath.
3. Vortex mixer.
4. Orbital shaker.
5. Black opaque bottom 96-well plates suitable for fluorometry.
6. Fluorometer capable of exciting at ~420 nm and measuring emission at ~460 nm.

3 Methods

While the coupled SIRT1 deacetylation/yPnc1 conversion reaction is typically carried out at 37 °C, all other steps may be performed at room temperature unless specified otherwise. In general, samples should be assayed in triplicate.

3.1 Nicotinamide (NAM) Standard Curve Preparation

1. Thaw the NAM standard solution on ice and perform serial dilutions in Eppendorf tubes to obtain concentrations that are 100× of the desired final concentration. Example final concentrations are 0, 5, 10, 20, 30, 40, and 50 μM NAM. Pipette 1 μL of each of these into the bottom of an Eppendorf tube (will be diluted 100-fold in the next step).
2. Thaw yPnc1 enzyme on ice and chill Reaction Buffer. Prepare a master-mix corresponding to the total number of sample reactions to be performed in a 15 mL BD Falcon tube (or similar tube). For each sample mix 1 μg of purified yPnc1 with 100 μL Reaction Buffer. Mix by pipetting up and down, and vortexing gently (~setting 4 for 5 s).
3. Next, pipette 100 μL of the master-mix into each NAM standard sample tube, mix up and down by pipetting, seal each tube, and incubate samples in a holder rack at 37 °C for 1 h. During this incubation period NAM will be converted into ammonia by yPnc1 (*see* **Note 4**).
4. During the incubation period, thaw the OPT Developer Reagent by incubating at 37 or 42 °C for about 15 min (keep covered in aluminum foil). Subsequently, vortex on the highest setting until completely mixed; ensure that no DTT precipitate is present.
5. Following the incubation period, remove the samples from the 37 °C incubator. Under dimly lit conditions (dark) (*see* **Note 5**) add 100 μL of OPT Developer Reagent to each tube as quickly

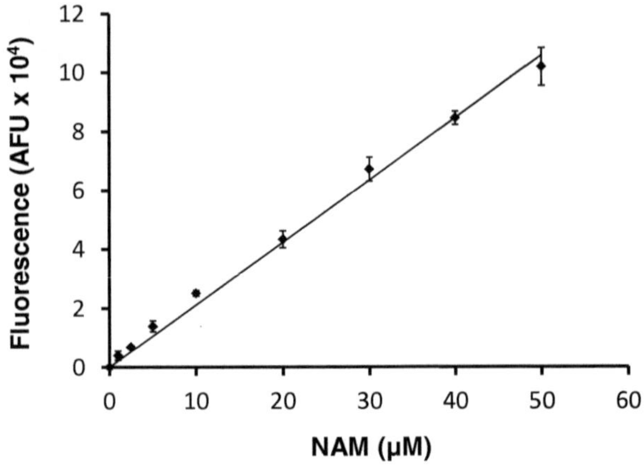

Fig. 3 Example nicotinamide standard curve produced using the PNC1-OPT assay. Each point represents mean ± s.d. ($n = 3$)

as possible to ensure consistency (*see* **Note 6**). Close tubes and vortex all samples immediately on the highest setting for 5 s.

6. Return the samples to the holder rack. Cover all the samples (the entire holder rack) with aluminum foil. Once covered, ambient lighting may be restored. Incubate the reactions on an orbital shaker for 1 h at room temperature (*see* **Note 7**).
7. Transfer samples to a 96-well plate for reading (do this under dim light). Read the fluorescence on a fluorometer with filters set to excitation at ~420 ± 10 nm and emission at ~460 ± 10 nm (*see* **Note 8**). A 0.1 or 1 s read is generally recommended.
8. A standard curve may be produced by plotting fluorescence intensity (*y*-axis) versus NAM concentration (*x*-axis), as depicted in Fig. 3.

3.2 Measurement of SIRT1 Deacetylase Activity

1. Thaw aliquoted solutions of β-NAD, acetylated peptide, and SIRT1 and yPnc1 enzyme preparations on ice. Prepare a master-mix in a 15 mL BD Falcon tube (or similar tube) corresponding to *twice* the total number of samples to be assayed. One set of reactions will be performed in the presence of β-NAD to measure SIRT1 deacetylase activity, while a second set of parallel reactions will be run in the absence of β-NAD to account for background fluorescence (*see* **Note 9**). For each single reaction, mix the following on ice in the order presented: Reaction Buffer (100 μL—volume of other components), native peptide substrate (typically 10–30 μM final concentration), purified yPnc1 enzyme (~1–2 μg), and purified SIRT1 enzyme (~1–2 μg) (*see* **Note 10**). Mix the master-mix by pipetting up and down, and gently vortex (setting 5 or 6 for 5 s).
2. Divide the master-mix into two tubes, and add either β-NAD (typically 75–200 μM final concentration), or an equal volume

of distilled water to each tube. Mix again by pipetting and gentle vortexing.

3. Array a series of labeled Eppendorf tubes in a tube holder rack, corresponding to each experimental reaction and its corresponding background control reaction. If test compounds (e.g., activators or inhibitors) are to be added, this should be performed at this stage (*see* **Note 11**).
4. Next, add 100 μL of master-mix into the appropriate Eppendorf tubes, close them, and return them to the tube holder rack. If a test compound is present, first mix by pipetting up and down. Incubate the reactions at 37 °C for 1 h (*see* **Note 12**).
5. During the incubation period, thaw the OPT Developer Reagent by incubating at 37 or 42 °C for about 15 min (keep covered in aluminum foil). Vortex on the highest setting until the solution is completely mixed; ensure that no DTT precipitate is present.
6. Following the incubation period, remove the samples from the 37 °C incubator, and under dimly lit or dark conditions, add 100 μL of OPT Developer Reagent to each tube as quickly as possible to ensure consistency (*see* **Note 6**). Close tubes and vortex all samples immediately on the highest setting for 5 s.
7. Return the samples to the holder rack and cover all the samples (the entire holder rack) with aluminum foil. Once covered, ambient lighting may be restored. Incubate the reactions on an orbital shaker for 1 h at room temperature (*see* **Note 7**).
8. Under dim light, transfer samples to a 96-well plate for reading. Read the fluorescence on a fluorometer with filters set to excitation at ~420 ± 10 nm and emission at ~460 ± 10 nm (*see* **Note 8**). A 0.1 or 1 s read is generally recommended.
9. Using the arbitrary fluorescence unit (AFU) readings for each reaction, calculate the net fluorescence for each sample by subtracting the fluorescence of the background control reaction (no NAD) from the experimental reaction (with NAD), $F_{\text{corrected}} = F_{\text{+NAD}} - F_{\text{-NADcontrol}}$. The resulting value is proportional to the amount of NAM produced during the deacetylation reaction.
10. Rates of deacetylation may be calculated using a linear equation obtained from the NAM standard curve.

4 Notes

1. The amount of background fluorescence observed in the assay is significantly affected by the choice of peptide substrate used. In general, shorter peptides elicit less background, as do peptides with fewer aromatic groups.

2. In situations where PBS is incompatible with the enzyme being studied, alternative buffers such as Tris may be used: 25 mM Tris pH = 8, 137 mM NaCl, 2.7 mM KCl, 1 mM $MgCl_2$. The pH of the reaction must be regulated since the subsequent development step is pH-sensitive; *see* ref. [33] for additional information on buffer selection.
3. Depending on the desired sensitivity and kinetics of the development reaction, the amount of ethanol in the developer solution may be adjusted. For example, a solution of 10 mM OPT and 10 mM DTT in 100 % ethanol may be used, although this will typically result in a decreased signal intensity/unit time.
4. If the doses of NAM selected for the standard curve are very high, the length of this incubation period may need to be increased.
5. This reaction is light-sensitive; *see* ref. [33] for more details.
6. If available, a multichannel pipetman may be used to simultaneously add the developer reagent to all samples. The timing and consistency of the addition of developer, and subsequent mixing, are critical to minimizing sample variance.
7. Depending on the concentrations of NAM present and the signal sensitivity desired, this time may need to be adjusted. A 1h incubation period has been chosen since it yields a substantial signal, and produces a linear readout of NAM over the ranges of concentrations suggested in this protocol. Importantly, the reaction kinetics are affected by both buffer composition and temperature. The development reaction may not reach completion within a 1-h time window, and the fluorometric signal may continue to increase in a linear fashion until a plateau signal is reached. The conditions outlined in [33] result in reaction completion in about 60–90 min. Therefore, it may be desirable to allow the incubation period to proceed for a longer time period. However, consistency in the length of this incubation period is critical in order to minimize sample variance.
8. Excitation and emission wavelengths of 420 and 460 nm are recommended for filter-based fluorometers. However, the optimal wavelengths for measurement are described in [33]. If a monochromator is available, excitation and emission wavelengths may be set to 413 and 476 nm, respectively.
9. In addition to inherent background from the plate/fluorometer, the buffer, peptide substrate, β-NAD^+ (at high doses), and resulting complexes formed in the presence of OPT and DTT may contribute to background fluorescence in the assay. In order to account for this, appropriate controls must be performed. When using concentrations of β-NAD less than the K_M value for SIRT1 (~200 μM), background fluorescence due to β-NAD is typically negligible. Therefore, performing parallel

reactions in the presence or absence of β-NAD and subtracting the two values yield a fluorescence value proportional to the actual amount of NAM produced ($F_{corrected} = F_{+NAD} - F_{-NADcontrol}$). This control also eliminates noise due to nonenzymatic deacetylation of the substrate, or deacetylation by contaminating, non-NAD⁺-dependent enzymes. For reactions involving high doses of β-NAD, performing parallel reactions in the presence or absence of SIRT1 enzyme may be more appropriate. Performing parallel reactions using either an acetylated peptide or the corresponding non-acetylated peptide is an alternative approach. Similarly, the quantity of fluorescence corresponding to NAM production may be obtained through calculation: $F_{corrected} = F_{acetyl\ substrate} - F_{non\text{-}acetyl\ substrate}$.

10. The amount of recombinant yPnc1 and SIRT1 enzymes which need to be added to obtain a strong signal will vary depending on the specific activity of the preparation.
11. Test compounds such as small molecule activators or inhibitors should first be tested in the context of a NAM standard curve to ensure that they do not alter yPnc1 activity, or quench the fluorescence signal. If they do, an alternative assay should be used. However, slight background fluorescence from test compounds is not prohibitive as this may be corrected for by using the corresponding reaction in the absence of β-NAD.
12. Longer incubation times may be used if lower amounts of SIRT1 enzyme are being used, or if activity on the substrate being assayed is weak.

Acknowledgements

This work was supported by an NSERC PGS-D Fellowship to B.P.H., the Paul F. Glenn Foundation for Medical Research, the United Mitochondrial Disease Foundation, The Juvenile Diabetes Research Foundation, and NIA/NIH grants to D.A.S. B.P.H. and D.A.S. are inventors on a patent licensed to Millipore. D.A.S. is a consultant to GlaxoSmithKline.

References

1. North BJ, Verdin E (2004) Sirtuins: Sir2-related NAD-dependent protein deacetylases. Genome Biol 5(5):224. doi:10.1186/gb-2004-5-5-224 gb-2004-5-5-224 [pii]
2. Gottlieb S, Esposito RE (1989) A new role for a yeast transcriptional silencer gene, SIR2, in regulation of recombination in ribosomal DNA. Cell 56(5):771–776. doi:0092-8674(89)90681-8 [pii]
3. Gotta M, Strahl-Bolsinger S, Renauld H, Laroche T, Kennedy BK, Grunstein M, Gasser SM (1997) Localization of Sir2p: the nucleolus as a compartment for silent information regulators. EMBO J 16(11):3243–3255. doi:10.1093/emboj/16.11.3243
4. Kaeberlein M, McVey M, Guarente L (1999) The SIR2/3/4 complex and SIR2 alone promote longevity in Saccharomyces cerevisiae by two different mechanisms. Genes Dev 13(19):2570–2580
5. Das C, Lucia MS, Hansen KC, Tyler JK (2009) CBP/p300-mediated acetylation of histone

H3 on lysine 56. Nature 459(7243):113–117. doi:nature07861 [pii] 10.1038/nature07861

6. Cohen HY, Miller C, Bitterman KJ, Wall NR, Hekking B, Kessler B, Howitz KT, Gorospe M, de Cabo R, Sinclair DA (2004) Calorie restriction promotes mammalian cell survival by inducing the SIRT1 deacetylase. Science 305(5682):390–392. doi:10.1126/science.1099196 1099196 [pii]
7. Rodgers JT, Lerin C, Gerhart-Hines Z, Puigserver P (2008) Metabolic adaptations through the PGC-1 alpha and SIRT1 pathways. FEBS Lett 582(1):46–53. doi:S0014-5793(07)01178-7 [pii] 10.1016/j.febslet.2007.11.034
8. Gerhart-Hines Z, Rodgers JT, Bare O, Lerin C, Kim SH, Mostoslavsky R, Alt FW, Wu Z, Puigserver P (2007) Metabolic control of muscle mitochondrial function and fatty acid oxidation through SIRT1/PGC-1alpha. EMBO J 26(7):1913–1923. doi:10.1038/sj.emboj.7601633
9. Bernier M, Paul RK, Martin-Montalvo A, Scheibye-Knudsen M, Song S, He HJ, Armour SM, Hubbard BP, Bohr VA, Wang L, Zong Y, Sinclair DA, de Cabo R (2011) Negative regulation of STAT3 protein-mediated cellular respiration by SIRT1 protein. J Biol Chem 286(22):19270–19279. doi:M110.200311 [pii] 10.1074/jbc.M110.200311
10. Dittenhafer-Reed KE, Feldman JL, Denu JM (2011) Catalysis and mechanistic insights into sirtuin activation. Chembiochem 12(2):281–289. doi:10.1002/cbic.201000434
11. Sauve AA, Celic I, Avalos J, Deng H, Boeke JD, Schramm VL (2001) Chemistry of gene silencing: the mechanism of NAD+-dependent deacetylation reactions. Biochemistry 40(51):15456–15463. doi:bi011858j [pii]
12. Smith BC, Denu JM (2006) Sir2 protein deacetylases: evidence for chemical intermediates and functions of a conserved histidine. Biochemistry 45(1):272–282. doi:10.1021/bi052014t
13. Jackson MD, Schmidt MT, Oppenheimer NJ, Denu JM (2003) Mechanism of nicotinamide inhibition and transglycosidation by Sir2 histone/protein deacetylases. J Biol Chem 278(51):50985–50998. doi:10.1074/jbc.M306552200 M306552200 [pii]
14. Smith BC, Denu JM (2007) Mechanism-based inhibition of Sir2 deacetylases by thioacetyl-lysine peptide. Biochemistry 46(50):14478–14486. doi:10.1021/bi7013294
15. Sauve AA, Wolberger C, Schramm VL, Boeke JD (2006) The biochemistry of sirtuins. Annu Rev Biochem 75:435–465. doi:10.1146/annurev.biochem.74.082803.133500
16. Jackson MD, Denu JM (2002) Structural identification of 2′- and 3′-O-acetyl-ADP-ribose as novel metabolites derived from the Sir2 family of beta -NAD+-dependent histone/protein deacetylases. J Biol Chem 277(21):18535–18544. doi:10.1074/jbc.M200671200 M200671200 [pii]
17. Borra MT, Langer MR, Slama JT, Denu JM (2004) Substrate specificity and kinetic mechanism of the Sir2 family of NAD+-dependent histone/protein deacetylases. Biochemistry 43(30):9877–9887. doi:10.1021/bi049592e
18. Smith BC, Denu JM (2007) Acetyl-lysine analog peptides as mechanistic probes of protein deacetylases. J Biol Chem 282(51):37256–37265. doi:M707878200 [pii] 10.1074/jbc.M707878200
19. Garrity J, Gardner JG, Hawse W, Wolberger C, Escalante-Semerena JC (2007) N-lysine propionylation controls the activity of propionyl-CoA synthetase. J Biol Chem 282(41):30239–30245. doi:M704409200 [pii] 10.1074/jbc.M704409200
20. Du J, Zhou Y, Su X, Yu JJ, Khan S, Jiang H, Kim J, Woo J, Kim JH, Choi BH, He B, Chen W, Zhang S, Cerione RA, Auwerx J, Hao Q, Lin H (2011) Sirt5 is a NAD-dependent protein lysine demalonylase and desuccinylase. Science 334(6057):806–809. doi:10.1126/science.1207861 334/6057/806 [pii]
21. Kowieski TM, Lee S, Denu JM (2008) Acetylation-dependent ADP-ribosylation by Trypanosoma brucei Sir2. J Biol Chem 283(9):5317–5326. doi:M707613200 [pii] 10.1074/jbc.M707613200
22. Hawse WF, Wolberger C (2009) Structure-based mechanism of ADP-ribosylation by sirtuins. J Biol Chem 284(48):33654–33661. doi:M109.024521 [pii] 10.1074/jbc.M109.024521
23. Kaeberlein M, McDonagh T, Heltweg B, Hixon J, Westman EA, Caldwell SD, Napper A, Curtis R, DiStefano PS, Fields S, Bedalov A, Kennedy BK (2005) Substrate-specific activation of sirtuins by resveratrol. J Biol Chem 280(17):17038–17045. doi:M500655200 [pii] 10.1074/jbc.M500655200
24. Howitz KT, Bitterman KJ, Cohen HY, Lamming DW, Lavu S, Wood JG, Zipkin RE, Chung P, Kisielewski A, Zhang LL, Scherer B, Sinclair DA (2003) Small molecule activators of sirtuins extend Saccharomyces cerevisiae lifespan. Nature 425(6954):191–196. doi:10.1038/nature01960 nature01960 [pii]
25. Borra MT, Smith BC, Denu JM (2005) Mechanism of human SIRT1 activation by resveratrol. J Biol Chem 280(17):17187–17195. doi:M501250200 [pii] 10.1074/jbc.M501250200
26. Fan Y, Hense M, Ludewig R, Weisgerber C, Scriba GK (2011) Capillary electrophoresis-based sirtuin assay using non-peptide substrates.

J Pharm Biomed Anal 54(4):772–778. doi:S0731-7085(10)00605-9 [pii] 10.1016/j.jpba.2010.10.010
27. Feng Y, Wu J, Chen L, Luo C, Shen X, Chen K, Jiang H, Liu D (2009) A fluorometric assay of SIRT1 deacetylation activity through quantification of nicotinamide adenine dinucleotide. Anal Biochem 395(2):205–210. doi:S0003-2697(09)00559-4 [pii] 10.1016/j.ab.2009.08.011
28. Milne JC, Lambert PD, Schenk S, Carney DP, Smith JJ, Gagne DJ, Jin L, Boss O, Perni RB, Vu CB, Bemis JE, Xie R, Disch JS, Ng PY, Nunes JJ, Lynch AV, Yang H, Galonek H, Israelian K, Choy W, Iffland A, Lavu S, Medvedik O, Sinclair DA, Olefsky JM, Jirousek MR, Elliott PJ, Westphal CH (2007) Small molecule activators of SIRT1 as therapeutics for the treatment of type 2 diabetes. Nature 450(7170):712–716. doi:nature06261 [pii] 10.1038/nature06261
29. Hubbard BP, Gomes AP, Dai H, Li J, Case AW, Considine T, Riera TV, Lee JE, E SY, Lamming DW, Pentelute BL, Schuman ER, Stevens LA, Ling AJ, Armour SM, Michan S, Zhao H, Jiang Y, Sweitzer SM, Blum CA, Disch JS, Ng PY, Howitz KT, Rolo AP, Hamuro Y, Moss J, Perni RB, Ellis JL, Vlasuk GP, Sinclair DA (2013) Evidence for a common mechanism of SIRT1 regulation by allosteric activators. Science 339(6124):1216–1219. doi:10.1126/science.1231097 339/6124/1216 [pii]
30. Smith BC, Hallows WC, Denu JM (2009) A continuous microplate assay for sirtuins and nicotinamide-producing enzymes. Anal Biochem 394(1):101–109. doi:S0003-2697(09)00493-X [pii] 10.1016/j.ab.2009.07.019
31. Ghislain M, Talla E, Francois JM (2002) Identification and functional analysis of the Saccharomyces cerevisiae nicotinamidase gene, PNC1. Yeast 19(3):215–224. doi:10.1002/yea.810 [pii] 10.1002/yea.810
32. Mondzac A, Ehrlich GE, Seegmiller JE (1965) An enzymatic determination of ammonia in biological fluids. J Lab Clin Med 66(3):526–531
33. Sugawara K, Oyama F (1981) Fluorogenic reaction and specific microdetermination of ammonia. J Biochem 89(3):771–774
34. Nishina H, Inageda K, Takahashi K, Hoshino S, Ikeda K, Katada T (1994) Cell surface antigen CD38 identified as ecto-enzyme of NAD glycohydrolase has hyaluronate-binding activity. Biochem Biophys Res Commun 203(2):1318–1323. doi:S0006-291X(84)72326-6 [pii] 10.1006/bbrc.1994.2326
35. Burkle A (2005) Poly(ADP-ribose). The most elaborate metabolite of NAD+. FEBS J 272(18):4576–4589. doi:EJB4864 [pii] 10.1111/j.1742-4658.2005.04864.x
36. Schneider B, Xu YW, Janin J, Veron M, Deville-Bonne D (1998) 3′-Phosphorylated nucleotides are tight binding inhibitors of nucleoside diphosphate kinase activity. J Biol Chem 273(44):28773–28778
37. Wood JG, Rogina B, Lavu S, Howitz K, Helfand SL, Tatar M, Sinclair D (2004) Sirtuin activators mimic caloric restriction and delay ageing in metazoans. Nature 430(7000):686–689. doi:10.1038/nature02789 nature 02789 [pii]

Chapter 12

Detecting Sirtuin-Catalyzed Deacylation Reactions Using ^{32}P-Labeled NAD and Thin-Layer Chromatography

Anita Zhu, Xiaoyang Su, and Hening Lin

Abstract

Sirtuins are a class of enzymes with important functions in regulation aging, metabolism, and genome stability. They were originally known as nicotinamide adenine dinucleotide (NAD)-dependent protein lysine deacetylases. However, recently it has been discovered that certain sirtuins with weak deacetylase activity also hydrolyze novel acyl lysine modifications. These findings indicate that other sirtuins with weak deacetylase activity may also possess novel activities on unknown protein posttranslational modifications. Analytical methods that can help to identify new activity of sirtuins and new acyl lysine modifications are thus needed. Here we describe a sensitive method that uses ^{32}P-labeled NAD and thin-layer chromatography to detect sirtuin-catalyzed deacylation reactions. This method can help to discover new acyl lysine modifications that can be removed by novel sirtuin activities.

Key words Sirtuin, Nicotinamide adenine dinucleotide, Thin-layer chromatography, Glutamate dehydrogenase, Sirt5, CobB, PfSir2a, ADP-ribose

1 Introduction

Sirtuins are a class of nicotinamide adenine dinucleotide (NAD)-dependent enzymes that regulate important biological pathways in bacteria, archaea, and eukaryotes [1, 2]. Previously, their ability to regulate aging, metabolism, and genomic stability was mainly attributed to deacetylation [1–5]. However, recent discoveries indicate that sirtuins with weak deacetylase activities actually prefer to hydrolyze novel acyl lysine modifications such as succinyl lysine (hydrolyzed by human Sirt5) and myristoyl lysine (hydrolyzed by Plasmodium falciparum Sir2A or PfSir2A) [6–8]. These findings suggest that other sirtuins with weak deacetylase activity may also prefer to hydrolyze acyl lysine modifications that are yet to be discovered. To aid in the discovery of novel posttranslation modifications as well as novel sirtuin activities, new analytical methods are needed.

Matthew D. Hirschey (ed.), *Sirtuins: Methods and Protocols*, Methods in Molecular Biology, vol. 1077,
DOI 10.1007/978-1-62703-637-5_12, © Springer Science+Business Media, LLC 2013

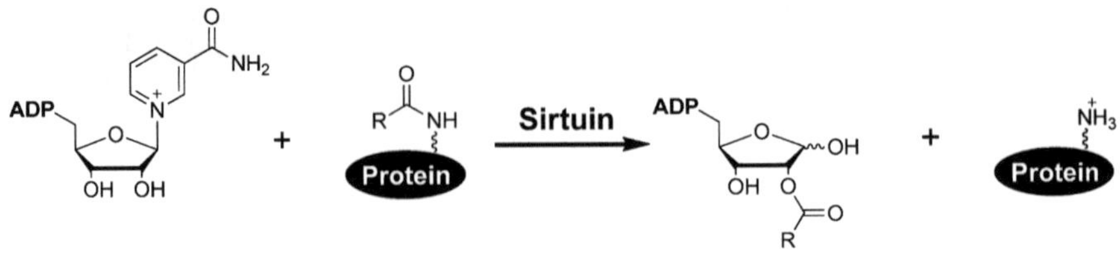

Fig. 1 Sirtuin-catalyzed NAD-dependent deacylation. The ADP in bold contains the ^{32}P label. The resulting acyl ADP-ribose is also ^{32}P-labeled, allowing the detection by TLC autoradiography

Currently, there are two major methods to detect acyl lysine modifications that are removed by sirtuins: western blots using acyl lysine antibodies and liquid chromatography-tandem mass spectrometry (LC-MS/MS). These methods are not suitable for discovery of new acyl lysine modifications. Antibodies for new acyl lysine modifications are not available during the initial discovery stage. LC-MS/MS can provide definitive proof for the discovery of new modifications, but the limitation is that typically it only works well to detect known modifications (MASCOT search only identifies modifications you specified in the search). Thus, alternative methods are needed in the initial discovery stage to detect acyl lysine modifications that are removed by sirtuins. The ^{32}P-labeled NAD (^{32}P-NAD) assay described here is specifically designed to fulfill this purpose.

^{32}P-NAD can be used to detect small molecules generated in sirtuin-catalyzed reactions [9]. One of the enzymatic products of sirtuin-catalyzed deacetylation reactions is O-acetyl ADP-ribose. When ^{32}P-NAD is used in the Sirt1-catalyzed deacetylation reaction, the resulting acetyl ADP-ribose will also be labeled by ^{32}P [10]. Similarly, in Sirt5-catalyzed demalonylation/desuccinylation reaction, the resulting malonyl ADP-ribose and succinyl ADP-ribose are also ^{32}P-labeled [7]. Other modifications removed by sirtuins will give the corresponding acyl ADP-ribose, too [6]. The acyl ADP-ribose can be resolved on a thin-layer chromatography (TLC) plate, allowing the detection and identification of the acyl ADP-ribose molecules by autoradiography. The principle of this assay is illustrated in Fig. 1.

The purpose of the ^{32}P-NAD assay is not to provide definitive proof, but rather serve as a screening step before venturing into more time-consuming and expensive experiments downstream, including antibody development and LC-MS/MS. Comparatively, ^{32}P-NAD assay is cheap and can provide insights into the structure of the modifications. The protein samples are usually trypsin-digested into peptides, and then treated with sirtuin proteins and ^{32}P-NAD. Alternatively, intact proteins can be used in the assay. In this protocol, we give examples of both. Using this assay, we will demonstrate that multiple posttranslational modifications can be removed by sirtuins and that commercially available glutamate dehydrogenase (GDH) contains succinyl lysine modifications.

2 Material

All solutions are prepared in deionized water with a resistivity of 18 MΩ/cm at 25 °C unless otherwise specified. All other reagents were of at least molecular biology grade. Solutions are prepared and stored at room temperature unless otherwise indicated. All peptides were custom synthesized by Genscript and are dissolved in deionized water unless otherwise specified. Institutional permit is required for safe handling of radiolabeled reagents, please refer to you institute'/university's protocol.

2.1 Trypsin Digestion

1. 1 M Tris pH 8.0: weigh out 121.14 g and dissolve in 500 mL of water. Use 6 M HCl to adjust pH to 8.0. Then transfer to a 1 L graduated cylinder and add water to make up 1 L.
2. 1 M 1, 4-dithio-D-threitol (DTT): Weigh out 15.4 mg and dissolve in 100 μL of water. Make it fresh every time.
3. Guanidine hydrochloride (solid, *see* **Note 1**).
4. Alkylation solution: 1 M iodoacetamide. Weigh out 9.25 mg of iodoacetamide and dissolve in 50 μL of water (*see* **Note 2**). *Caution: wear necessary protective gear when handling.*
5. 1 M Tris pH 7.4: Weigh out 121.14 g and dissolve in 500 mL of water. Use 6 M HCl to adjust pH to 7.4 and add water to a final volume of 1 L.
6. 1 M $CaCl_2$: Dissolve 147 mg of $CaCl_2 \cdot 2H_2O$ into 1 mL of water.
7. Trypsin digestion buffer (50 mM Tris, pH 7.4, 1 mM $CaCl_2$): Combine 157.5 μL of 1 M Tris pH 7.4, 3.6 μL of 1 M $CaCl_2$, and 3 mL of water (*see* **Note 3**).
8. Trypsin solution (0.1 μg/μL): Dissolve 20 μg of trypsin in 0.2 mL of 50 mM acetic acid.
9. Quenching Solution (10 % trifluoroacetic acid or TFA): Pipet 1 mL of TFA with a graduated glass pipet into 9 mL of water in a hood (*see* **Note 4**).

2.2 Desalting

1. Wetting solution (90 % methanol with 0.1 % TFA): Combine 9 mL of methanol and 100 μL of quenching solution and 900 μL of water.
2. Equilibration solution (0.1 % TFA): Add 100 μL of quenching solution to 9.9 mL of water.

3. Washing solution (5 % methanol in 0.1 % TFA): Combine 500 μL of methanol and 100 μL of quenching solution in 9.4 mL of water.
4. Eluting solution (50 % Acetonitrile or ACN in 0.1 % TFA): Mix 5 mL of ACN and 100 μL of quenching solution in 4.9 mL of water.
5. Sep-Pak Vac 1 cc (50 mg) C18 cartridge.

2.3 ^{32}P-NAD Assay

1. H3K9 peptide solutions (1 mM acylated H3K9 peptides): Acetyl and succinyl H3K9 peptides are dissolved in water to a final concentration of 1 mM. Butyryl and octanoyl H3K9 peptides are dissolved in 50:50 (*v*:*v*) DMSO: water. Myristoyl H3K9 peptide was dissolved in DMSO.
2. 5 M NaCl: Dissolve 58.44 g in 200 mL of water.
3. 10× reaction buffer (1.5 M NaCl and 500 mM Tris buffer pH 8.0): Combine 300 μL of 5 M NaCl, 500 μL of 1 M Tris pH 8.0, and 200 μL water for a total of 1 mL.
4. 50 mM DTT. Dilute 5 μL of 1 M DTT into 95 μL of water.
5. 1 M Ammonium bicarbonate: Dissolve 79 g of ammonium bicarbonate into 1 L of water.
6. TLC running buffers (70 % ethanol and 30 % 1 M ammonium bicarbonate by volume): Combine 70 mL of 95 % ethanol and 30 mL of 1 M ammonium bicarbonate.
7. ^{32}P-NAD (1 μCi/μL): Dilute the 5 μCi/μL stock five times with water before use. *Caution: Use proper protect gear and follow university protocol in regard to the use of radioactive compounds.*
8. Sirt5, CobB, and PfSir2a are all 6xHis-tagged proteins purified according to the QIAexpressionist manual by gravity flow with nickel-NTA resins.
9. TLC silica gel 60, 20×20 cm.
10. TLC developing chamber with lid.
11. Cellophane membrane.
12. Storage Phosphor Screen and holder.
13. Image eraser model 810-unv.
14. Storm Image scanner 860.
15. Imagequant TL v2005.

2.4 ^{32}P-NAD Assay of Intact GDH Protein

1. GDH (24 mg/mL) (from bovine liver, buffered aqueous glycerol solution).
2. 6.8 mM α-ketoglutarate. Weigh out 1 mg and dissolve in 1 mL of water.
3. 10 mM Ammonia chloride. Weigh out 5.3 mg and dissolve in 10 mL of water.

4. ^{32}P-NAD assay solution. For a 10 μL reaction volume (150 mM of NaCl, 50 mM Tris pH 8.0, 1 mM DTT, 0.1 μCi ^{32}P-NAD, 1 μM Sirt5, 48 μg of GDH), add the following into a 1.5 mL microfuge tube:

1 μL	10× reaction buffer
2 μL	50 mM DTT
2 μL	24 mg/mL GDH
1 μL	1 μCi of ^{32}P-NAD
1 μL	10 μM Sirt5
1 μL	6.8 mM α-ketoglutarate
1 μL	10 mM ammonia chloride
1 μL	Water

3 Methods

3.1 Trypsin Digestion of the GDH

1. Mix the following in a 1.5 mL microfuge tube and incubate at room temperature for 1 h to allow proteins to denature (*see* **Notes 1 and 5**).

62.5 μL	GDH (24 mg/mL, 1.5 mg)
258 mg	Guanidine hydrochloride
6.75 μL	1 M DTT
27 μL	1 M Tris pH 8.0

Add water to a final volume of 450 μL.

2. Add 22.5 μL 1 M iodoacetamide (final concentration of iodoacetamide is 50 mM), wrap the tube with aluminum foil (see **Note 2**), and gently mix for 1 h at room temperature.
3. Dilute mixture with 3.15 mL of trypsin digestion buffer to 3.6 mL (*see* **Note 6**). Vortex to mix.
4. Add 150 μL of trypsin solution (*see* **Note 5**).
5. Incubate at 37 °C overnight.
6. Add 67.5 μL of quenching solution (10 % TFA) to stop the reaction. The pH should be around 2–3.

3.2 Desalting

Do not let the Sep-Pak cartridge run dry after wetting unless otherwise stated (*see* **Notes 7** and **8**).

1. Wet Sep-Pak C18 cartridge with 1 mL of wetting solution. Repeat two more times.
2. Equilibrate the cartridge with 1 mL of equilibration solution 3 times.

3. Load the sample of trypsin-digested proteins slowly (no faster than one drop per second).
4. Wash the cartridge with 0.5 mL of equilibration buffer.
5. Wash the cartridge with 1 mL of wash solution twice. For the second wash, blow the cartridge dry to get rid of residual wash solution.
6. Elute the peptides into a clean tube with 1 mL of elution buffer and blow the cartridge dry.
7. Freeze the elution with liquid nitrogen and lyophilize. The lyophilized sample can be stored for months at −80 °C.

3.3 ^{32}P-NAD Assay

1. Reconstitute the trypsin-digested peptides in water to a concentration of 10 mg/mL.
2. ^{32}P-NAD labeling assays are carried out in 10 μL reaction volume (150 mM of NaCl, 50 mM Tris pH 8.0, 1 mM DTT, 0.1 μCi ^{32}P-NAD, 1 μM enzyme, 100 μM peptide). For each reaction, add the following into a 1.5 mL microfuge tube:

1 μL	10× reaction buffer
2 μL	50 mM DTT
1 μL	100 μM H3K9 peptides or 1 mg/mL of trypsin-digested peptides
1 μL	1 μCi of ^{32}P-NAD
1 μL	10 μM Sirt5, CobB or PfSir2a protein
4 μL	Water

The ^{32}P-NAD is added second to last and the sirtuin enzymes are always added last (*see* **Notes 9**, **10**, and **11**).

Results for a typical experiment with standard acyl peptides are shown in Fig. 2.

3. Incubate the reaction at 37 °C for 15 min. At the same time, the TLC plate and running buffer can be prepared (*see* **Note 12**).
4. Cut the TLC plate according to Fig. 3. The width of the plate can be changed according to the number of samples. Add 100 mL of TLC running buffer into the tank and cover with lid (*see* **Note 13**).
5. Briefly vortex and spin the sample tubes at 12,000× *g* for 15 s. Load 0.5 μL of each sample onto the TLC plate.
6. Wait 2 min for the spots to dry completely and then put the TLC plate in the chamber slowly and evenly. Cover the chamber with a lid. Take out the TLC plate when the solvent line reaches the finishing line. The development takes 40–60 min (*see* **Note 14**).

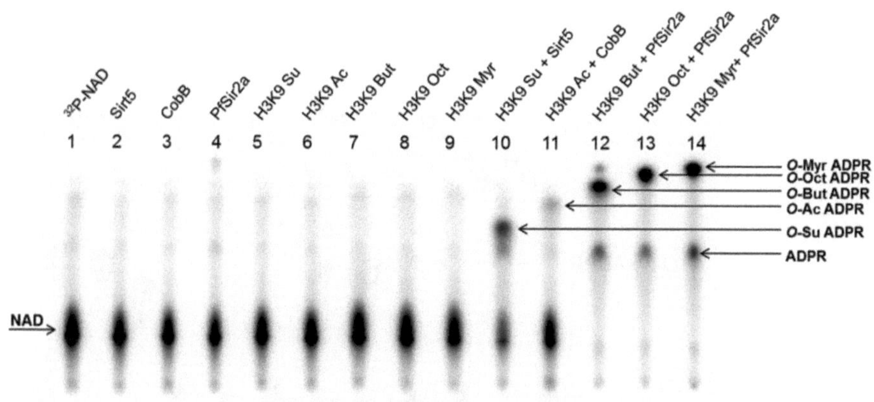

Fig. 2 ^{32}P-NAD activity assay of Sirt5, CobB, and PfSir2a with succinylated, acetylated, butylylated, octanoylated, and myristoylated H3K9 peptide. The reactions were incubated at 37 °C for 15 min and 0.5 μL were spotted on a TLC. The lower spots in *lanes 1–11* are NAD. The identities of other spots are indicated on the right side. *Lane 1* contains reaction buffer without peptide or protein. *Lanes 2–4* contain reaction buffers with only proteins. PfSir2a in *lane 4* has a weak spot that is likely myristoyl ADPR, suggesting that it may be fatty acylated. This spot can be removed by pre-incubating PfSir2A with NAD followed by buffer exchange to remove small molecules. *Lanes 5–9* contain reaction buffer and peptide only. *Lanes 10–14* are reactions that contain both acyl peptides and sirtuins

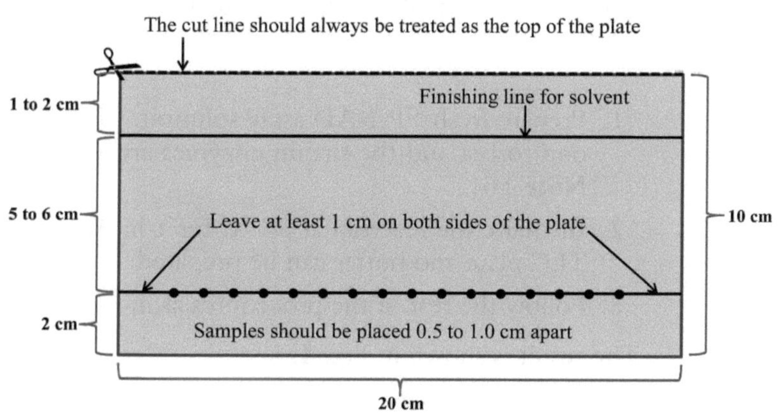

Fig. 3 Diagram for cutting TLC plate (*see* **Note 15**)

7. Leave plate for 20–30 min to dry. Wrap it with cellophane membrane (to avoid contaminating the phosphor screen) and put it into the storage phosphor screen cassette. Make sure the front side of the TLC plate is facing to the phosphor screen. Expose overnight.
8. Remove the plate, scan the phosphor screen using the Storm Image scanner, and process the image using Imagequant TL.
9. After imaging, erase screen with the Image Eraser for 5 min.
10. Discard the ^{32}P-NAD TLC plate appropriately.

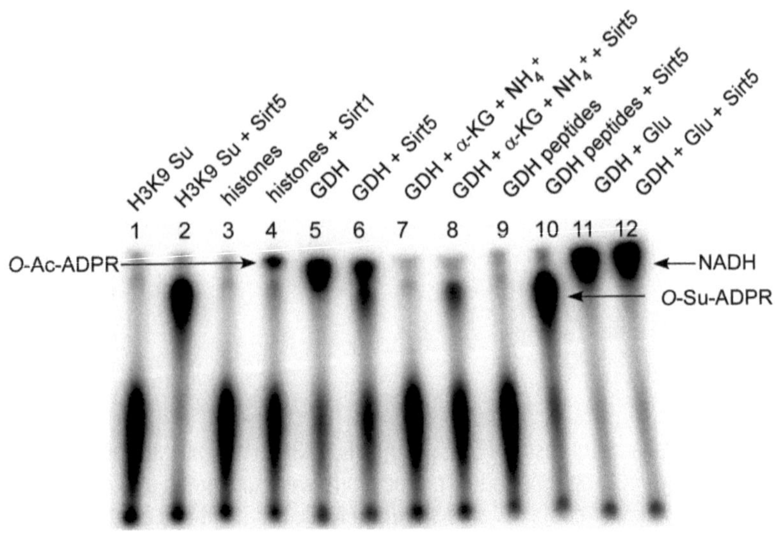

Fig. 4 Bovine GDH contains succinyl lysine residues that can be removed by Sirt5. *Lanes 2, 4, and 11* are control reactions to indicate the positions of *O-Su-ADPR*, *O*-Ac-ADPR, and NADH, respectively. When GDH was incubated with Sirt5 and ^{32}P-NAD (*lane 6*), two new spots were formed, one is NADH, and another is *O*-Su-ADPR. NADH is formed from GDH catalyzed glutamate dehydrogenation reaction. The formation of NADH by GDH can be prevented by adding α-ketoglutarate and ammonium in the reaction to force the reaction to go backward (*lanes 7 and 8*)

3.4 ^{32}P-NAD Assay of Intact GDH Protein

1. Prepare fresh ^{32}P-NAD assay solution. ^{32}P-NAD is added second to last and the sirtuin enzymes are always added last (*see* **Note 16**).
2. Incubate the reaction at 37 °C for 1 h. At the same time, the TLC plate and buffer can be prepared.
3. Follow the rest of the procedures as in Subheading 3.3.

The result is shown in Fig. 4.

4 Notes

1. Alternatively, a 10 M stock solution of guanidine hydrochloride can be made and added to denaturing buffer for a final concentration of 6 M. Denaturing solution containing guanidine hydrochloride can be heated to 95 °C to facilitate the denaturation. An alternative to guanidine hydrochloride is 10 M urea (6 M final concentration), but the denaturing solution containing urea should not be heated above 37 °C because it will carbamylate peptides.
2. Iodoacetamide is sensitive to light and toxic. Stock solution must be made fresh, used immediately, and be shielded from light.

During alkylation, cover tubes with aluminum foil or incubate in a dark place such as a closed drawer at room temperature.

3. Dilute the entire solution after alkylation into the prepared trypsin digestion buffer. The concentration of Tris is calculated to take into account the existing Tris buffer in the denatured protein solution while the $CaCl_2$ was calculated according to final volume.
4. All solutions containing TFA should not be stored for more than a week. It is best to make solutions containing TFA the day of use.
5. The trypsin digestion can be scaled down if less protein needs to be digested. The trypsin to protein ratio is typically 1:20–1:100.
6. The purpose of this step is to lower the concentration of urea or guanidine hydrochloride to less than 1 M.
7. Compressed air or nitrogen gas can be used to push the solution through the cartridge for a faster flow. If air pressure is not applied, the step can take up to several hours. Pressure should not be applied in the sample binding step. For the sample binding step, it is important that the sample rate should be less than one drop per second.
8. During the Sep-Pak desalting step, the wash and elution conditions can be further optimized based on the polarity of acyl modification. In the case of hydrophilic modifications like succinylation, a lower percentage of methanol can be used in the solutions. If working with more hydrophobic modifications, a higher percentage of acetonitrile (up to 90 %) in elution buffer would be helpful. In such cases it would be best to first wash the Sep-Pak column with 100 % acetonitrile to wash away polymers that may be soluble in acetonitrile.
9. The solution mixture is prepared in a nonradioactive space to minimize exposure to radioactive material. For multiple samples, it is better to prepare a mixture of the reaction buffer to minimize pipetting error. The mixture of reaction buffer usually includes the 10× reaction buffer, DTT, and water. As a general rule of thumb always prepare the mixture for an extra sample. For example, when preparing for 6 samples, make a buffer mixture for 7 samples.
10. In order to detect the enzymatic product resulting from ^{32}P-NAD, the protein and peptide samples should not contain nonradioactive NAD. Otherwise, the nonradioactive NAD may be consumed to remove the acyl groups, leaving the ^{32}P-NAD unreacted. NAD and other small molecules can be removed by buffer exchange or dialysis for larger peptides or for protein samples.

11. A higher concentration of the sirtuin proteins can be used to achieve better sensitivity.
12. For some acyl ADP-ribose molecules, they may decompose into ADP-ribose readily. For such modifications, the ^{32}P-NAD reaction incubation time can be reduced to preserve the acyl ADP-ribose. The best reaction time should be optimized. The activity of the enzyme should also be considered when determining the incubation time at 37 °C
13. On the TLC plate, samples should be at least 0.5 cm apart from each other. The edge of the plate that is cut with scissors should not be used as the bottom/starting side. When the plate is cut, the silica near the cutting line tends to loosen and if this side is used as the starting side, the running front will be jagged. For similar reasons, leave a 1–2 cm empty space on both the left and right sides of the plate. Samples loaded too close to the sides may not run evenly with the rest of the samples.
14. The run time may vary significantly depending on the maker and the condition of the silica plates as well as the solvent used. The development should be closely monitored. The development tank should always be covered with either aluminum foil or a lid to speed up the development. When there is no cover, depending on the ventilation of the space for the run, the top of the plate may dry out and development may be too slow.
15. Depending on the tool used to cut the TLC, the edges may be in better or worse condition. Flexible plates are usually cut with scissors, but paper cutters work very well on the plates with flexible backings. The diagram is not to scale.
16. When intact protein is used as sirtuin substrate, the intrinsic enzymatic activity of the substrate must be taken into consideration. For instance, in the case of GDH, when glutamate is present, GDH can convert NAD and glutamate to NADH and α-ketoglutarate. Because the GDH solution naturally contains a small amount of glutamate, the ^{32}P-NAD (which is very little in terms of molar concentration) can be consumed significantly by the GDH activity rather than being used in the Sirt5 desuccinylation reaction. Other NAD utilizing enzymes may give similar problems. One way to eliminate the substrate's NAD utilizing activity is to drive the reaction in the NAD-forming direction thus leaving the ^{32}P-NAD available to the sirtuins. As we have shown in the GDH case, the reaction buffer contained 0.68 mM α-ketoglutarate and 1 mM ammonia chloride, which would strongly suppress the NADH formation so that the succinyl ADP-ribose can be better detected.

References

1. S-i I, Guarente L (2010) Ten years of NAD-dependent SIR2 family deacetylases: implications for metabolic diseases. Trends Pharmacol Sci 31(5):212–220
2. Haigis MC, Sinclair DA (2010) Mammalian sirtuins: biological insights and disease relevance. Annu Rev Pathol 5(1):253–295. doi:10.1146/annurev.pathol.4.110807.092250
3. Sauve AA, Wolberger C, Schramm VL, Boeke JD (2006) The biochemistry of sirtuins. Annu Rev Biochem 75:435–465
4. S-i I, Armstrong CM, Kaeberlein M, Guarente L (2000) Transcriptional silencing and longevity protein Sir2 is an NAD-dependent histone deacetylase. Nature 403(6771):795–800
5. Tanner KG, Landry J, Sternglanz R, Denu JM (2000) Silent information regulator 2 family of NAD-dependent histone/protein deacetylases generates a unique product, 1-*O*-acetyl-ADP-ribose. Proc Natl Acad Sci U S A 97(26):14178–14182
6. Zhu AY, Zhou Y, Khan S, Deitsch KW, Hao Q, Lin H (2012) Plasmodium falciparum Sir2A preferentially hydrolyzes medium and long chain fatty acyl lysine. ACS Chem Biol 7(1):155–159. doi:10.1021/cb200230x
7. Du J, Zhou Y, Su X, Yu JJ, Khan S, Jiang H, Kim J, Woo J, Kim JH, Choi BH, He B, Chen W, Zhang S, Cerione RA, Auwerx J, Hao Q, Lin H (2011) Sirt5 is a NAD-dependent protein lysine demalonylase and desuccinylase. Science 334(6057):806–809. doi:10.1126/science.1207861
8. Peng C, Lu Z, Xie Z, Cheng Z, Chen Y, Tan M, Luo H, Zhang Y, He W, Yang K, Zwaans BMM, Tishkoff D, Ho L, Lombard D, He T-C, Dai J, Verdin E, Ye Y, Zhao Y (2011) The first identification of lysine malonylation substrates and its regulatory enzyme. Mol Cell Proteomics 10(12):M111.012658. doi:10.1074/mcp.M111.012658
9. Du J, Jiang H, Lin H (2009) Investigating the ADP-ribosyltransferase activity of sirtuins with NAD analogues and ^{32}P-NAD. Biochemistry 48(13):2878–2890. doi:10.1021/bi802093g
10. Alvarez Y, Rodriguez M, Municio C, Hugo E, Alonso S, Ibarrola N, Fernandez N, Sanchez Crespo M (2012) Sirtuin 1 is a key regulator of the IL-12 p70/IL-23 balance in human dendritic cells. J Biol Chem. doi:10.1074/jbc.M112.391839

Chapter 13

Chemical Acetylation and Deacetylation

Kristofer S. Fritz

Abstract

Lysine acetylation is an important posttranslational modification that alters protein structure and function. Understanding the mechanisms involved in regulating protein acetylation remains a key factor in elucidating what role this modification plays in numerous disease pathologies. Here, we describe an in vitro strategy to examine the site-specific deacetylation of proteins utilizing the chemical acetylation of protein lysine residues via acetic anhydride. The impact of chemical acetylation on protein lysine residues is characterized by native gel electrophoresis and Western blotting. Acetyl-Lys modifications are then examined for deacetylation using a SIRT3 deacetylase activity assay and followed by stable isotope dilution mass spectrometry.

Key words Acetylation, Acetic anhydride, Native gel electrophoresis, Mass spectrometry, Sirtuin, SIRT3, Deacetylase

1 Introduction

Protein acetylation is recognized to play a critical role in the regulation of numerous cellular processes, including lipid and amino acid metabolism, energy production, oxidative stress response, and intracellular signaling. The acetylation of lysine residues is regulated through two main protein families; protein acetyltransferases and protein deacetylases [1]. These two classes of proteins regulate homeostatic processes, nutrient response, cell signaling, and stress-induced changes throughout subcellular compartments and involves major functions of the nucleus, cytosol, and mitochondria [2]. Mitochondrial dysfunction is linked to a number of metabolic disorders and provides an important therapeutic target for ameliorating metabolic diseases [3]. Interestingly, loss of mitochondrial deacetylase activity induces mitochondrial protein hyperacetylation and mitochondrial dysfunction, contributing to metabolic syndrome [4]. The Sirtuin (SIRT) family of proteins are classified as class III histone deacetylases (HDACs)

Matthew D. Hirschey (ed.), *Sirtuins: Methods and Protocols*, Methods in Molecular Biology, vol. 1077,
DOI 10.1007/978-1-62703-637-5_13, © Springer Science+Business Media, LLC 2013

and SIRT3 is the mitochondrial resident deacetylase known to regulate global mitochondrial protein deacetylation [5].

Understanding biochemical mechanisms of Sirtuin-regulated lysine deacetylation and how this modification alters protein structure and function is the primary focus of research involving this mitochondrial posttranslational modification. A number of chemical and enzymatic strategies are available to characterize the impact of lysine acetylation on protein activity. This particular strategy employs "light" (h_6) and "heavy" (d_6) acetic anhydride to chemically acetylate proteins followed by a SIRT3 deacetylation assay. Additionally, native gel electrophoresis [6] and Western blotting [7] will be utilized to visualize protein acetylation. Finally, the quantification of lysine deacetylation will be demonstrated by stable isotope dilution MALDI-TOF-TOF mass spectrometry.

2 Materials

Prepare all solutions using HPLC grade water and analytical grade reagents. Prepare and store all reagents at room temperature in acid-washed (2 % HCl solution) glass bottles. Dispose of all toxic, acid, and organic waste in accordance with institutional regulations.

2.1 Acetylation

1. Acetic anhydride ≥99 % (light h_6) (Sigma-Aldrich, St. Louis, MO.). Deuterated acetic anhydride (heavy d_6) (d_6, ≥98 %) (Cambridge Isotope Laboratories, Inc., Andover, MA) (*see* **Note 1**). Store acetic anhydride away from light and moisture (*see* **Note 2**).
2. 100 % Methanol (*see* **Note 3**).
3. Stock acetic anhydride: 500 mM in methanol (*see* **Note 4**). Store at 4 °C.
4. Phosphate buffered saline (1× PBS) (*see* **Note 5**): 8 g NaCl, 0.2 g KCl, 1.44 g Na_2HPO_4, 0.24 g of KH_2PO_4, pH 7.4. Prepared in 1 L water.
5. Target protein for treatment: Human recombinant ALDH2 (>95 % purity) (*see* **Note 6**).

2.2 Native Gel Electrophoresis

1. Native running buffer: 250 mM glycine and 25 mM Tris base, pH 8.3: Dissolve 18.75 g of glycine and 3 g of Tris base in 1 L water (*see* **Note** 7).
2. 5× native loading buffer: 0.25 M Tris base (pH 6.8), 50 % glycerol, and 0.5 % bromophenol blue (*see* **Note 8**).
3. Stock 1.5 M Tris base, pH 8.8: Dissolve 36.3 g Tris base into 150 mL water and adjust to pH 8.8 with HCl. Add water to a final volume of 200 mL.

4. Stock 1.25 M Tris base, pH 6.8: Dissolve 15.1 g Tris base into 50 mL water and adjust to pH 6.8 with HCl. Add water to a final volume of 100 mL.
5. 10 % ammonium persulfate (APS) solution: Dissolve 1 g of ammonium persulfate in 10 ml of water.
6. 30 % acrylamide/bis solution (Bio-Rad) (*see* **Note 9**).
7. Stock tetramethylethylenediamine (TEMED) (*see* **Note 10**).

2.3 Western Blotting

1. Transfer buffer: 28.8 g glycine, 6.05 g Tris base, 400 mL methanol, and add water to 1 L.
2. Tris buffered saline (TBS)–Tween (TBS/T): 10 mM Tris base, 0.9 % NaCl, 0.05 % Tween-20, pH 7.4: Add 1.21 g Tris base and 9 g NaCl in 800 mL water and adjust the pH to 7.4 with HCl. Add 0.5 mL Tween-20 and add water to 1 L.
3. TBS/T w/5 % Milk (TBS/T/M): Dissolve 10 g powdered milk in 200 mL TBS/T (*see* **Note 11**).
4. Filter paper (Bio-Rad) and Hybond-P PVDF Membrane (GE Healthcare).
5. Anti-acetyl-Lys primary antibody (Cell Signaling Technology, #9441) and anti-rabbit-HRP secondary (Jackson Immuno Research Laboratories, Inc.).

2.4 Enzymatic Deacetylation Assay

1. Active form of human recombinant SIRT3 (0.3 μg/μL stock concentration) (Cayman Chemical, Ann Arbor, MI) (*see* **Note 12**).
2. Stock 50 mM NAD^+ in water, prepared fresh.
3. SIRT3 assay buffer: 50 mM Tris base, pH 8.0, 137 mM NaCl, 2.7 mM KCl, and 1 mM $MgCl_2$. Store at 4 °C. Stable for at least 4 months.

2.5 Mass Spectrometry

1. Reagents for in-gel proteolytic digestion: 50 mM ammonium bicarbonate, pH 8.0 (Add 0.0396 g of ammonium bicarbonate to 10 mL of water), 100 % acetonitrile, 50 % acetonitrile in 50 mM ammonium bicarbonate (Add 0.0396 g of ammonium bicarbonate to 5 mL of water and 5 mL of acetonitrile), 5 % formic acid in acetonitrile, 1.0 M dithiothreitol (DTT) (0.15 g in 1 mL water), 1.0 M iodoacetamide (IAM) (0.185 g iodoacetamide in 1 mL of water), and trypsin (0.1 μg/μL, Promega) (*see* **Note 13**).
2. C18 ZipTip reagents: acetonitrile, 0.1 % formic acid in water, 0.1 % formic acid in 60 % acetonitrile, and C18 ZipTips (Millipore).
3. SpeedVac (centrifugal vacuum apparatus).
4. Mass Spectrometer: ABI 4,800 Plus MALDI-TOF-TOF (AB/SCIEX), Opti-TOF 96 well insert.

3 Methods

3.1 Chemical Acetylation

1. Using both light (h_6) and heavy (d_6) acetic anhydride, prepare fresh stocks of 2 mM, 4 mM, 6 mM, 8 mM, 10 mM, 15 mM, 30 mM, 60 mM, 100 mM, and 500 mM in methanol (*see* **Note 14**). Keep acetic anhydride solutions at 4 °C on ice to reduce solvent evaporation.
2. Aliquot 5.6 μg of ALDH2 into a final volume of 18 μL of 1× PBS for each of the desired acetic anhydride treatment concentrations. In this case, a total of 11 aliquots will be used for light (h_6) acetic anhydride and 11 aliquots for heavy (d_6) acetic anhydride incubations (*see* **Note 15**).
3. To each of the ALDH2 aliquots prepared above, add 2 μL of acetic anhydride to the ALDH2 solution (1:10 dilution to each stock). This yields a final concentration of 5 μM ALDH2 and acetic anhydride concentrations ranging from 0 to 50 mM. Perform treatments identically for both light (h_6) and heavy (d_6) acetic anhydride. Incubate the reaction for 60 min at room temp (RT) (*see* **Note 16**).
4. Upon completion, perform 3.2 Native Gel Electrophoresis and 3.3 Western blotting as detailed below to visualize changes to protein charge state due to increased acetylation (Fig. 1).
5. To quantify SIRT3 specificity for lysine deacetylation, perform 3.4 Enzymatic Deacetylation Assay and 3.5 Mass Spectrometry.

3.2 Native Gel Electrophoresis

1. Using native gel electrophoresis conditions, the charge of each of the proteins will depend on the primary amino acid sequence of the protein (isoelectric point), posttranslational modifications, and the pH during electrophoresis. Each of these factors will alter the mobility of the protein during electrophoresis [6].

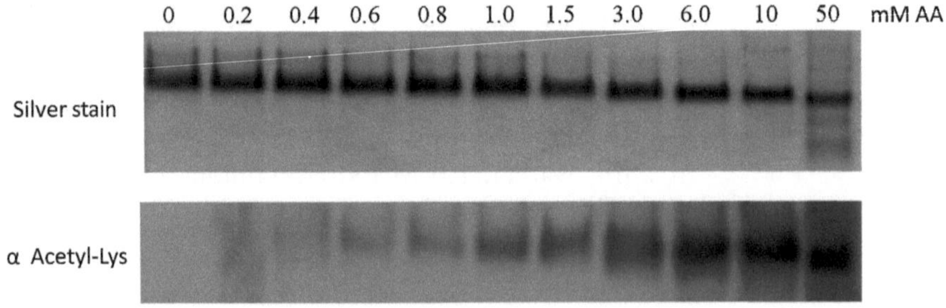

Fig. 1 ALDH2 Silver stain and anti-acetyl-Lys Western analysis. The chemical acetylation of lysine residues by acetic anhydride (AA) alters the migration of ALDH2 through the native PAGE. This is likely observed due to the acetylation of positively charged lysine residues, which in turn leaves ALDH2 less positive. Here, the combination of using a Silver-stained gel and an anti-acetyl-Lys Western provides a thorough illustration of ALDH2 acetylation

2. Thoroughly clean and dry glass plates, spacers, and comb. Assemble glass plates, spacers, and the comb as described by the manufacturer.
3. To prepare the 10 % resolving gel, mix 4.0 mL water, 3.3 mL of acrylamide solution (30 %), 2.5 mL 1.5 M Tris base (pH 8.8), 0.1 mL 10 % APS, and 4 μL TEMED. The addition of TEMED will begin polymerization of the gel (*see* **Note 17**). Mix gently prior to pouring into the gel mould (glass plates). Allow the gel to polymerize for approximately 30 min.
4. Gently cover the top of the resolving gel with water. Once the gel is solidified, remove the water.
5. To prepare the 5 % stacking gel, mix 2.1 mL water, 0.5 mL of acrylamide solution (30 %), 2.5 mL 1.25 M Tris base (pH 6.8), 30 μL 10 % APS, and 3 μL TEMED. Gently mix the solution and pour onto the top of the polymerized resolving gel.
6. Add a 15-well comb and allow the gel to polymerize for 20 min.
7. Once polymerized, place the gel into the running apparatus and pour native running buffer into the inner and outer chambers to the appropriate fill line.
8. For native gel electrophoresis, mix the sample protein solution with native loading buffer at a ratio of 5:1. Do not boil the samples (*see* **Note 18**).
9. Load the samples and standard molecular weight marker into the gel lanes.
10. Migrate the samples using a constant voltage of 175 V. Run the gel so the lowest molecular weight marker approaches the bottom of the gel.
11. Continue to Subheading 3.3 for Western blotting procedure to visualize protein acetylation. Alternatively, protein migration can be visualized by a Silver stain of the native gel (Fig. 1).

3.3 Western Blotting

1. Carefully remove the gel from the glass plates and rinse in water. Place the gel in transfer buffer for 5 min. Mark the gel by cutting the corner indicating the top side of the molecular weight loading control.
2. Wearing gloves, cut two pieces of filter paper (Bio-Rad) for each gel to be transferred. Cut to the same size as the gel and soak them in transfer buffer for 5 min at RT. Cut a piece of PVDF membrane also to the same size of the gel. Wet the membrane by soaking in 100 % methanol for 5 min. Once wetted, place the membrane in transfer buffer.
3. For a semi-dry transfer (*see* **Note 19**), first place one of the filter papers on the transfer apparatus followed by placing the PVDF membrane exactly on top of the filter paper. Then lay the gel on top of the membrane and finally place the second

piece of filter paper on top of the gel. After placing each of the four items, it is critical to roll out any air bubbles with a glass pipette.

4. Cover the stack with the top transfer plate and transfer the gel at 20 V (fixed voltage) for 60 min.
5. Once complete, switch off the blotting apparatus and disassemble the transfer apparatus.
6. Carefully remove the membrane with clean metal tweezers and incubate the membrane in 15 mL TBS/T/M by gentle shaking for 1 h. This is the "blocking" step.
7. Wash the membrane by shaking in TBS/T 3× for 5 min each.
8. Incubate the membranes with an anti-acetyl-Lys antibody, diluted in TBS/T/M at 1:1,000 (it is best to optimize antibody performance by experimentation).
9. Wash the membrane by shaking in TBS/T 3× for 5 min each.
10. Incubate each membrane separately for 1 h in appropriately conjugated secondary antibodies suspended in TBS/T/M at an appropriate dilution.
11. Wash the membrane by shaking in TBS/T 3× for 5 min each.
12. Develop and expose the membrane using the preferred method available (i.e., film developer or Storm instrument) (Acetylation of ALDH2 is shown via Western blot in Fig. 1).

3.4 Enzymatic Deacetylation Assay

1. Using the 10 mM acetic anhydride-treated ALDH2, apply the acetylated ALDH2 to the SIRT3 deacetylase assay (*see* **Note 20**).
2. To a clean 0.5 mL eppendorf tube, add 25 μL SIRT3 assay buffer, 15 μL NAD^+ (50 mm stock), 5 μL SIRT3, and 5 μL of ALDH2 protein (d_3 acetylated).
3. As a control, an identical assay should be used for 5 μL of ALDH2 protein (h_3 acetylated) in the absence of SIRT3. In this case, add 25 μL SIRT3 assay buffer, 15 μL NAD^+ (50 mm stock), 5 μL water, and 5 μL of ALDH2 protein (h_3 acetylated).
4. Allow the deacetylation assay to proceed for 60 min at 37 °C.
5. Upon completion, add 10 μL of 5× SDS-PAGE loading buffer (reducing) to each sample. This will halt the deacetylase assay.
6. In a clean 0.5 mL eppendorf tube add each reaction mixture 1:1; 10 μL of h3-acetyl-ALDH2 and 10 μL of d3-acetyl-ALDH2. Here, equivalent mixtures of h3-acetyl-ALDH2 and deacetylated d3-acetyl-ALDH2 will be added 1:1 which will allow for the quantification of site-specific lysine deacetylation by SIRT3 in Subheading 3.5 Mass Spectrometry (*see* Fig. 2 for experimental scheme).

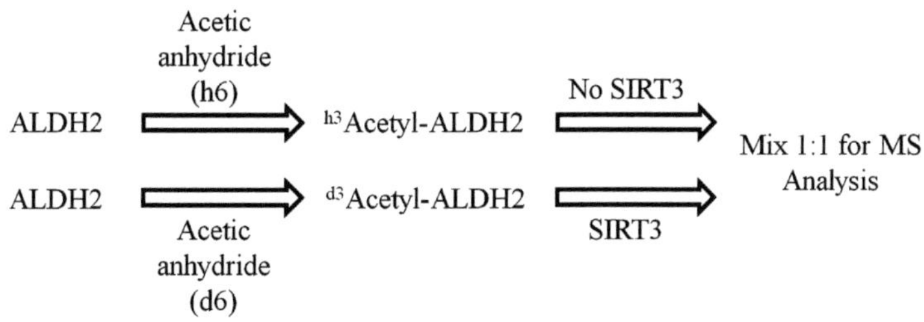

Fig. 2 Experimental workflow for the chemical acetylation and SIRT3-mediated deacetylation of ALDH2. Stable isotope dilution is used to quantify changes in lysine acetylation that are directly related to SIRT3 deacetylase activity

7. Boil the samples at 95 °C for 5 min and resolve using reducing SDS-PAGE conditions followed by coomassie staining and de-staining. *See* ref. [8] for complete reducing SDS-PAGE protocol.

3.5 Mass Spectrometry

1. Once the SDS-PAGE gel is destained, remove the ALDH2 protein band with a new razor blade (*see* **Note 21**). When cutting out the protein band, cut into 1 mm squares. Always stain and destain gels in glassware washed in 10 % acetic acid (*see* **Note 22**).
2. Incubate each spot with 100 μL of 50 mM ammonium bicarbonate (pH 8.0) for 15 min at RT. Make sure the gel pieces are completely covered by the solution. Remove the buffer and discard.
3. To destain the gel pieces, add 100 μL of 50 % acetonitrile in 50 mM ammonium bicarbonate and shake at RT until the gel pieces are no longer stained. This may take overnight.
4. Once destained, remove the buffer and add 100 μL of 100 % acetonitrile and incubate for 15 min at RT. Remove the buffer and allow the gel pieces to air-dry for 20 min.
5. Reduce the in-gel protein with 50 μL of 10 mM DTT (dilute stock DTT 1:100) in 50 mM ammonium bicarbonate for 45 min at 60 °C. Make sure the gel pieces are completely covered by the solution, as the gel pieces will swell. Discard the solution.
6. Alkylate the in-gel protein with 50 μL of 40 mM IAM (dilute stock IAM 1:25) in 50 mM ammonium bicarbonate for 30 min at RT in the dark. Make sure the gel pieces are completely covered by the solution. Discard the solution.
7. Wash the gel pieces with 100 μL of 50 mM ammonium bicarbonate for 15 min at RT and discard.
8. Add 25 μL of 50 mM ammonium bicarbonate and 0.3 μg trypsin (3 μL of a 0.1 μg/μL stock). Digest overnight at 37 °C in a water bath or temperature controlled oven. The gel pieces should be completely covered by the protease solution.
9. Remove the solution from the tube and transfer it to a clean 1.5 mL eppendorf tube. This solution now contains the digested peptides.

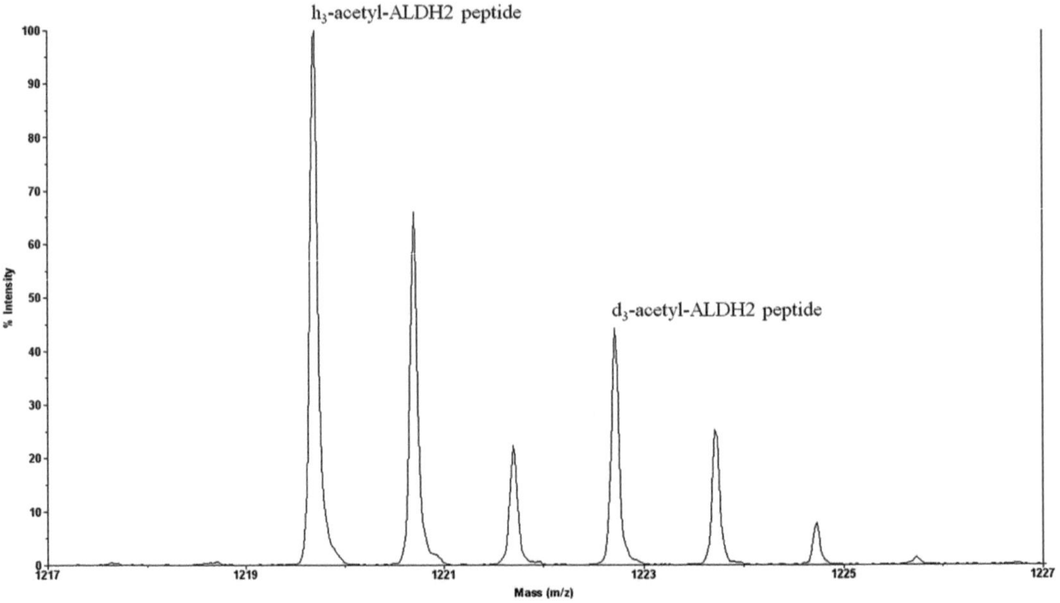

Fig. 3 This stable isotope dilution strategy employs the analysis of light (h_3) and heavy (d_3) acetylation of ALDH2 tryptic peptides by MALDI-TOF mass spectrometry. By analyzing the area under the curve (AUC) for both the light (no SIRT3) and heavy (active SIRT3) acetylated peptides, the deacetylation of certain lysine residues can be quantified. In this instance, by comparing a h_3-acetylated peptide (21,017 AUC) and a d_3-acetylated peptide (8,578 AUC) one can determine that SIRT3 reduced the specific lysine acetylation by 60 %

10. Extract the peptides from the gel pieces with 50 μL of 50 % acetonitrile in 50 mM ammonium bicarbonate solution for 20 min in a sonicating water bath. Add the extracts to the tube from the previous step.
11. Extract the peptides once more with 5 % formic acid in acetonitrile the same as the previous step. Add the extracts to the tube from the previous step.
12. Dry down the resulting protein digest extract to completion in a speedvac and resuspend in 20 μL of 0.1 % TFA in water.
13. Aliquot 10 μL for the C18-ZipTip procedure for spotting on a MALDI-TOF plate. For each peptide sample, wet the ZipTip 2× in 10 μL 100 % acetonitrile, rinse 3× in 10 μL 0.1 % TFA in water, bind sample to ZipTip using 7× passes of 10 μL sample over resin, wash 5× in 10 μL 0.1 % TFA in water, and finally elute the sample by taking up 10 μL 0.1 % TFA in 60 % acetonitrile and eluting into a clean 0.5 mL Eppendorf tube (*see* **Note 23**).
14. The purified peptides should be spotted onto the MALDI plate using 0.8 μL sample and 0.8 μL α-Cyano-4-hydroxycinnamic acid (CHCA) in 0.1 % formic acid in 60 % acetonitrile. Allow to dry and analyze using optimized conditions on a MALDI-TOF-TOF mass spectrometer. (*See* Fig. 3 for illustration of stable isotope dilution MS quantification of SIRT3 deacetylation).

4 Notes

1. While this procedure uses normal and stable isotope-labeled acetic anhydride to monitor the acetylation and deacetylation of lysine residues, other chemical reagents may be used to examine a variety of lysine acyl modifications employing this same procedure (i.e., propionic anhydride for lysine propionylation).
2. Acetic anhydride is hygroscopic and should be stored away from moisture, preferably under nitrogen or with desiccant. In the atmosphere, it is converted to Acetic acid which is subject to photooxidative degradation.
3. Acetic anhydride is highly reactive and stock solutions should be prepared in 100 % methanol.
4. It is best to prepare stock acetic anhydride solutions fresh.
5. PBS is used in this experiment because the acetylation reaction is followed by an enzymatic assay and gel electrophoresis separation. In certain instances, when only determining the sites of acetylation, 50 mM Ammonium Bicarbonate (pH 8.0) may be used so the reaction can be directly followed by standard in-solution trypsin digestion and MS/MS analysis.
6. Here, human recombinant ALDH2 is used to demonstrate the acetylation reaction, however, any protein of interest may be applied to this method.
7. For native gel electrophoresis, prepare all reagents fresh, just before use.
8. SDS and reducing agents should not be used for native gel electrophoresis. Proteins separated by this technique resolve not only by molecular mass and intrinsic charge, but also the cross-sectional area, and thus experience different electrophoretic forces which are dependent on the secondary and tertiary structure.
9. Acrylamide is a toxic chemical and great care should be used when handling and disposing of this reagent. Follow all MSDS guidelines for handling and disposal.
10. TEMED is toxic and flammable. Follow all MSDS guidelines for handling and disposal.
11. Stir thoroughly by vortexing or magnetic stir bar to ensure the milk is completely solubilized.
12. Other active sirtuin enzymes may be used. For example, SIRT 3–5 may be used in this assay when examining mitochondrial resident proteins.
13. If desired, other proteolytic enzymes may be used, such as chymotrypsin or Asp-N.

14. The range of acetic anhydride concentrations should be optimized for the protein of interest. In this case, a range of 200 μM to 50 mM acetic anhydride was used for ALDH2 acetylation.
15. While only the 10 mM treatment will be used for the SIRT3 assay and MS analysis, the entire range of acetic anhydride concentrations provides an in-depth analysis of how lysine acetylation alters protein migration through native gel electrophoresis (Fig. 1).
16. While 60 min was used for this experiment, the duration of treatment should be optimized for other target proteins. Once the acetylation treatment is complete, an activity assay may be employed to determine the impact of in vitro chemical acetylation on enzymatic activity or protein function.
17. Take care to avoid air bubbles in the gel after polymerization.
18. Boiling the sample will destroy native protein conformation.
19. Alternative transfer techniques may be used, such as a "wet transfer."
20. It is important to verify that the protein acetylation reaction worked properly by performing a Western blot with anti-acetyl-Lys antibody. Once verified, proceed to the deacetylase activity assay.
21. A new razor blade is required to ensure that no other protein contamination is added to the purified acetylated protein in the gel.
22. Always use acid-washed glassware. Plastic containers may have protein residue from previous experiments or may contain soap residues.
23. Alternatively, the dried sample may be suspended in 10 μL of 0.2 % formic acid in water for LC-MS analysis. Consult your proteomics core for specific solvent requirements.

References

1. Vassilopoulos A, Fritz KS, Petersen DR, Gius D (2011) The human sirtuin family: evolutionary divergences and functions. Hum Genomics 5(5):485–496
2. Dali-Youcef N, Lagouge M, Froelich S, Koehl C, Schoonjans K, Auwerx J (2007) Sirtuins: the 'magnificent seven', function, metabolism and longevity. Ann Med 39(5):335–345
3. Schemies J, Uciechowska U, Sippl W, Jung M (2010) NAD(+) -dependent histone deacetylases (sirtuins) as novel therapeutic targets. Med Res Rev 30(6):861–889. doi:10.1002/med.20178
4. Hirschey MD, Shimazu T, Jing E, Grueter CA, Collins AM, Aouizerat B, Stancakova A, Goetzman E, Lam MM, Schwer B, Stevens RD, Muehlbauer MJ, Kakar S, Bass NM, Kuusisto J, Laakso M, Alt FW, Newgard CB, Farese RV Jr, Kahn CR, Verdin E (2011) SIRT3 deficiency and mitochondrial protein hyperacetylation accelerate the development of the metabolic syndrome. Mol Cell 44(2):177–190. doi:10.1016/j.molcel.2011.07.019
5. He W, Newman JC, Wang MZ, Ho L, Verdin E (2012) Mitochondrial sirtuins: regulators of protein acylation and metabolism. Trends

Endocrinol Metab 23(9):467–476. doi:10.1016/j.tem.2012.07.004
6. Arndt C, Koristka S, Bartsch H, Bachmann M (2012) Native polyacrylamide gels. Methods Mol Biol 869:49–53. doi:10.1007/978-1-61779-821-4_5
7. Manoussopoulos IN, Tsagris M (2009) Native electrophoresis and western blot analysis: method and applications. Methods Mol Biol 536:277–287. doi:10.1007/978-1-59745-542-8_29
8. Blancher C, Cormick RM (2012) Sodium dodecyl sulphate-polyacrylamide denaturing gel electrophoresis and Western blotting techniques. Methods Mol Biol 878:89–110. doi:10.1007/978-1-61779-854-2_6

Chapter 14

Accurate Measurement of Nicotinamide Adenine Dinucleotide (NAD^+) with High-Performance Liquid Chromatography

Jun Yoshino and Shin-ichiro Imai

Abstract

Nicotinamide adenine dinucleotide (NAD^+) plays a critical role in regulating numerous biological and physiological pathways including metabolism, inflammation, cancer, and aging in mammals. Here we describe a highly quantitative method with reverse-phase high-performance liquid chromatography (HPLC) for the determination of NAD^+ levels in cells and tissues. This methodology provides accurate, reliable, and reproducible results of NAD^+ measurement, which enables us to analyze various pathophysiological changes in NAD^+ levels in vitro and in vivo.

Key words Nicotinamide adenine dinucleotide (NAD^+), HPLC, NAMPT

1 Introduction

Nicotinamide adenine dinucleotide (NAD^+) is a classic coenzyme known to play an important role in cellular redox reactions in a wide variety of organisms [1]. In mammals, an accumulating body of evidence suggests that NAD^+ is involved in numerous biological pathways including metabolism, cancer, stress response, inflammation, and aging [1–4]. In recent years, the emerging importance of NAD^+ biology has fueled enthusiasm to investigate both NAD^+ biosynthetic pathways and NAD^+-dependent mediators in diverse model organisms. These studies have uncovered a new layer of functional roles for NAD^+ biosynthetic enzymes and NAD^+-consuming regulators.

To synthesize NAD^+, nicotinamide and nicotinic acid (two different forms of vitamin B_3), tryptophan, and nicotinamide riboside (NR) are used as precursors [3, 5, 6]. NAD^+ biosynthetic pathways from these precursors are summarized in Fig. 1. In mammals, nicotinamide is a predominant precursor. Nicotinamide is converted into nicotinamide mononucleotide (NMN) by the rate-limiting

Matthew D. Hirschey (ed.), *Sirtuins: Methods and Protocols*, Methods in Molecular Biology, vol. 1077,
DOI 10.1007/978-1-62703-637-5_14,

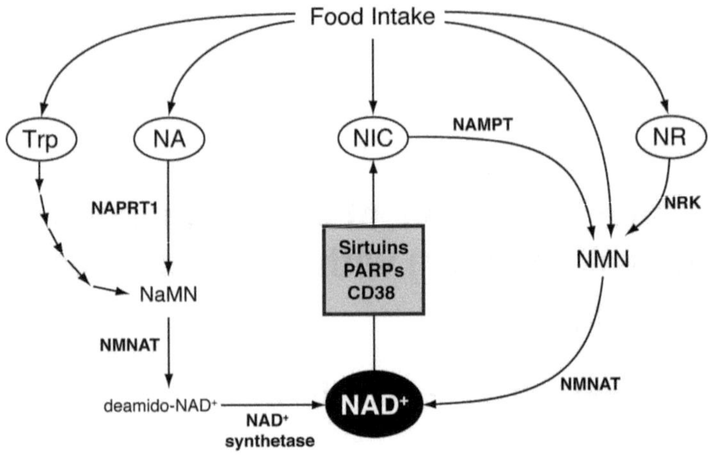

Fig. 1 Mammalian NAD+ biosynthetic pathways. In mammals, NAD+ can be generated from four precursors including tryptophan (Trp), nicotinic acid (NA), nicotinamide (NIC), and nicotinamide riboside (NR). The de novo pathway starts from Trp and is converged into the Preiss–Handler pathway through nicotinic acid mononucleotide (NaMN). In the Preiss–Handler pathway, nicotinic acid phosphoribosyltransferase (NAPRT1) produces NaMN from NA, and NAD+ is generated from NaMN via nicotinamide/nicotinic acid mononucleotide adenylyltransferase (NMNAT) and NAD+ synthetase. In the salvage pathway, NIC is converted to nicotinamide mononucleotide (NMN) by the rate-limiting enzyme, nicotinamide phosphoribosyltransferase (NAMPT). Our recent data suggest that NMN is also found in our daily food (Yoshino and Imai, unpublished data). NAD+ is synthesized from NMN by NMNAT. NR is converted to NMN by nioctinamide riboside kinase (NRK) and utilized for NAD+ biosynthesis. NAD+ is consumed by various enzymes such as sirtuins, poly(ADP-ribose) polymerases (PARPs), and cyclic ADP ribose hydrolase (CD38)

enzyme nicotinamide phosphoribosyltransferase (NAMPT) [7], and NMN is then converted into NAD+ by the second enzyme, nicotinamide/nicotinic acid mononucleotide adenylyltransferase (NMNAT) (Fig. 1). NAD+ exerts its pleiotropic functions through NAD+-consuming enzymes, such as poly (ADP-ribose) polymerases (PARPs) [8, 9], mono-ADP ribosyltransferases [10], cyclic ADP ribose hydrolase (e.g., CD38) [11], and sirtuins [12, 13]. Recent studies show that alteration in NAD+ levels has a profound impact on various physiological functions, particularly metabolic functions, through these NAD+-consuming enzymes. For example, we and other group have reported that both NAMPT protein and NAD+ levels show circadian oscillatory patterns that are regulated by the core clock machinery in peripheral tissues, such as the liver and white adipose tissue [14, 15]. This oscillation of NAD+ modulates the activity of SIRT1, one of the mammalian sirtuin family members, and thereby regulates the circadian transcriptional regulation of clock genes, including the *Nampt* gene itself, comprising a novel feedback loop of circadian rhythm [16]. Recently, we have also found that NAMPT-mediated NAD^+ biosynthesis is compromised by high-fat diet feeding and aging, contributing to the pathogenesis of diet- and age-induced type 2 diabetes [17]. Haplodeficiency of *Nampt* also causes defects in glucose-stimulated insulin secretion in pancreatic islets [18]. Furthermore, genetically engineered mouse models show that inhibition of PARP-1 or CD38 significantly increases NAD^+ content, SIRT1 activity, and mitochondrial functions in skeletal muscle and prevents diet-induced obesity by enhancing energy expenditure [9, 19, 20].

Given that mammalian sirtuins (SIRT1-7) [12, 13] are implicated in many age-associated diseases such as type 2 diabetes [17, 21], Alzheimer's disease [22, 23], osteoporosis [24, 25], and cancer [26, 27], it is conceivable that the regulation of NAD^+ biosynthesis plays a crucial role in the pathophysiology of those diseases. NAD^+ biosynthesis can also be an important therapeutic target. Indeed, enhancing NAD^+ biosynthesis by key NAD^+ intermediates, namely NMN [17] and NR [28], or by pharmacological inhibition of CD38 [29], improves glucose tolerance, insulin secretion and action, plasma lipid panel, and energy expenditure through activating SIRT1 and SIRT3 in high-fat diet-fed mice. These findings provide novel insight into the development of nutriceutical intervention against age-associated metabolic complications, such as obesity and type 2 diabetes [30]. Therefore, accurate and reliable methodology of NAD^+ measurement is now getting higher demands in various scientific fields. Here we describe a method with high-performance liquid chromatography (HPLC) to accurately measure the levels of NAD^+ in cells and tissues. We will also show a detailed validation of this HPLC-driven methodology with tandem mass spectrometry.

2 Materials

Prepare all solutions and buffers using HPLC-grade water.

2.1 NAD^+ Extraction

1. Perchloric acid ($HClO_4$): Dilute 70 % $HClO_4$ in HPLC-grade water and prepare 10 % solution (*see* **Note 1**). Store at 4 °C.
2. Potassium Carbonate (K_2CO_3): Weight and transfer 20.7 g of K_2CO_3 into a 50 mL conical tube. Make to 50 mL water and prepare 3 M solution (*see* **Note 2**).

2.2 HPLC

1. 0.5 M Potassium Phosphate Monobasic (KH_2PO_4): Transfer 34.0 g KH_2PO_4 into a glass beaker. Add about 400 mL of HPLC-grade water and stir until dissolved. Make up to 500 mL with water in a graduated cylinder. Filtrate solution with 0.22 μm filter.

2. 0.5 M Potassium Phosphate Dibasic (K_2HPO_4): Weight 43.5 g K_2HPO_4 and prepare a 500 mL solution, as described in the previous step.
3. 0.05 M Phosphate Buffer (pH 7.0): Add 38.5 mL of 0.5 M KH_2PO_4 and 61.5 mL of 0.5 M K_2HPO_4 to a 1 L graduated cylinder and make to 1 L with water. Filtrate with 0.22 μm filter and de-gas the buffer (*see* **Note 3**).
4. Methanol: HPLC-grade 100 % Methanol.
5. NAD^+ standard solution: Prepare 50 mM stock solution in water from the pure NAD^+ compound (*see* **Note 4**). Dilute the stock solution in water and make 100 μM working stock solution (*see* **Note 5**).
6. Columns: SUPELCOSIL™ LC-18-T HPLC column, 3 μm particle size, (15 cm × 4.6 cm; Sigma) (*see* **Note 6**). SUPELCOSIL™ LC-18-T Supelguard™ Cartridge (*see* **Note 7**).
7. HPLC instrument (Shimadzu Scientific Instruments, Columbia, MD): LC-20AT Quaternary Solvent Delivery Unit, SPD-20A Ultraviolet (UV) Detector, SIL-20A Autosampler, DGU-20A5 Degasser (*see* **Note 8**).

3 Methods

3.1 Acid Extraction on a Tissue Sample

1. Collect tissue samples and rinse in cold PBS (Phosphate buffered saline) (*see* **Note 9**).
2. Remove excess PBS with Kim wipe or diaper and snap-freeze the tissue samples in liquid nitrogen (*see* **Note 10**). Store at −80 °C. If a tissue sample is very small (e.g., hippocampus and hypothalamus), measure tissue weight before freezing and skip **step 3** (*see* **Note 11**).
3. Weight frozen tissue samples when ready to measure NAD^+ by HPLC (*see* **Note 12**).
4. Calculate the necessary amount of $HClO_4$ for each sample (*see* **Note 13**).
5. Add a necessary volume of pre-chilled $HClO_4$ to a 10 mL tube on ice. Transfer a frozen tissue sample into the tube and keep on ice.
6. Homogenize a frozen tissue sample in cold $HClO_4$ using a Polytron Homogenizer or Tissue Lyser II (Qiagen, Valencia, CA) (*see* **Note 14**).
7. Keep the tissue $HClO_4$ homogenate on ice for 15 min (*see* **Note 15**).
8. Centrifuge the homogenate at maximal speed (~22,000 ×*g*) at 4 °C for 5 min.

9. Carefully take the supernatant and transfer it to a new sample tube (*see* **Note 16**).
10. Add a one-third volume of 3 M K_2CO_3 and vortex rigorously (*see* **Note 17**).
11. Keep the sample on ice for 10 min (*see* **Note 18**).
12. Centrifuge the sample at maximal speed (~22,000 ×*g*) at 4 °C for 5 min.
13. Carefully take the supernatant and transfer it to a new sample tube.

3.2 Acid Extraction on cultured cells

1. Grow cells and give necessary treatments (*see* **Note 19**).
2. Aspirate culture media and wash cultured cells in cold PBS.
3. Aspirate PBS and add pre-chilled $HClO_4$ (*see* **Note 20**).
4. Harvest the cells using a cell scraper and transfer the cell lysate into a 1.7 mL tube on ice.
5. Cells are thoroughly homogenized using a syringe with a 23–26 gage needle or extensive pipetting.
6. Vortex the sample rigorously and keep on ice.
7. Prepare the sample, as described in the previous procedure (**steps 7–13**).

3.3 Reverse-Phase HPLC

1. To set up the sample, dilute the tissue/cell extract in 0.05 M Phosphate Buffer as needed, and transfer it into a HPLC glass vial.
2. Place the vials in the autosampler of the HPLC instrument.
3. Inject 50–100 μL of samples into the HPLC.
4. NAD^+ standards should also be measured with biological samples (*see* **Note 21**).
5. To set up the mobile phase, the HPLC is run at a flow rate of 1 mL/min with 100 % buffer A (0.05 M Phosphate Buffer) from 0 to 5 min, a linear gradient to 95 % buffer A/5 % buffer B (100 % methanol) from 5 to 6 min, 95 % buffer A/5 % buffer B from 6 to 11 min, a linear gradient to 85 % buffer A/15 % buffer B from 11 to 13 min, 85 % buffer A/15 % buffer B from 13 to 23 min, a linear gradient to 100 % buffer A from 23 to 24 min, and 100 % buffer A from 24 to 30 min.
6. The pressure through the HPLC system is also carefully monitored during the measurement (*see* **Note 22**).
7. Detection of the peak for NAD^+: NAD^+ is monitored by absorbance at 261 nm.
8. The peak for NAD^+ is usually eluted as a sharp peak at 11 min and completely separable from peaks for other NAD^+-related metabolites (Fig. 2) (*see* **Note 23**).

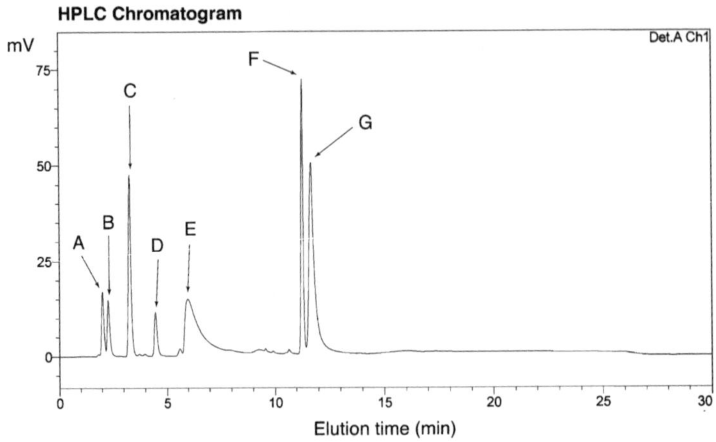

	NAD$^+$-related metabolite	elution time (min)
A	Nicotinic acid mononucleotide (NaMN)	2.02+/-0.00
B	Nicotinamide mononucleotide (NMN)	2.29+/-0.00
C	ATP	3.29+/-0.01
D	Nicotinic acid	4.46+/-0.01
E	Nicotinamide riboside (NR)	6.02+/-0.03
F	Nicotinamide adenine dinucleotide (NAD$^+$)	11.25+/-0.03
G	Nicotinamide	11.66+/-0.03

Fig. 2 Separation of NAD$^+$-related metabolites with a LC-18-T column. Separation of NAD$^+$-related metabolites using high-end HPLC with a LC-18-T column. A representative chromatogram and elution times of each compound are shown

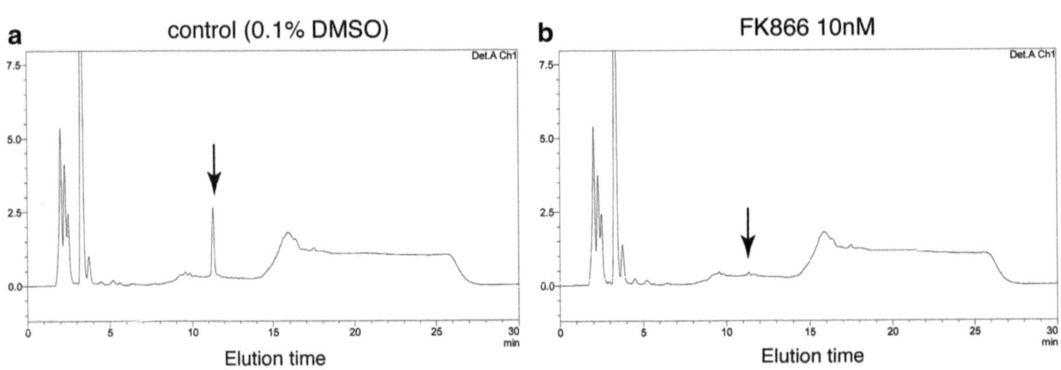

Fig. 3 NAD$^+$ detection in mammalian extract using HPLC system. Representative chromatograms of control (**a**) and FK866 (a potent specific NAMPT inhibitor)-treated (**b**) NIH3T3 cells from our HPLC system. Cells were incubated with 0.1 % DMSO (control) or 10 nM FK866 for 24 h. The peak for NAD$^+$ is indicated in each chromatogram (*arrow*)

9. We confirmed that the NAD$^+$ peak detected in mammalian extract, such as NIH3T3 fibroblasts, contained NAD$^+$ >99 % by using the NAMPT-specific chemical inhibitor FK866 [31] (Fig. 3).

10. Quantification of NAD^+: NAD^+ levels are quantitated based on the peak area compared to a standard curve and normalized to weights of tissues or protein content of cultured cells (*see* **Note 24**).

3.4 Validation by a Tandem Mass Spectrometry

To further validate our HPLC-driven NAD^+ measurement (*see* **Note 25**), the NAD^+ peak from the liver extract of a regular chow-fed, NMN-injected C57BL/6 mouse was fractionated and analyzed by liquid chromatography-tandem mass spectrometry (LC/MS/MS) [a Shimadzu 10A HPLC system (Shimadzu) coupled to a TSQ Quantum Ultra triple quadrupole mass spectrometer (Thermo Finnigan, San Jose, USA)] in the positive mode.

1. For Ql ion scanning, the NAD^+-containing HPLC fraction was first examined by Ql ion scanning in LC/MS/MS with a C18 column.
2. The results from Ql ion scanning clearly show that the ion species with 664 and 702 of m/z form only specific peaks, except for solvent peaks generated from the buffer (m/z = 391.4, 429.3, and 819.5), in this HPLC fraction (Fig. 4).
3. Comparing to the Ql spectra of standard NAD^+ and the buffer only, these two ion species were identified as $[M+H]^+$ (m/z = 664) and $[M+K]^+$ (m/z = 702) of NAD^+.
4. To further confirm the identity of NAD^+ peak, we compared the MS^2 scan for the Q1 spectrum (m/z = 664) and found that MS^2 product ions completely matched between the HPLC fraction and standard NAD^+ (Fig. 5a).
5. Using the selected reaction monitoring (SRM) mode in this LC/MS/MS method, we also measured relative changes in NAD^+ levels in liver extracts from C57BL/6 mice injected with NMN at the dose of 500 mg/kg body weight.
6. We successfully confirmed that the time course of relative NAD^+ changes measured by LC/MS/MS was highly consistent with those measured by our HPLC-driven methodology [17] (Fig. 5b).
7. These results provide a convincing validation for the accuracy of the HPLC-driven NAD^+ measurement method described in this chapter.

4 Notes

1. $HClO_4$ is a very strong acid and corrosive solution and needs to be very carefully handled. Therefore, it is recommended that 10 % solution is prepared in a fume hood.
2. K_2CO_3 solution is a strong base and should be carefully prepared and stored in a plastic conical tube (NOT in glass products).

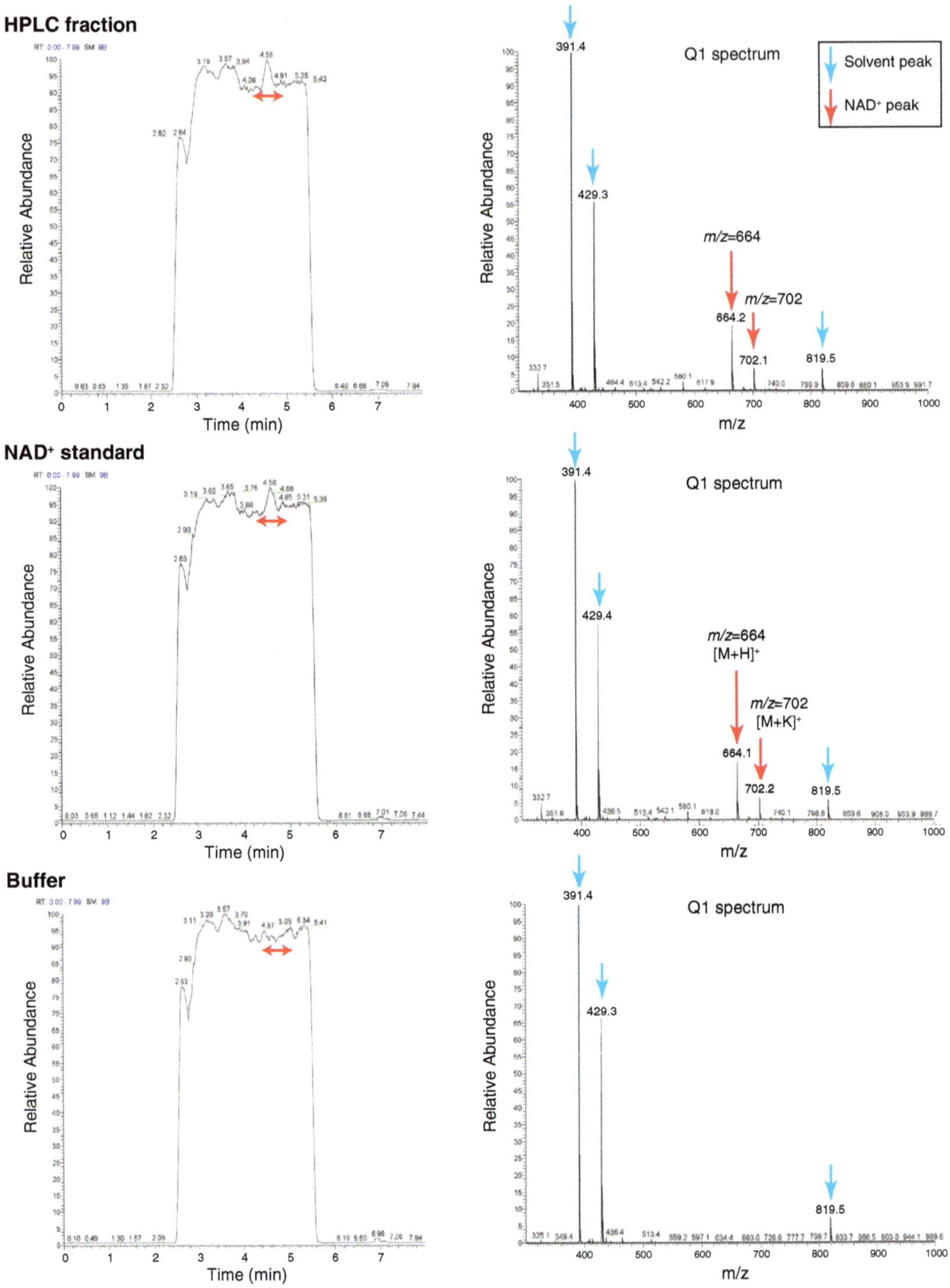

Fig. 4 Q1 ion scanning of the NAD$^+$ fraction isolated by HPLC. Liquid chromatography-tandem mass spectrometry (LC/MS/MS) was performed with a Shimadzu 10A HPLC system coupled to a TSQ Quantum ultra triple quadrupole mass spectrometer in the positive mode. The NAD$^+$ peak was fractionated from a regular chow-fed liver extract by HPLC with a C18 column and analyzed by LC/MS/MS. NAD$^+$ standard (Molecular weight = 663.4) and blank (buffer) samples were also analyzed. Total ion chromatograms and Q1 scan spectra of indicated peaks are shown

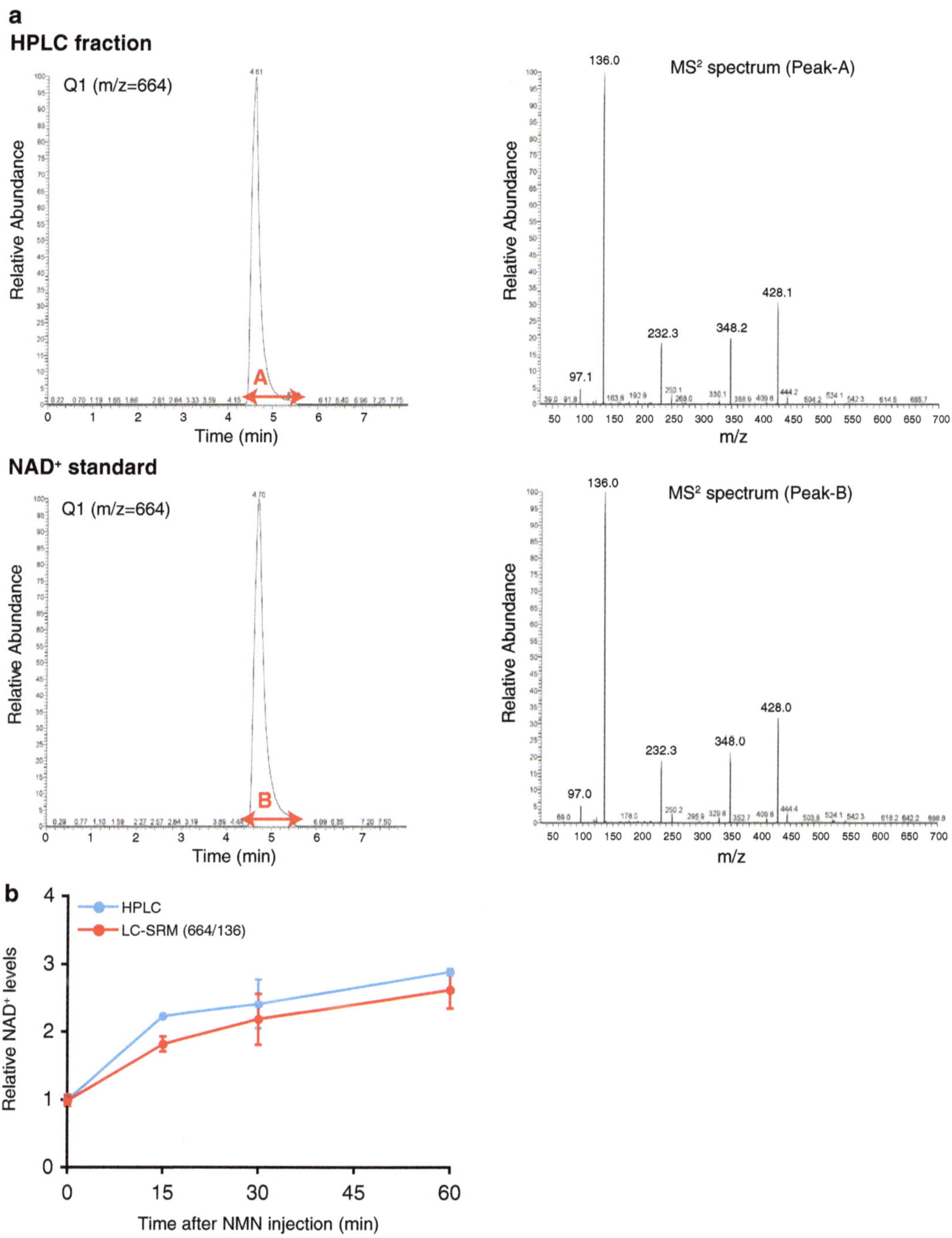

Fig. 5 MS^2 analysis and quantitation of NAD^+ by LC/MS/MS. (**a**) Extracted ion chromatograms of the NAD^+ fraction and standard using LC/MS/MS. The ion $[M+H]^+$ ($m/z=664$) was selected in Q1. MS^2 spectra of the indicated peaks are shown. (**b**) Relative changes in NAD^+ levels in liver extracts prepared from NMN-injected B6 female mice. Relative changes in NAD^+ levels were assessed by HPLC and LC/MS/MS and normalized to those at 0 min time point. NAD^+ quantitation was conducted using the selected reaction monitoring (SRM) mode (m/z 664 → 136)

3. We usually check the pH of prepared 0.05 M Phosphate Buffer using Litmus paper. The proper control of mobile phase pH is not only important for the reproducibility in the elution time but also critical for the shape and the separation of the peak.
4. We typically use "β-Nicotinamide adenine dinucleotide hydrate" (Sigma, N1511) to make NAD^+ standard solution. NAD^+ powder may be difficult to dissolve in water and need to be vortex rigorously until dissolved. Aliquots of 50 mM NAD^+ stock solution should be stored at the deep freezer (−80 °C). Avoid multiple freeze/thaw cycles with stock solution. Frozen NAD^+ stock solution is stable at −80 °C at least for 1 year.
5. 100 μM NAD^+ working stock solution should be stored in the sample box at the refrigerator (4 °C). Working stock solution needs to be kept on ice when preparing standard solutions for the HPLC measurement. Working stock solution is stable at 4 °C at least for a couple of weeks.
6. Column selection is the most important part for the HPLC methodology. We find that the LC-18-T HPLC column is the best for NAD^+ measurement in terms of sensitivity, reproducibility, and convenience.
7. A HPLC guard column can extend the lifetime of the main column by preventing the accumulation of impurities on the column. A guard column should be connected with the HPLC instrument when biological samples are measured.
8. Sensitivity of UV detector is particularly important when examining the tissue sample that contains low amounts of NAD^+, such as adipose tissue.
9. Tissue collection should be done as quickly and smoothly as possible to avoid the degradation of NAD^+.
10. Liquid nitrogen, but not dry ice, needs to be used to snap-freeze the tissue samples. Frozen samples should be stored at the deep freezer (−80 °C) without being thawed until NAD^+ measurement.
11. Small frozen tissue samples tend to be thawed very easily and quickly so that NAD^+ is degraded when measuring frozen tissue weight. To avoid this issue, we usually measure tissue weight quickly before freezing it and calculate the volume of $HClO_4$ (*see* also **Note 13**).
12. Frozen tissue samples have to be kept in the dry ice box and immediately be weighted without being thawed. In most cases, no more than 100 mg of tissue is required for the determination of NAD^+ levels. Frozen tissue samples can be cut into small pieces on dry ice, as needed.
13. Typically, a tissue sample is extracted in 10 % $HClO_4$ at a 1:10–50 ratio (tissue weight: $HClO_4$ volume).

14. Extra cautions should be taken when dealing with $HClO_4$ (*see* also **Note 1**). Use a Polytron Homogenizer in a fume hood and wear a safety goggle as needed. The Tissue Lyser II is operated for 5 min with the frequency of 30 shakes/s using the pre-chilled adaptor plates.
15. We usually transfer the $HClO_4$ homogenate into a 1.7 mL tube during incubation on ice. For NAD^+ quantitation, no more than 1 mL of tissue lysate is required. Excess amount of the homogenate can be discarded.
16. If the volume of the supernatant is more than 0.5 mL, it is recommended to transfer it into a 5–10 mL tube in order not to spill the solution due to the CO_2 production in the next step (*see* also **Note 18**).
17. The precipitation of white crystals ($KClO_4$) is observed after adding K_2CO_3 solution.
18. CO_2 gas is produced in this neutralization and desalting step. Therefore, a cap of the tube should be opened at least once during incubation on ice to remove the gas.
19. We usually grow cells in a 6-well plate for NAD^+ measurement. We use 3 wells for NAD^+ and 3 wells for protein concentration measurement in each experimental condition. The average of protein concentrations through 3 wells is used for the normalization of NAD^+ levels.
20. Typically 0.3–0.5 mL of $HClO_4$ is added to each well on 6-well plate. Extra caution should be taken when dealing with $HClO_4$ (*see* also **Note 1**).
21. We prepare NAD^+ standard solution in 0.05 M Phosphate Buffer at four different concentrations (e.g., 0.2, 0.5, 1.0, and 2.0 μM). The linearity and reproducibility for NAD^+ standard solutions should be evaluated at each measurement.
22. Make sure there is no leak in the instrument. Salt precipitation can result in the high pressure. Pump failure can cause a fluctuation in pressure.
23. Elution time must be always consistent in the same experimental condition. Wrong buffer preparation or HPLC pump problems can cause significant changes in the elution time.
24. By using this HPLC-driven methodology, we successfully detected the pathophysiological changes in NAD^+ levels in multiple tissues induced by circadian oscillation [14], high-fat diet feeding [17], and aging [17].
25. Although the NAD^+ peak is completely separable from other NAD^+-related metabolites in this method (Fig. 2), there is still a possibility that NAD^+ fraction extracted from biological samples could contain compounds other than NAD^+. Therefore, we addressed this issue by using LC/MS/MS as described in this chapter.

Acknowledgments

We thank Drs. Xuntian Jiang and Daniel Ory for mass spec analysis in the Metabolomics Facility at Washington University School of Medicine. We also thank members of the Imai lab for their critical discussions. This work was supported in part by the National Institute on Aging (AG02150), the Ellison Medical Foundation, and the Longer Life Foundation to S.I. and by institutional support from the Washington University Nutrition Obesity Research Center (P30DK056341) and the Washington University Diabetes Research and Training Center (P60DK020579). S.I. serves as a scientific advisory board member for Sirtris, a GSK company.

References

1. Belenky P, Bogan KL, Brenner C (2007) NAD+metabolism in health and disease. Trends Biochem Sci 32(1):12–19. doi:10.1016/j.tibs.2006.11.006
2. Stein LR, Imai S (2012) The dynamic regulation of NAD metabolism in mitochondria. Trends Endocrinol Metab 23(9):420–428. doi:10.1016/j.tem.2012.06.005
3. Revollo JR, Grimm AA, Imai S (2007) The regulation of nicotinamide adenine dinucleotide biosynthesis by Nampt/PBEF/visfatin in mammals. Curr Opin Gastroenterol 23(2):164–170. doi:10.1097/MOG.0b013e32801b3c8f
4. Imai S (2009) The NAD world: a new systemic regulatory network for metabolism and aging–Sirt1, systemic NAD biosynthesis, and their importance. Cell Biochem Biophys 53(2):65–74. doi:10.1007/s12013-008-9041-4
5. Magni G, Amici A, Emanuelli M, Raffaelli N, Ruggieri S (1999) Enzymology of NAD+synthesis. Adv Enzymol Relat Areas Mol Biol 73:135–182
6. Bogan KL, Brenner C (2008) Nicotinic acid, nicotinamide, and nicotinamide riboside: a molecular evaluation of NAD+precursor vitamins in human nutrition. Annu Rev Nutr 28:115–130. doi:10.1146/annurev.nutr.28.061807.155443
7. Revollo JR, Grimm AA, Imai S (2004) The NAD biosynthesis pathway mediated by nicotinamide phosphoribosyltransferase regulates Sir2 activity in mammalian cells. J Biol Chem 279(49):50754–50763. doi:10.1074/jbc.M408388200
8. Krishnakumar R, Kraus WL (2010) The PARP side of the nucleus: molecular actions, physiological outcomes, and clinical targets. Mol Cell 39(1):8–24. doi:10.1016/j.molcel.2010.06.017
9. Bai P, Canto C (2012) The role of PARP-1 and PARP-2 enzymes in metabolic regulation and disease. Cell Metab 16(3):290–295. doi:10.1016/j.cmet.2012.06.016
10. Corda D, Di Girolamo M (2003) Functional aspects of protein mono-ADP-ribosylation. EMBO J 22(9):1953–1958. doi:10.1093/emboj/cdg209
11. Lee HC (2001) Physiological functions of cyclic ADP-ribose and NAADP as calcium messengers. Annu Rev Pharmacol Toxicol 41:317–345. doi:10.1146/annurev.pharmtox.41.1.317
12. Imai S, Guarente L (2010) Ten years of NAD-dependent SIR2 family deacetylases: implications for metabolic diseases. Trends Pharmacol Sci 31(5):212–220.doi:10.1016/j.tips.2010.02.003
13. Guarente L (2011) Franklin H. Epstein lecture: sirtuins, aging, and medicine. N Eng J Med 364(23):2235–2244. doi:10.1056/NEJMra1100831
14. Ramsey KM, Yoshino J, Brace CS, Abrassart D, Kobayashi Y, Marcheva B, Hong HK, Chong JL, Buhr ED, Lee C, Takahashi JS, Imai S, Bass J (2009) Circadian clock feedback cycle through NAMPT-mediated NAD+biosynthesis. Science 324(5927):651–654. doi:10.1126/science.1171641
15. Nakahata Y, Sahar S, Astarita G, Kaluzova M, Sassone-Corsi P (2009) Circadian control of the NAD+salvage pathway by CLOCK-SIRT1. Science 324(5927):654–657. doi:10.1126/science.1170803
16. Imai S (2010) "Clocks" in the NAD world: NAD as a metabolic oscillator for the regulation of metabolism and aging. Biochim Biophys Acta 1804(8):1584–1590. doi:10.1016/j.bbapap.2009.10.024
17. Yoshino J, Mills KF, Yoon MJ, Imai S (2011) Nicotinamide mononucleotide, a key NAD(+) intermediate, treats the pathophysiology of

diet- and age-induced diabetes in mice. Cell Metab 14(4):528–536. doi:10.1016/j.cmet.2011.08.014
18. Revollo JR, Korner A, Mills KF, Satoh A, Wang T, Garten A, Dasgupta B, Sasaki Y, Wolberger C, Townsend RR, Milbrandt J, Kiess W, Imai S (2007) Nampt/PBEF/visfatin regulates insulin secretion in beta cells as a systemic NAD biosynthetic enzyme. Cell Metab 6(5):363–375. doi:10.1016/j.cmet.2007.09.003
19. Bai P, Canto C, Oudart H, Brunyanszki A, Cen Y, Thomas C, Yamamoto H, Huber A, Kiss B, Houtkooper RH, Schoonjans K, Schreiber V, Sauve AA, Menissier-de Murcia J, Auwerx J (2011) PARP-1 inhibition increases mitochondrial metabolism through SIRT1 activation. Cell Metab 13(4):461–468. doi:10.1016/j.cmet.2011.03.004
20. Barbosa MT, Soares SM, Novak CM, Sinclair D, Levine JA, Aksoy P, Chini EN (2007) The enzyme CD38 (a NAD glycohydrolase, EC 3.2.2.5) is necessary for the development of diet-induced obesity. FASEB J 21(13):3629–3639. doi:10.1096/fj.07-8290com
21. Gillum MP, Erion DM, Shulman GI (2010) Sirtuin-1 regulation of mammalian metabolism. Trends Mol Med. doi:10.1016/j.molmed.2010.09.005
22. Braidy N, Jayasena T, Poljak A, Sachdev PS (2012) Sirtuins in cognitive ageing and Alzheimer's disease. Curr Opin Psychiatry 25(3):226–230. doi:10.1097/YCO.0b013e32835112c1
23. Donmez G, Wang D, Cohen DE, Guarente L (2010) SIRT1 Suppresses beta-amyloid production by activating the alpha-secretase gene ADAM10. Cell 142(2):320–332. doi:10.1016/j.cell.2010.06.020
24. Cohen-Kfir E, Artsi H, Levin A, Abramowitz E, Bajayo A, Gurt I, Zhong L, D'Urso A, Toiber D, Mostoslavsky R, Dresner-Pollak R (2011) Sirt1 is a regulator of bone mass and a repressor of sost encoding for sclerostin, a bone formation inhibitor. Endocrinology 152(12):4514–4524. doi:10.1210/en.2011-1128
25. Li Y, He X, Li Y, He J, Anderstam B, Andersson G, Lindgren U (2011) Nicotinamide phosphoribosyltransferase (Nampt) affects the lineage fate determination of mesenchymal stem cells: a possible cause for reduced osteogenesis and increased adipogenesis in older individuals. J Bone Miner Res 26(11):2656–2664. doi:10.1002/jbmr.480
26. Finley LW, Carracedo A, Lee J, Souza A, Egia A, Zhang J, Teruya-Feldstein J, Moreira PI, Cardoso SM, Clish CB, Pandolfi PP, Haigis MC (2011) SIRT3 opposes reprogramming of cancer cell metabolism through HIF1alpha destabilization. Cancer Cell 19(3):416–428. doi:10.1016/j.ccr.2011.02.014
27. Sebastian C, Zwaans BM, Silberman DM, Gymrek M, Goren A, Zhong L, Ram O, Truelove J, Guimaraes AR, Toiber D, Cosentino C, Greenson JK, Macdonald AI, McGlynn L, Maxwell F, Edwards J, Giacosa S, Guccione E, Weissleder R, Bernstein BE, Regev A, Shiels PG, Lombard DB, Mostoslavsky R (2012) The histone deacetylase SIRT6 is a tumor suppressor that controls cancer metabolism. Cell 151(6):1185–1199. doi:10.1016/j.cell.2012.10.047
28. Canto C, Houtkooper RH, Pirinen E, Youn DY, Oosterveer MH, Cen Y, Fernandez-Marcos PJ, Yamamoto H, Andreux PA, Cettour-Rose P, Gademann K, Rinsch C, Schoonjans K, Sauve AA, Auwerx J (2012) The NAD(+) precursor nicotinamide riboside enhances oxidative metabolism and protects against high-fat diet-induced obesity. Cell Metab 15(6):838–847. doi:10.1016/j.cmet.2012.04.022
29. Escande C, Nin V, Price NL, Capellini V, Gomes AP, Barbosa MT, O'Neil L, White TA, Sinclair DA, Chini EN (2012) Flavonoid apigenin is an inhibitor of the NAD+ase CD38: implications for cellular NAD+metabolism, protein acetylation, and treatment of metabolic syndrome. Diabetes. doi:10.2337/db12-1139
30. Imai S (2010) A possibility of nutriceuticals as an anti-aging intervention: activation of sirtuins by promoting mammalian NAD biosynthesis. Pharmacol Res 62(1):42–47. doi:10.1016/j.phrs.2010.01.006
31. Hasmann M, Schemainda I (2003) FK866, a highly specific noncompetitive inhibitor of nicotinamide phosphoribosyltransferase, represents a novel mechanism for induction of tumor cell apoptosis. Cancer Res 63(21):7436–7442

Chapter 15

In Vivo Measurement of the Acetylation State of Sirtuin Substrates as a Proxy for Sirtuin Activity

John Dominy, Pere Puigserver, and Carles Cantó

Abstract

Evaluating the precise catalytic activity of sirtuin proteins in vivo is a challenging endeavor. Enzymological methods, including those employed in commercially available kits, require the isolation of immunopurified protein from cells or tissues, which can perturb regulatory protein–protein interactions as well as remove the enzyme from the reaction-altering effects of intracellular NAD^+, nicotinamide, and *O*-acetyl-ADP ribose concentrations. As such, the measurement of the steady state acetylation status of select sirtuin substrates in vivo remains an important tool for evaluating changes in sirtuin activity. Here, we describe how to perform the analysis of the acetylation status of key SIRT1 and SIRT3 targets in rodent tissues and cultured cells.

Key words Acetylation, Sirtuins, PGC-1α, Ndufa9, Immunoprecipitation, Western blot, Transfection, Homogenization

1 Introduction

SIRT1 and SIRT3 have been demonstrated to be key metabolic regulators through the deacetylation of a wide range of substrates [1]. SIRT1, for example, deacetylates key nuclear transcriptional regulators of energy metabolism. In contrast, SIRT3, due to its mitochondrial localization, directly deacetylates key mitochondrial enzymes for oxidative respiration. In both cases, the activation of SIRT1 and SIRT3 generally promotes adaptations aimed to improve energy utilization in situations of nutrient scarcity.

While the relevance of both enzymes for general physiology has been largely reported via genetic approaches, the analysis of physiological changes in endogenous SIRT1 and SIRT3 activities is still challenging. To this date, a key proof for demonstrating increased sirtuin activity, at least for SIRT1 and SIRT3, is to measure the acetylation status of well-known substrates. Of course, this method has its limitations: (1) It does not rule out the possibility of co-regulation of acetyltransferase reactions, and (2) the existence

Matthew D. Hirschey (ed.), *Sirtuins: Methods and Protocols*, Methods in Molecular Biology, vol. 1077,
DOI 10.1007/978-1-62703-637-5_15, © Springer Science+Business Media, LLC 2013

of substrate overlap among different deacetylases might lead to indefinite conclusions about enzyme identity. However, it presents advantages over many commercial kits, which directly measure sirtuin activities by mixing cellular extracts with externally provided substrates or use protein immunopurified from the cell. These methodologies are useful for evaluating intrinsic changes in sirtuin activities due to covalent posttranslational modifications (i.e., phosphorylation [2]) which are rarely lost upon extraction. However, the physiological regulation of sirtuins sometimes relies on more labile parameters in situ, such as allosteric regulation, NAD^+ bioavailability, or nicotinamide (NAM) levels [3], and this perspective is lost in extraction-based kits. Therefore, the evaluation of endogenous acetylation status of sirtuin substrates is always a key piece of data when exploring their activity.

Here, we provide an overview on how to measure SIRT1 and SIRT3 activities through the evaluation of the acetylation status of key substrates in cultured cells and in murine tissues. Needless to say, both deacetylases can target a large number of substrates. For this reason, we will focus on a well-described "gold standard" substrate for each of them: the peroxisome proliferator-activating receptor coactivator 1α (PGC-1α) for SIRT1 [4] and Ndufa9 for SIRT3 [5]. In both cases, the method relies on the immunoprecipitation of the substrate and consecutive analysis of acetylated lysine content through western blotting techniques.

SIRT1 and PGC-1α: The biological activity of PGC-1α, a transcriptional co-activator capable of modulating the expression of many genes involved in central metabolism, is controlled in part by its acetylation state [4]. High levels of PGC-1α acetylation are correlated with low levels of activity, whereas low levels of acetylation are associated with high levels of activity. Steady state levels of PGC-1α acetylation are determined by the relative rates of PGC-1α's acetyltransferase, GCN5, and its deacetylase, SIRT1 [4, 6]. As such, PGC-1α acetylation can be used as a surrogate biological marker of SIRT1 activity in the cell, assuming that there is no significant perturbation of GCN5 activity. Indeed, under conditions wherein SIRT1 is most active, such as when intracellular concentrations of NAD^+ are elevated and/or when the protein kinase A pathway is stimulated, PGC-1α becomes deacetylated [7–9]. Conversely, when SIRT1 catalytic activity is at its nadir, such as that obtained by lowering NAD^+ or by specific pharmacological inhibition of the enzyme, the acetylation of PGC-1α is significantly elevated [8].

A major hurdle in the detection of PGC-1α acetylation, particularly in cultured cells, is the issue of absolute protein abundance. Compared to the levels found in some animal tissues, cultured primary cells, such as those derived from skeletal muscle, brown fat, and liver, contain low, albeit detectable, quantities of

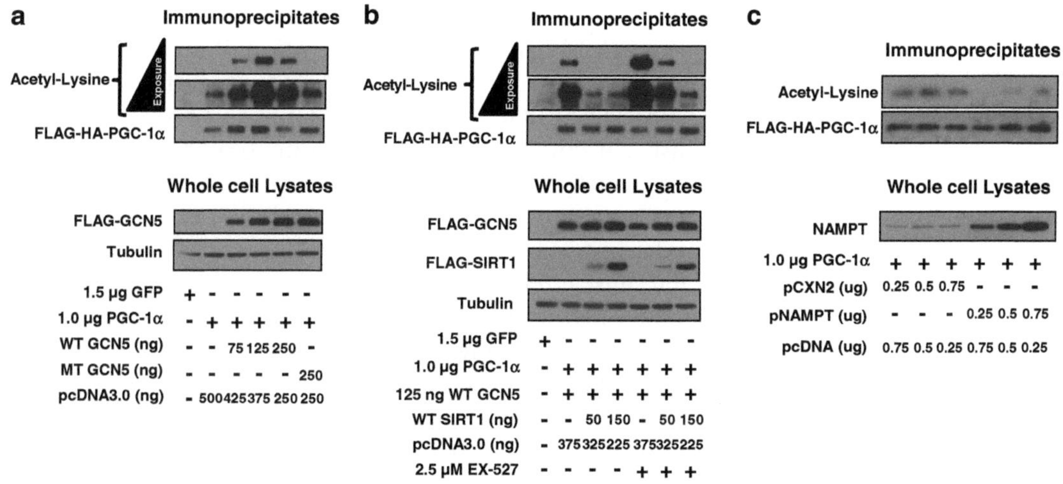

Fig. 1 Use of the U-2 OS cell system for the expression and immunopurification of acetylated PGC-1α: a surrogate marker of endogenous Sirt1 activity. (**a**) Illustration of the effects of GCN5 co-expression to boost the acetylation of ectopically expressed PGC-1α. Cells were transfected with the indicated quantities of plasmid using the same method outlined in this chapter. MT GCN5 (Y621A/F622A) is a catalytically incompetent allele of human GCN5. (**b**) Inhibition of endogenous Sirt1 with the pharmacological inhibitor EX-527 boosts PGC-1α acetylation; inhibition is overcome by the ectopic expression of Sirt1. Cells were transfected with the indicated amount of plasmids and were cultured in 2.5 μM EX-527 for 12 h prior to harvesting. (**c**) Overexpression of the NAD^+ biosynthetic enzyme nicotinamide phosphoribosyltransferase (NAMPT) causes deacetylation of PGC-1α relative to a control plasmid (pCXN2). Quantities of plasmids used in transfection are indicated. In this particular experiment, 2 μg of total DNA and 4 μL of Polyfect were used per well—a slight deviation from the procedure described in this chapter

endogenous PGC-1α protein, whereas most transformed cell lines contain vanishingly small amounts of protein. This is particularly problematic as significant quantities of PGC-1α must be immunoprecipitated from cells in order to reliably measure acetylation by standard Western blotting methods. Efforts to evaluate the effects of various treatments on PGC-1α acetylation in cell lines are therefore greatly facilitated by overexpression of an epitope-tagged form of PGC-1α protein. It is further enhanced by overexpression of the acetyltransferase GCN5 (Fig. 1a). This method has been routinely employed for exploring factors that may affect the activity of SIRT1 in cell culture systems. Our cell line of preference is the human osteosarcoma line, U-2 OS, which has a high growth rate and is easy to transfect—although other cell lines such as HEK 293 or C2C12 muscle cells have also been employed with success. Overall, the system is both reliable and efficient as a first-pass screening tool for measuring various treatment effects on acetylation before proceeding to primary cell culture and animal models. Evidence of this can be seen by pharmacological inhibition of SIRT1, as shown in Fig. 1b, which yields a very large increase in PGC-1α acetylation—an effect that can nevertheless be overcome by overexpression of additional SIRT1 protein. On the other hand, promotion of endogenous SIRT1 activity by overexpression of an enzyme in the NAD+ biosynthetic pathway, NAMPT, yields a significant decrease in the acetylation of PGC-1α (Fig. 1c). Thus, this system can serve as a robust cell culture tool for exploring biological pathways/ compounds which modulate SIRT1 activity.

In the case of mouse tissues, PGC-1α is strongly expressed in highly oxidative tissues, such as brown adipose tissue (BAT), heart, and skeletal muscle. However, the detection at the protein level by western blot on total homogenates is, in our experience, rather limited with current commercial antibodies. In order to favor the immunoprecipitation of PGC-1α, we generally begin with rather large amounts of protein or with nuclear fractions of the tissue. In big, nonfibrous tissues, such as liver, the use of nuclear fractions is recommended, and 300–500 μg of nuclear protein can be used as input. However, in tissues such as skeletal muscle or heart, we favor the use of 1–2 mg of total protein input, as the fractionation procedure is more complicated and leads to poor yields. When testing BAT or small muscles (such as soleus or EDL), total homogenates are also used as input material. A published comparison of immunoprecipitations performed on 1 mg of total homogenates from different tissues clearly highlighted the higher abundance of PGC-1α in heart, while reflected how the detection in brain or white adipose tissue would require still higher amounts of input material [10]. Given the dynamic nature of PGC-1α acetylation, it is of utmost importance to carefully choose the sample and conditions to be tested. For example, PGC-1α acetylation in liver and muscle is higher in the fed state, and further boosted by chronic high-fat feeding. In some tissues, like heart or oxidative skeletal muscle, the basal levels of PGC-1α acetylation are extremely low, making it difficult to detect any further deacetylation by nutritional or lifestyle interventions [7].

SIRT3 and Ndufa9 immunoprecipitates. SIRT3 deacetylates a plethora of mitochondrial proteins, and its activation is linked to the enhancement of fat oxidation and procurement of ATP production [11]. SIRT3 was identified as the main mitochondrial deacetylase, and defective SIRT3 function has been linked to mitochondrial protein hyperacetylation [12, 13]. In 2008, multiple components of the respiratory complex I were identified as acetylated proteins, whose acetylation status was critically controlled by SIRT3 [5]. More precisely, SIRT3 could directly interact with the subunit Ndufa9, deacetylate complex I subunits and, through this mechanism, increase Complex I activity [5]. The impact of SIRT3 on Complex I acetylation has been confirmed thereafter in later studies [13].

An advantage of measuring the acetylation of Ndufa9 immunoprecipitates over measuring PGC-1α acetylation is that Ndufa9 is an abundant protein, present at detectable levels in most cell types and at very high levels in most animal tissues. In addition, there are a number of monoclonal commercial antibodies for Ndufa9 that render excellent blotting and immunoprecipitation performances. Below, we will describe how to evaluate complex I acetylation through Ndufa9 immunoprecipitation in mouse tissues. A similar procedure can be applied to most cultured cell lines

2 Materials

2.1 Cell Culture Maintenance

1. Human osteosarcoma cell line, U-2 OS (American Type Culture Collection).
2. Dulbecco's Modified Eagle's Medium containing 10 % FBS and 1× penicillin G/streptomycin sulfate solution.
3. 15 cm mammalian cell culture dishes.
4. Tabletop centrifuge.

2.2 Cell Transfections

1. Serum-free Dulbecco's Modified Eagle's medium.
2. Dulbecco's Modified Eagle's Medium containing 10 % fetal bovine serum and 1× penicillin G/streptomycin sulfate solution.
3. Phosphate buffered saline.
4. Buffered 0.05 % trypsin solution with EDTA (Gibco).
5. Sterile 50 mL conical tubes (Corning).
6. Sterile 6-well mammalian culture plates (Falcon).
7. Sterile 1.5 mL eppendorf tubes.
8. Polyfect transfection reagent (Qiagen).
9. Expression plasmids: Empty pcDNA 3.0; enhanced green fluorescence protein in pcDNA; FLAG-HA-murine PGC-1α in pcDNA 3.0; FLAG-human GCN5 in pcDNA 3.0 (these plasmids are available from the authors upon request).
10. Tabletop vortexer.

2.3 Harvesting Cells and Performing Immunoprecipitation

1. Lysis buffer with protease inhibitor cocktail: We usually use a RIPA-based buffer (0.5 %, w/v, sodium deoxycholate, 0.01 %, w/v, SDS, 1 %, v/v, nonidet-P40 dissolved in phosphate buffered saline), buffer containing 10 mM nicotinamde (to inhibit class III HDAC activity), 1 μM trichostatin A (to inhibit class I/II HDAC activity), and 1× mammalian protease inhibitor cocktail.
2. Anti-FLAG M2 antibody resin (Sigma).
3. Cell scrapers (Fisher).
4. Rotator.

5. Cold room.
6. 70 % Ethanol.
7. Gel loading pipette tips.
8. 2× Laemmeli SDS-PAGE loading buffer.

2.4 Tissue Obtention, Protein Obtention, and Immunoprecipitation

1. Anesthesia—preferably isofluorane, linked to a respiratory device adapted for mouse use.
2. General dissection material, including small and curved precission scissors and forceps.
3. Saline.
4. Absorvent paper.
5. A Petri dish.
6. Liquid nitrogen.
7. Porcelain grinder.
8. Eppendorf tubes.
9. Eppendorf homogenizer (motor pestle).
10. Ice.
11. Phosphate buffered saline.
12. Immunoprecipitaton buffer with protease inhibitor cocktail (IP100): 25 mM Tris–HCl pH 7.9, 5 mM $MgCl_2$, 10 % glycerol, 100 mM KCl, 1 % NP40; 0.3 mM dithiothreitol, and freshly added protease inhibitor cocktail (Calbiochem). In order to prevent deacetylation during the extraction, 5 mM NAM and 0.1 mM sodium butyrate (as a general deacetylase inhibitor) were added to the buffer. During immunoprecipitation we use an additional buffer, similar to IP100, but containing 500 mM KCl. For Ndufa9 immunoprecipitation tests RIPA-based buffers containing 5 mM nicotinamde (to inhibit class III HDAC activity), 0.1 mM sodium butyrate also works nicely.
13. Protein A Sepharose.
14. Anti-PGC-1α polyclonal antibody (Millipore) or Anti-Ndufa9 monoclonal antibody (Abcam).
15. Tabletop centrifuge.
16. Rotator.
17. Cold room.
18. Laemmli sample buffer (1× and 3×).
19. Thermomixer.

2.5 SDS-PAGE

1. Apparatus for one-dimensional electrophoresis.
2. Electrode buffer for Lammeli method.

3. SDS-polyacrylamide gels (8 %, w/v, acrylamide). [We have obtained the best resolution of the full-length PGC-1α (120 kDa) and GCN5 (90 kDa) proteins when we use manually prepared Tris-Glycine SDS-PAGE gels. The gel should be poured so the stacking layer has at least 0.5 cm between the bottom of the well and the top of the resolving portion of the gel. The stacking layer consists of 5.1 % (w/v) acrylamide with a final pH of 6.8. The resolving layer consists of 8 % (w/v) acrylamide with a final pH of 8.8. To test the acetylation of Ndufa9 immunoprecipitates, the resolving layer should consist of 12.5 % (w/v)].
4. Protein molecular weight markers.

2.6 Electrotransfer

1. Polyvinylidine difluoride (PVDF) membranes [such as 0.45 Immobilon-P PVDF membranes (Millipore)].
2. Adjustable speed rocking platform.
3. Electroblotting apparatus including cassette and sponges.
4. Whatman filter paper.
5. Transfer buffer: 25 mM Tris base, 193 mM glycine, and 10 % methanol with PVDF membranes.

2.7 Immunoblotting

1. 1× TBST buffer.
2. Blocking buffers: 1× TBST containing 5 %, w/v, non-fat dry milk; 1× TBST containing 5 %, w/v, bovine serum albumin.
3. Primary antibodies for cell culture analysis: rabbit polyclonal anti-pan acetylated-lysine antibody (Cell Signaling Technology, 9441); rabbit polyclonal anti-PGC-1α antibody (Santa Crux, sc-13067); rabbit anti-GCN5L2 antibody (Cell Signaling Technology, 3305); rabbit anti-β-tubulin (Abcam, Ab6046). For tissue analysis of PGC-1α or Complex I acetylation, we use rabbit polyclonal anti-pan acetylated-lysine antibody (Cell Signaling Technology, 9441); rabbit polyclonal anti-PGC-1α antibody (Millipore, AB3242); mouse anti-Ndufa9 antibody (Abcam, ab14713); mouse anti α-tubulin (Sigma, T9026).
4. Secondary antibody: horseradish peroxidase-conjugated goat anti-rabbit secondary antibody (Pierce, 1858415) and chemiluminescent substrate (e.g., Supersignal West Pico or Supersignal West Dura from Thermo Scientific). For tissue analysis, we used Jackson peroxidase-conjugated donkey anti-mouse and anti-rabbit antibodies (711-035-150 and 711-035-152, respectively).
5. Restore Western blot stripping buffer (Pierce).
6. X-ray film and developer.

3 Methods

3.1 Cell Culture Maintenance

1. U-2 OS cells are to be grown and maintained in 15 cm dishes to provide material for seeding 6 well plates.
2. Seed a vial of these cells into a 15 cm dish and exchange the medium every other day with 25 mL DMEM + 10 % FBS and 1× penicillin G/streptomycin sulfate solution.
3. Cells should be passaged at a ratio of 1:3 when they are 80 % confluent.
4. For passaging, aspirate medium. Add 10 mL of PBS to wash the cell monolayer and then aspirate.
5. Add 3 mL of 0.05 % trypsin solution, warmed to 37 °C, and place plate into the 37 °C cell culture incubator for 6 min, or until cells begin to dissociate.
6. Add 3 mL of 37 °C culture medium to the trypsinized monolayer and pipette up and down to collect all cells.
7. Transfer 2 mL of collected cells to each 15 cm dish to be seeded.
8. Each 15 cm plate should contain 25 mL of fresh, 37 °C culture medium.

3.2 Transfection

1. Calculate the number of transfection reactions needed and set aside the appropriate number of 15 cm plates for this purpose (*see* **Note 1**).
2. Trypsinize the monolayers of each plate as described in Subheading 3.1.
3. To each 15 cm dish, add 3 mL of 37 °C culture medium.
4. Collect all of the dissociated cells by pipetting up and down and then transfer to a sterile 50 mL conical (one plate per conical).
5. Centrifuge at 300 × *g* for 3 min to pellet cells.
6. Aspirate supernatant and then add an additional 15 mL 37 °C culture medium.
7. Resuspend cells by gentle trituration and repeat centrifugation.
8. Aspirate supernatant and resuspend cells in a volume of medium that is twice the number of wells that can be transfected per 15 cm plate; e.g., 36 mL of medium for 18 possible wells.
9. Once all cells have been appropriately resuspended, aliquot 2 mL to each well of the 6 well plates.
10. Put 6 well plates on the shelf of a humidified cell culture incubator, shaking orthogonally to assure even dispersal of the cells.
11. Allow cells to attach in the incubator for 3 h prior to transfection.

12. Also, check by microscopic evaluation (40× magnification) to verify that cells are 40 % confluent prior to transfection.
13. The transfection of each well will require a total of 1.5 μg of DNA and 3 μL of Polyfect that are diluted in room temperature serum-free DMEM and delivered in a 100 μL aliquot to the cells (*see* **Notes 2** and **3**).
14. Taking into account pipetting losses, 1.95 μg of total DNA and 3.9 μL Polyfect should be used in a final volume of 130 μL.
15. Reagents are to be added and mixed in sterile eppendorf tubes; the order of addition is serum-free DMEM, DNA, and then Polyfect.
16. For GFP transfected wells, add an appropriate volume of plasmid for 1.95 μg to serum-free DMEM (volume of DMEM is equal to 130 μL-(volume of plasmid)-3.9 μL polyfect).
17. Add 3 μL Polyfect and pulse vortex to mix.
18. For PGC-1α/GCN5 transfections, add sufficient volume of plasmids for 1.3 μg PGC-1α, 0.163 μg GCN5, 0.163 μg pcDNA3.0 to serum-free DMEM (volume of DMEM is equal to 130 μL-(volume of plasmids)-3.9 μL Polyfect).
19. Add 3 μL Polyfect and pulse vortex to mix.
20. Allow transfection mixes to sit at room temperature for 20 min prior to addition to the cells.
21. During this incubation, aspirate medium off of cells and replace with 2 mL of 37 °C culture medium.
22. After the 20 min incubation period, add 100 μL of transfection mix to each well in a drop-wise fashion.
23. Swirl the plates to ensure complete mixing and return to the incubator overnight.
24. 24 h after transfection, aspirate medium, wash with PBS, and then replace with new culture medium warmed to 37 °C.
25. Return plate to incubator.
26. If screening for the effects of compounds on PGC-1α acetylation, compounds can be added at any time point after this medium change.
27. 45 h after transfection, aspirate medium and then replace with new culture medium warmed to 37 °C (*see* **Note 4**).
28. Return plate to incubator. Cells will be harvested 3 h later.

3.3 Cell Harvesting and Immunoprecipitation

1. Chill PBS and lysis buffer to 4 °C prior to harvesting. All harvesting steps are to be conducted on ice.
2. Remove plate from incubator and place on ice.
3. Aspirate medium and add 2 mL PBS.

4. Aspirate and add 200 μL lysis buffer to each well.
5. Using a cell scraper, collect monolayer and pipette into a prechilled eppendorf tube. Before proceeding to the next well, clean the cell scraper with 70 % ethanol and wipe dry.
6. After harvesting, rotate tubes for 1 h in a 4 °C cold room.
7. Vortex tubes for 10 s at maximum speed and then centrifuge at 4 °C for 20 min at 16,000 × *g*.
8. Transfer supernatants to a new, prechilled eppendorf tube.
9. Remove 20 μL of the supernatant and save in a separate tube for inputs analysis.
10. To the remaining supernatant, add 12 μL of a 50 % slurry of M2 affinity resin, equilibrated in lysis buffer.
11. Rotate at 4 °C for at least 3 h to immunoprecipitate PGC-1α.
12. Upon completion of incubation with the M2 resin, the samples must be washed to remove extraneous protein.
13. Centrifuge tubes at 1,000 × *g* for 1.5 min.
14. Carefully aspirate supernatant and add 1 mL of lysis buffer.
15. Tap tubes to mix resin in lysis buffer and centrifuge at 1,000 × *g* for 1.5 min.
16. Repeat washing the samples for a total of 4 times.
17. After the final wash, carefully remove all supernatant with a gel loading pipette tip crimped at the end.
18. Add 14 μL of 2× Laemmli SDS-PAGE loading buffer and heat at 95 °C for 5 min.
19. To the input samples, add 4 μL of 6× Laemmeli SDS-PAGE loading buffer and heat at 95 °C for 5 min.
20. Cool samples to room temperature and centrifuge at 1,000 × *g* to collect condensation.
21. Samples are ready to be run by SDS-PAGE.

3.4 SDS-PAGE Resolution, Electrotransfer, and Western Blotting on Cellular Extracts

1. Load samples and protein ladder onto an 8 % polyacrylamide resolving gel.
2. Run the gel at a constant 160 V until the dye front reaches the bottom of the gel.
3. Disassemble gel and place into transfer buffer on a rocking platform to equilibrate for 30 min.
4. Wet a precut PVDF membrane with methanol, rinse with water, and then equilibrate in transfer buffer for 15 min.
5. Upon completion of the equilibration, assemble the gel and PVDF membranes into a transfer cassette, sandwiched between two pieces of Whatman filter paper and porous sponges.

6. Place into transfer apparatus and transfer overnight at a constant voltage of 30 V.
7. Following transfer, remove membranes and place in 5 % w/v milk blocking buffer for 1 h.
8. Wash with TBST 3 times, 5 min per wash and then incubate with primary antibodies.
9. For immunoprecipitated samples, the acetylated lysine signal should be analyzed first. The acetylated lysine antibody should be diluted 1:1,000 in blocking buffer containing 5 % w/v bovine serum albumin.
10. Incubations with this antibody should be conducted overnight at 4 °C with gentle rocking.
11. The following day, wash the blot 3 times with TBST, 10 min per wash on a rocking platform at room temperature.
12. Incubate blot for 1 h with gentle rocking in a 1:50,000 dilution of secondary antibody in 5 % w/v milk blocking buffer.
13. Wash the blot 3 times with TBST buffer, 10 min per wash, on a rocking platform.
14. Incubate with West Dura chemiluminescent substrate for 5 min and then develop the blot using X-ray film. The apparent molecular weight of acetylated PGC-1α is ~120 kDa when compared against a standard protein ladder. Acetylated GCN5, which can sometimes be visualized with highly sensitive chemiluminescent substrates, runs at ~90 kDa.
15. Total levels of PGC-1α can be assessed using the anti-PGC-1α primary antibody. Blots must first be stripped with the user's preferred method (we use Restore stripping buffer from Pierce).
16. Repeat steps above, starting with **step 8**, Subheading 3.4, with the exception that the primary antibody can be incubated for 1 h at room temperature. The apparent molecular weight of full-length PGC-1α is ~120 kDa.
17. Total levels of GCN5 can be assessed using the anti-GCN5L2 primary antibody. Blots must first be stripped and steps repeated starting with **step 8**, Subheading 3.4. The apparent molecular weight of full-length GCN5 is ~90 kDa
18. As a loading control on input blots, users can probe their samples with the anti-β tubulin primary antibody. Blots must first be stripped and steps repeated above beginning with **step 8**, Subheading 3.4. The apparent molecular weight of full-length tubulin is ~50 kDa.

3.5 Homogenization and Immunoprecipitation on Animal Tissues

1. Anesthetize mice and carefully extract the tissues of interest. Whenever possible, try to keep the mice alive during the extraction procedure. In some tissues, such as skeletal muscle, PGC-1α acetylation is gradually lost after the death of the animal.

2. Immediately after extraction, quickly place the tissue on a surface (for example, an inverted Petri dish) with saline solution and make sure to remove any extraneous fat, tendons, and connective tissue out of your sample of interest. In the case of liver, wash it with excess saline solution, make sure to remove the gall bladder, and cut the tissue into small pieces to facilitate later processing. For the heart, make sure to remove blood as much as possible with an excess saline solution. In all cases, once the tissue is ready to be frozen (*see* **Note 5**), quickly remove the excess saline with absorbent paper.
3. Flash freeze the tissue in liquid nitrogen. Not more than a few seconds should take place between the extraction and freezing steps. The tissue can be then stored in liquid nitrogen or at −80 °C.
4. To homogenize the tissue (*see* **Note 6** for specifications on tissue size), prepare eppendorf tubes with 300 μL of ice-cold IP100 buffer and keep them on ice.
5. In parallel precool a porcelain grinder with liquid nitrogen.
6. Wait until the excess of liquid nitrogen is almost gone.
7. Then, quickly add your tissue of interest and pulverize it.
8. During this process, the tissue/powder should remain frozen in liquid nitrogen quality. Therefore, always keep the grinder with some remaining liquid nitrogen while pulverizing the tissue.
9. Additional liquid nitrogen can be added if necessary.
10. Once the tissue is pulverized, quickly transfer the powder to the eppendorf tube containing IP100 buffer, with the help of a spatula precooled in liquid nitrogen (*see* **Note 7**).
11. Once the powder is in the eppendorf tube with the lysis buffer, use an eppendorf homogenizer (motor/pestle) to fully homogenize the tissue.
12. Homogenize for a minute, making sure the tubes are always kept on ice.
13. In case clumps of tissue are formed at the bottom of the tube, carefully move the pestle up and down until the clump is removed from the bottom and is mixed properly.
14. After homogenization, place the tube on rotation for 1 h at 4 °C.
15. Centrifuge the sample for 10 min at 13,000 RCF in a tabletop centrifuge at 4 °C.
16. Then, transfer the supernatant to another prechilled eppendorf tube and keep it on ice.
17. Discard the pellet, containing non-soluble materials.
18. The protein content of the lysate can be measured by any of the most common available methods, even though we have generally used Lowry or BCA. We strongly recommend to

dilute the lysate at least 5 times for the protein measurement (*see* **Note 8**).

19. At this stage, the sample can be frozen at −80 °C. However, it is always recommended that the immunoprecipitation of PGC-1α or Ndufa9 is performed immediately after protein extraction.
20. Prepare the Protein-A Agarose (for PGC-1α immunoprecipitation) or Protein G-Sepharose (for Ndufa9 immunoprecipitation) beads. There are commercially available resins, as well as powder for resuspension. Both work equally well, but we recommend fast-flow resins for ease of use. In the case of beads that require resuspension, resuspend at a proportion of 100 mg of beads in 1 mL of MilliQ water (*see* **Note 9**).
21. In order to estimate the amount of beads required for immunoprecipitation, take into account that each immunoprecipitation tube should contain 20–30 μL of beads. Beads in suspension are generally at a 3:2 proportion with the suspension milieu. Therefore, it is recommended to place in each tube 50 μL of your suspension solution. This number should be corrected in case the bead:milieu proportion is different.
22. The number of tubes to be prepared with beads should take into account, at least (*see* Fig. 3): tubes to preclear samples, tubes for unspecific antibody binding, tubes for specific antibody binding.
23. Centrifuge the suspension solution for 1 min at 200 RCF in a tabletop centrifuge (*see* **Note 10**).
24. Aspirate the supernatant and wash the beads 3 times with 10 volumes of IP100 buffer.
25. Sample preclearing: for PGC-1α immunoprecipitation on total homogenates, we generally use, at least, 1 mg of sample. Since the sample will be immunoprecipitated against specific and unspecific antibodies, this means that we will preclear twice the amount of sample to start with. In the case of Ndufa9, immunoprecipitations can be performed with 500 μg of sample.
26. Add the appropriate amount of sample to the preclearing tubes and bring the volume up to 200 μL with IP100. In very dilute samples, the volume of sample required might surpass 200 μL. In this case equilibrate all the samples to a similar volume (*see* **Note 11**).
27. Incubate the sample with the beads for 1 h at 4 °C with gentle shaking.
28. Then, centrifuge the sample for 1 min at 200 RCF in a table top centrifuge and keep the supernatant on ice.
29. Transfer the supernatant to the antibody bound resin as soon as possible.

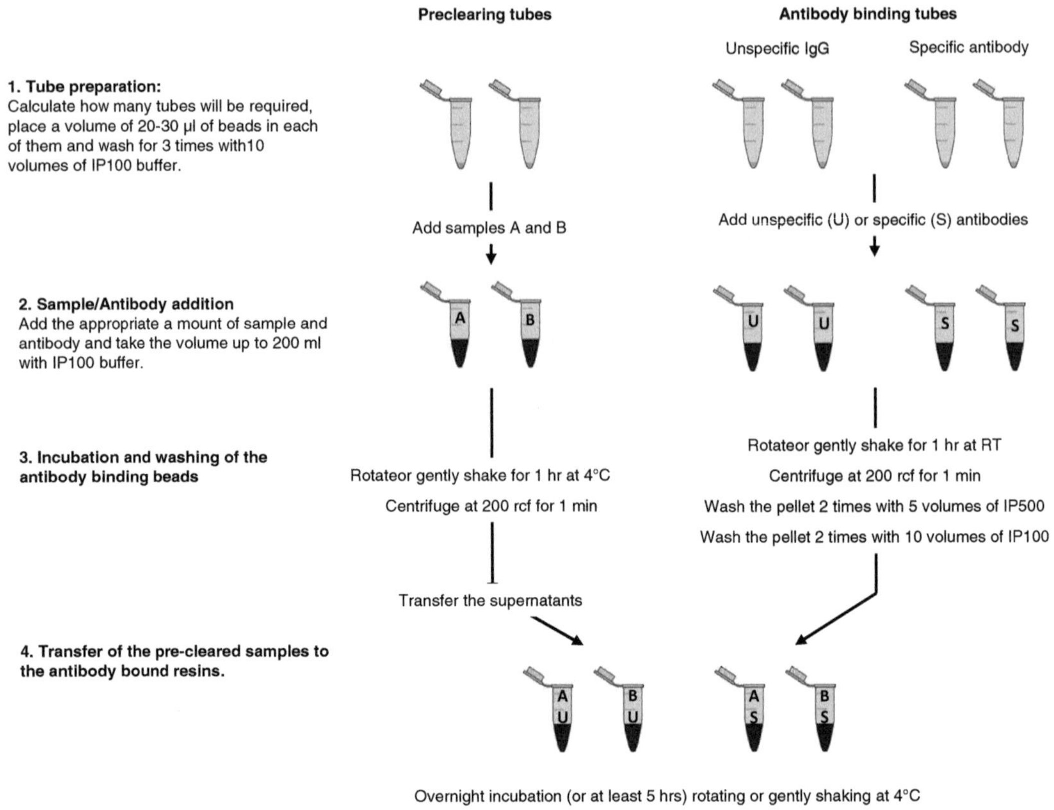

Fig. 2 Schematic diagram of the initial immunoprecipitation procedures. The graph depicts the tube preparation steps, sample preclearing, antibody binding, antibody washing, and sample/antibody conjugation steps

30. Antibody binding: Add 5 µg of PGC-1α antibody to each tubes. In the case of Ndufa9, 2 µg of antibody are enough. Do not forget to add the corresponding amount of unspecific IgGs to the control tubes.
31. Bring the total volume of the tube to 200 µL with IP100 buffer.
32. Then, incubate for one 1 h at room temperature with gentle shaking.
33. Then, centrifuge the tube at 200 RCF for 1 min and keep the pellet.
34. Wash the pellet 2 times with at least 5 volumes of IP500 buffer and wash again 2 times with IP100 buffer.
35. Mix the antibody bound resins with the precleared fractions and incubate overnight, or at least 5 h, at 4 °C with gentle shaking (Fig. 2).
36. Centrifuge for 1 min at 4 °C at 200 RCF. Keep the supernatant for input testing.

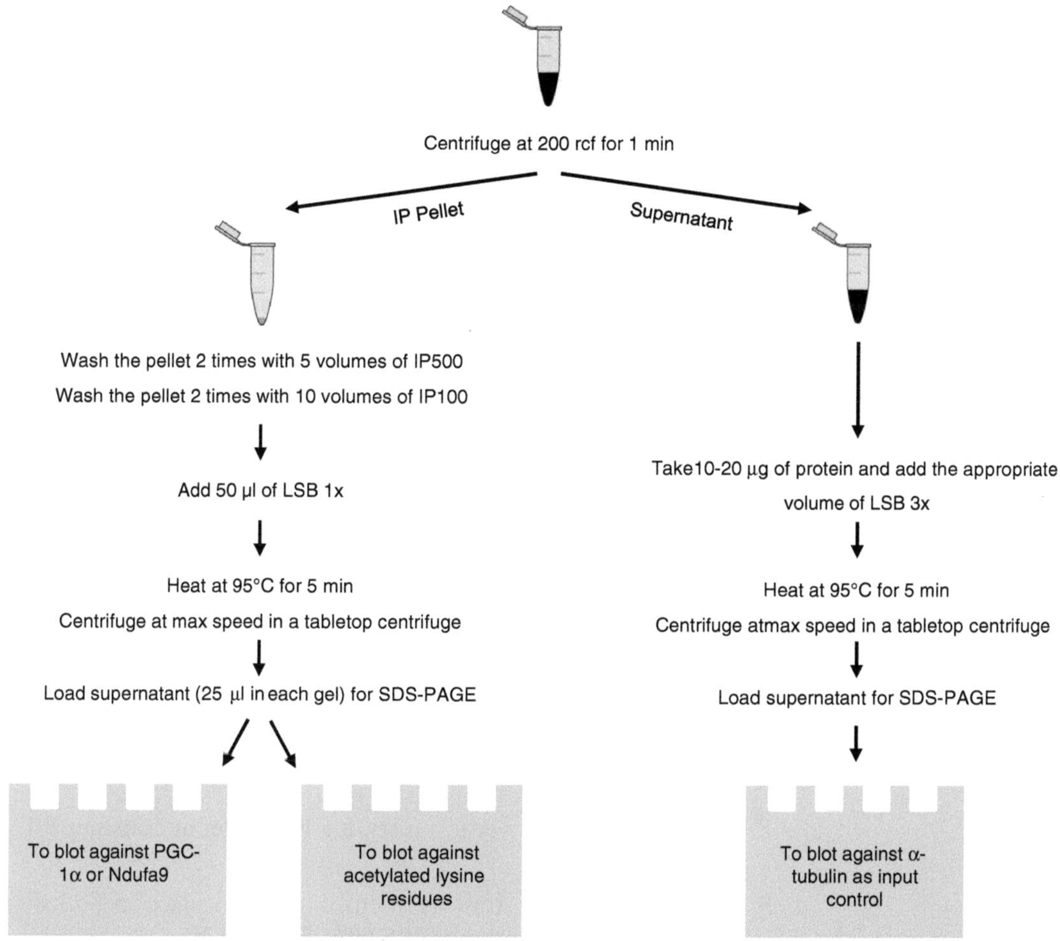

Fig. 3 Schematic diagram of the final immunoprecipitation procedures. The graph depicts the clearing of the sample and recommended gel loading strategy for the analysis of PGC-1α and Ndufa9 immunoprecipitates from tissue extracts

37. Wash the beads 3 times with 10 volumes of IP500 and 2 times with 10 volumes of IP100.
38. Aspire as much as possible the IP100 supernatant and add 50 μL of Laemmli Sample Buffer 1×, containing 5 % β-mercaptoethanol.
39. Take the equivalent of 20 mg of protein from the supernatants collected above, add the appropriate volume of LSB 3×, containing 15 % β-mercaptoethanol.
40. Boil the samples at 95 °C for 5 min.
41. Centrifuge the samples at maximum speed in a table top centrifuge.
42. Transfer the supernatant to another tube or load them for SDS-PAGE (Fig. 3).

3.6 SDS-PAGE Resolution, Electrotransfer, and Western Blotting on Tissue Samples

There are a few key differences on SDS-PAGE planning when using tissue samples or cellular extracts. With cellular extracts, the overexpression of PGC-1α makes it possible to detect PGC-1α by stripping the membrane after blotting for acetylated lysine residues. The very weak detection of endogenous PGC-1α in tissues compromises this possibility. Therefore, it is advised to test in parallel total PGC-1α levels and acetylated lysine content (Fig. 3). A similar approach is used for Complex I acetylation tests.

1. Load samples and protein ladder onto an 8 % polyacrylamide resolving gel for PGC-1α tests or a 12.5 % polyacrylamide gel to resolve Ndufa9 immunoprecipitates.
2. Run the gel at a constant 130 V until the dye front reaches the bottom of the gel.
3. Disassemble gel and place into transfer buffer on a rocking platform to equilibrate for 30 min.
4. Wet a precut PVDF membrane with methanol, rinse with water, and then equilibrate in transfer buffer for 15 min.
5. Upon completion of the equilibration, assemble the gel and PVDF membranes into a transfer cassette, sandwiched between two pieces of Whatman filter paper and porous sponges.
6. Place into transfer apparatus and transfer overnight at a constant voltage of 30 V.
7. For Complex I acetylation tests, a 1 h transfer at constant 80 V is enough.
8. Following transfer, remove membranes and place in 5 % w/v bovine serum albumin in TBST buffer for 1 h.
9. Wash with TBST to eliminate the excess of bovine serum albumin.
10. In order to avoid future cross-reactions, at this step we place the membrane into plastic and cut out all the molecular ranges that are beyond our interest.
11. PGC-1α is generally detected around 120 kDa (*see* **Note 12** and Fig. 4a). We therefore cut the membrane at the 100 kDa marker and will only keep using the top part. In the case of Complex I acetylation tests, we only use the part of the membrane below 50 kDa.
12. Prepare your antibody dilutions: the acetylated lysine and PGC-1α antibodies should be diluted 1:1,000 in TBST buffer containing 1 % w/v bovine serum albumin. The Ndufa9 antibody should be diluted 1:2,000. For input evaluation, we generally use α-tubulin. The α-tubulin antibody should be diluted 1:5,000.
13. Incubations with this antibody should be conducted overnight at 4 °C with gentle rocking.

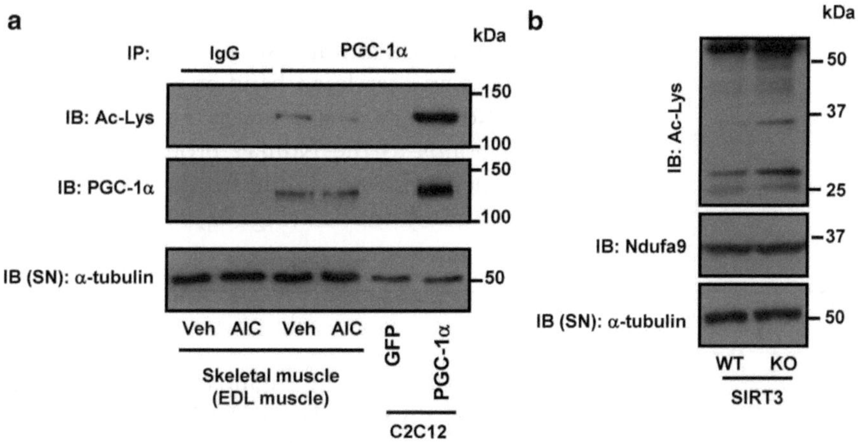

Fig. 4 Examples of PGC-1α acetylation and Complex I acetylation tests in total homogenates from mouse skeletal muscle. (**a**) Mice were treated with either vehicle (Veh, saline) or AICAR (250 mg/kg) for 6 h immediately after the feeding period without access to food after the injection. Then, mice were sacrificed and the two EDL muscles were extracted. The two muscles of a mice were pooled together to extract protein. Then, total 1 mg of total homogenate was immunoprecipitated against unspecific IgGs or against PGC-1α. In parallel, 500 μg of protein from C2C12 myotubes infected with adenovirus encoding for either GFP of FLAG-PGC-1α were immunoprecipitated against PGC-1α in order to further validate the specificity of the band. As specified in Subheading 3, and based on previous experience, only the 100–150 kDa range of the membranes blotted for PGC-1α or acetyl-lysine are used in order to gain focus and lose cross-reaction with unspecific bands. As shown by the control test in FLAG-PGC-1α overexpressing C2C12 myotubes, full-length PGC-1α appears as a highly acetylated band around the 120 kDa range. C2C12 myotubes are an excellent negative control, as they display no detectable endogenous full-length PGC-1α protein levels. A band can be detected in skeletal muscle when the samples are immunoprecipitated against PGC-1α, but not against unspecific IgGs. As expected, AICAR treatment promotes PGC-1α deacetylation, demonstrating that these conditions allow detection of changes in acetylation levels. (**b**) Total protein homogenates of gastrocnemius muscles from either WT or muscle-specific SIRT3 knock-out mice were obtained. Then, 500 μg of protein were used to immunoprecipitate against Ndufa9. As expected, the SIRT3 knock-out mice display a global higher acetylation profile on Ndufa9 immunoprecipitates, with some bands very clearly defined within the 25–50 kDa range. Please notice that band profile and intensity might be different between tissues

14. The following day, wash the blot 4 times with TBST, 15 min per wash on a rocking platform at room temperature.
15. Incubate blot for 1 h with gentle rocking in a 1:25,000 dilution of secondary antibody in TBST containing 1 % w/v of bovine serum albumin.
16. Wash the blot 4 times with TBST buffer, 15 min per wash, on a rocking platform.
17. Incubate with ECL chemiluminescent substrate for 5 min and then develop the blot using X-ray film. The apparent molecular weight of acetylated PGC-1α is ~120 kDa when compared against a standard protein ladder (Fig. 4a). Ndufa9 is a prominent band close to the 37 kDa mark. In the case of acetyl-lysine blots on Ndufa9 immunoprecipitates, multiple bands can be

observed between 50 and 25 kDa, probably corresponding to acetylation events on multiple subunits of Complex I (Fig. 4b). While there is an acetylated band at a very similar molecular weight of Ndufa9, we have never tested whether this band truly corresponds to the Ndufa9 protein itself. In the input gels a clear defined band should be observed around 50 kDa, corresponding to α-tubulin (Fig. 4a, b)

4 Notes

1. The relative surface area of a standard 15 cm culture dish and a single well of a 6 well plate is 18.3:1. The number of wells that a given 15 cm dish of U-2 OS cells can yield for transfection is dependent upon this ratio as well as the confluency at the time of trypsinization. Because cells are to be seeded for transfection into each 6 well at a confluency of 40 %, the total number of yieldable wells is 18.3·(confluency of the 15 cm dish)/0.4. Thus, for instance, an 80 % confluent 15 cm dish will yield a maximum of 36.6 transfectable wells (18.3·0.8/0.4).
2. The total amount of DNA to transfect into cells is at the discretion of the user, but we have found 1.5 μg DNA is optimal with the Polyfect system. Using this system, we have determined that the ideal amount constructs to use for acetylation manipulations are 1 μg PGC-1α, 0.125 μg GCN5, and 0.125 μg pcDNA. The pcDNA can be conveniently substituted with other expression constructs when testing the effects of a query protein's overexpression in PGC-1α acetylation. GFP should be transfected into separate wells for each experiment to evaluate transfection efficiency—visualized by fluorescence microscopy—as well as to use as a negative control for Western blotting (no detectable acetylation should be seen in lanes containing GFP control). In general, healthy U-2 OS cultures should display a transfection efficiency of between 50 and 80 %.
3. We prefer to use Polyfect for the transfection of cells, where possible. We have found that lipid-based transfection reagents, such as Lipofectamine 2000, result in significantly less acetylation of PGC-1α on a per unit protein basis than dendrimer-based transfection reagents such as Polyfect.
4. Exchanging spent culture medium for fresh culture medium 3 h prior to harvesting significantly improves the acetylation signal for PGC-1α. We assume that this may be due to the depletion of glucose levels in the medium with prolonged culturing. Indeed, the culturing of cells in glucose-free medium or the addition of the non-metabolizable glucose analog, 2-deoxyglucose, results in a significant reduction of PGC-1α acetylation.

5. When aiming to evaluate PGC-1α acetalyation in some tissues where the sample amount is generally not limiting, like in liver, it is strongly recommended to perform nuclear isolation and test PGC-1α in nuclear extracts. In this case, perform the nuclear extracts in fresh tissue instead of frozen samples. For the preparation of nuclear extracts from liver or big muscle pieces, we refer the reader to references [14, 15], respectively. When using nuclear extracts, the starting amount of protein necessary for the immunoprecipitaton of PGC-1α is 500 μg for liver and 300 μg for muscle.
6. Generally, an ideal proportion of 10 mg of tissue per 100 μL of buffer. This generally renders a concentration ~10 μg/μL of protein in muscle, and up to 50 μg/μL in liver. Considering that we generally aim to have at least 2 mg of protein for the assay, we recommend using around 30 mg of muscle (in 300 μL of buffer). In the case that only lower amounts of muscle are available, the amount of buffer should also be scaled down, but never to less than 200 μL, since this would lead to inefficient homogenization. In the case of small muscles, such as soleus or EDL, it is recommended to pool the two muscles of the same mouse for 300 μL of buffer. In the case of the heart or big muscles, such as the gastrocnemius, which can weight more than 100 mg, it is recommended to use less than half o it. For BAT, generally we can homogenate one of the two patches of a young lean animal in 300 μL of buffer, making sure to avoid the lipidic phase when collecting the supernatant after the homogenate centrifugation.
7. The pulverized tissue can also be stored once more in liquid nitrogen or −80 °C. The transfer to the eppendorf tube should be as fast as possible, preventing thawing of the sample and placing the powder on the liquid, rather than on the walls of the tube. In the case of Ndufa9 immunoprecipitation, we generally use RIPA-based buffers all along the homogenization and immunoprecipitation procedure, rather than IP100 or IP500, but following similar steps.
8. Both the Lowry and BCA method lose linearity at high protein concentrations, hence the need for dilution. In addition, sodium butyrate shows a mild cross-reaction with the BCA protein determination, which is lost when diluting the sample 5–10 times with water.
9. The resuspension of 100 mg of beads in 1 mL of MilliQ water will provide a volume ~600 μL of rehydrated beads.
10. Centrifugation at higher speeds can compromise the integrity of the beads.
11. In general, whenever higher volumes are required, we recommend to proportionally increase the volume of beads.

However, we have efficiently immunoprecipitated PGC-1α with combinations of 30 μL of beads with up to 400 μL of lysate. At higher volumes of lysate we have observed losses in immunoprecipitation efficiency unless the amount of beads is also augmented.

12. As mentioned in Subheading 3.4, PGC-1α is generally detected around 110–130 kDa, with slight variations depending on the tissue. Of note, many commercial antibodies detect a band ~85–95 kDa. This band likely does not correspond to full-length PGC-1α, as manifested by overexpression experiments and often appears even in samples with close to undetectable PGC-1α expression at the mRNA level (such as immortalized cells). We strongly recommend being cautious when interpreting results based on this band's signal.

Acknowledgments

We are thankful to the members of the Johan Auwerx and Pere Puigserver labs for fruitful discussions regarding the set up of these techniques. C.C. would like to also thank the Laszlo Tora lab at the Institute de Génétique et de Biologie Moleculaire et Céllulaire for useful advice on immunoprecipitations. This work was supported by funds from the Dana-Farber Cancer Institute and with grants from the American Diabetes Association, Department of Defense and NIH/NIDDK (R01 069966) awarded to P.P. J.E.D. was supported in part by an NRSA Kirschstein fellowship from the National Institutes of Health. C.C. is an employee of the Nestlé Institute of Health Sciences S.A. and declares no financial interest in the data and technologies discussed in this manuscript.

References

1. Houtkooper RH, Pirinen E, Auwerx J (2012) Sirtuins as regulators of metabolism and healthspan. Nat Rev Mol Cell Biol 13(4):225–238. doi:10.1038/nrm3293, pii: nrm3293
2. Gerhart-Hines Z, Dominy JE Jr, Blattler SM, Jedrychowski MP, Banks AS, Lim JH, Chim H, Gygi SP, Puigserver P (2011) The cAMP/PKA pathway rapidly activates SIRT1 to promote fatty acid oxidation independently of changes in NAD(+). Mol Cell 44(6):851–863. doi:10.1016/j.molcel.2011.12.005,pii:S1097-2765(11)00944-0
3. Canto C, Auwerx J (2012) Targeting sirtuin 1 to improve metabolism: all you need is NAD+? Pharmacol Rev 64(1):166–187. doi:10.1124/pr.110.003905, pii: pr.110.003905
4. Rodgers JT, Lerin C, Haas W, Gygi SP, Spiegelman BM, Puigserver P (2005) Nutrient control of glucose homeostasis through a complex of PGC-1alpha and SIRT1. Nature 434(7029):113–118. doi:10.1038/nature03354, pii: nature03354
5. Ahn BH, Kim HS, Song S, Lee IH, Liu J, Vassilopoulos A, Deng CX, Finkel T (2008) A role for the mitochondrial deacetylase Sirt3 in regulating energy homeostasis. Proc Natl Acad Sci U S A 105(38):14447–14452. doi:10.1073/pnas.0803790105, pii: 0803790105
6. Lerin C, Rodgers JT, Kalume DE, Kim SH, Pandey A, Puigserver P (2006) GCN5 acetyltransferase complex controls glucose metabolism

through transcriptional repression of PGC-1alpha. Cell Metab 3(6):429–438. doi:10.1016/j.cmet.2006.04.013

7. Canto C, Gerhart-Hines Z, Feige JN, Lagouge M, Noriega L, Milne JC, Elliott PJ, Puigserver P, Auwerx J (2009) AMPK regulates energy expenditure by modulating NAD+ metabolism and SIRT1 activity. Nature 458(7241):1056–1060. doi:10.1038/nature07813, pii: nature07813
8. Gerhart-Hines Z, Dominy JE Jr, Blattler SM, Jedrychowski MP, Banks AS, Lim JH, Chim H, Gygi SP, Puigserver P (2011) The cAMP/PKA pathway rapidly activates SIRT1 to promote fatty acid oxidation independently of changes in NAD(+). Mol Cell 44(6):851–863. doi:10.1016/j.molcel.2011.12.005
9. Nin V, Escande C, Chini CC, Giri S, Camacho-Pereira J, Matalonga J, Lou Z, Chini EN (2012) Role of deleted in breast cancer 1 (DBC1) protein in SIRT1 deacetylase activation induced by protein kinase A and AMP-activated protein kinase. J Biol Chem 287(28):23489–23501. doi:10.1074/jbc.M112.365874
10. Canto C, Houtkooper RH, Pirinen E, Youn DY, Oosterveer MH, Cen Y, Fernandez-Marcos PJ, Yamamoto H, Andreux PA, Cettour-Rose P, Gademann K, Rinsch C, Schoonjans K, Sauve AA, Auwerx J (2012) The NAD(+) precursor nicotinamide riboside enhances oxidative metabolism and protects against high-fat diet-induced obesity. Cell Metab 15(6):838–847. doi:10.1016/j.cmet.2012.04.022, pii: S1550-4131(12)00192-1
11. Verdin E, Hirschey MD, Finley LW, Haigis MC (2010) Sirtuin regulation of mitochondria: energy production, apoptosis, and signaling. Trends Biochem Sci 35(12):669–675. doi:10.1016/j.tibs.2010.07.003, pii: S0968-0004(10)00135-0
12. Lombard DB, Alt FW, Cheng HL, Bunkenborg J, Streeper RS, Mostoslavsky R, Kim J, Yancopoulos G, Valenzuela D, Murphy A, Yang Y, Chen Y, Hirschey MD, Bronson RT, Haigis M, Guarente LP, Farese RV Jr, Weissman S, Verdin E, Schwer B (2007) Mammalian Sir2 homolog SIRT3 regulates global mitochondrial lysine acetylation. Mol Cell Biol 27(24):8807–8814. doi:10.1128/MCB.01636-07, pii: MCB.01636-07
13. Fernandez-Marcos PJ, Jeninga EH, Canto C, Harach T, de Boer VC, Andreux P, Moullan N, Pirinen E, Yamamoto H, Houten SM, Schoonjans K, Auwerx J (2012) Muscle or liver-specific Sirt3 deficiency induces hyperacetylation of mitochondrial proteins without affecting global metabolic homeostasis. Sci Rep 2:425. doi:10.1038/srep00425
14. Timchenko NA, Wilde M, Darlington GJ (1999) C/EBPalpha regulates formation of S-phase-specific E2F-p107 complexes in livers of newborn mice. Mol Cell Biol 19(4):2936–2945
15. Lagouge M, Argmann C, Gerhart-Hines Z, Meziane H, Lerin C, Daussin F, Messadeq N, Milne J, Lambert P, Elliott P, Geny B, Laakso M, Puigserver P, Auwerx J (2006) Resveratrol improves mitochondrial function and protects against metabolic disease by activating SIRT1 and PGC-1alpha. Cell 127(6):1109–1122. doi:10.1016/j.cell.2006.11.013, pii: S0092-8674(06)01428-0

Part IV

Methods to Study Sirtuin Biology

Chapter 16

Oxygen Flux Analysis to Understand the Biological Function of Sirtuins

Dongning Wang, Michelle F. Green, Eoin McDonnell, and Matthew D. Hirschey

Abstract

The sirtuins are a family of highly conserved NAD$^+$-dependent lysine deacylases with important roles in metabolic regulation. Of the seven mammalian sirtuins, three localize to the mitochondria: SIRT3, SIRT4, and SIRT5. Mitochondrial sirtuins are crucial regulators of the metabolic network that controls energy homeostasis and impacts cancer, obesity, diabetes, mitochondrial diseases, metabolic disorders, and many other human diseases of aging. To best study the mitochondrial function of the sirtuins, we have employed an oxygen flux analyzer as a tool to track and record the extracellular oxygen consumption rate and acidification rate that reflects mitochondrial respiration and glycolysis, respectfully. Here we described the methods using this assay to study the substrate utilization and mitochondrial function in a human hepatocellular carcinoma cell line, Huh7. Additionally, we have generated a stable SIRT4 knocked-down Huh7 cell line. With this cell line, we evaluated how the absence of SIRT4 affects mitochondrial function, glucose utilization, glutamine oxidation, and fatty acid oxidation in these cells.

Key words Mitochondrial sirtuins, Seahorse XF extracellular flux analyzer, Oxygen consumption rate, Seahorse assay, Substrate utilization, Mitochondrial function, SIRT4

1 Introduction

The major energy-producing pathway in cells when oxygen is present is oxidative phosphorylation. Nutrients derived from different metabolic pathways are oxidized to produce cellular energy in the form of ATP and macromolecules for cellular function. The mitochondrion is the central organelle where all of these processes take place.

Currently, the most common method to study substrate oxidation in vitro is to provide the cells, or isolated mitochondria, with radiolabeled substrate and then track and quantify the radiolabeled intermediates and end-products. This method has several challenges including the use of toxic radioactive materials, as well as limited sensitivity. For example, the end-products of some reactions are the

Matthew D. Hirschey (ed.), *Sirtuins: Methods and Protocols*, Methods in Molecular Biology, vol. 1077,
DOI 10.1007/978-1-62703-637-5_16, © Springer Science+Business Media, LLC 2013

substrates of others, and therefore will be challenging to monitor pathway flux.

Alternatively, oxygen dissolved in the reaction vessel of liquid-phase systems, or that accumulates in the sample chamber of gas-phase systems, can be directly detected polarographically by Clark type electrodes [1]. Isolated mitochondria are often used for this measurement. Although this system has had some success with a suspension of cultured cells, it does not work for adherent cells due to low oxygen diffusion. Additionally, setup of the Clark electrode system is tedious and requires considerable experience.

Seahorse XF extracellular flux analyzer is an alternative to the Clark type electrode. It was first introduced in 2006. By using the optical sensors, it simultaneously measures the proton and oxygen level in a very small volume of media above a monolayer of cultured cells. Valuable insight into the physiological state of cells, and the alteration of the state of those cells, can be gained through measuring the rate of oxygen consumed by the cells (OCR), an indicator of mitochondrial respiration. The cells also generate ATP through glycolysis, the conversion of glucose to lactate, independent of oxygen. The measurement of lactic acid produced is indirectly via the protons released into the extracellular medium surrounding the cells. Therefore, the extracellular acidification rate (ECAR) obtained from a "seahorse assay" reflects the glycolytic function of the cells [2]. Together, OCR and ECAR can provide important insight into the metabolic role of mitochondrial proteins.

The mitochondrial sirtuins remove acyl groups from lysine residues on proteins and control the levels of these posttranslational modifications in the mitochondria [3]. SIRT3 is the best studied of the three mitochondrial sirtuins and functions to remove acetyl groups from target lysine residues. Acetylation is emerging as an important regulatory mechanism in mitochondria where over one third of proteins contain acetylation sites and metabolic enzymes are preferentially modified [4]. Furthermore, acetylation levels and sirtuin activity change depending on the nutritional status of the cell [5–9]. In the case of SIRT3 a pattern is emerging where nutrient starvation leads to activation of the sirtuin followed by deacetylation of substrates and return to metabolic homeostasis. This is illustrated by the regulation of LCAD, an enzyme important in fatty acid oxidation, by SIRT3 [5]. During metabolic stress brought on by fasting acetylation levels of LCAD increase leading to decreased activity of the enzyme and lower levels of fatty acid oxidation. However, when SIRT3 is present LCAD is deacetylated during fasting leading to increased activity of the enzyme and increased fatty acid oxidation, an important source of fuel during fasting. Thus, SIRT3 contributes to shifts in substrate use during times of metabolic stress.

Less is known about the functions of the other mitochondrial sirtuins, SIRT4 and SIRT5, which lack robust deacetylase activity.

SIRT5 has recently been described as a protein lysine desuccinylase and demalonylase [10, 11] and regulates the urea cycle enzyme CPS1 [12]. Less is known about the function of SIRT4. Interestingly SIRT4 has weak ADP-ribosylase activity [13] and has a potent role in regulating lipid oxidation [14]. Thus, more work is required to obtain a better understanding of how the mitochondrial sirtuins function to regulate metabolism in the mitochondria.

In this chapter, we describe how the Seahorse extracellular flux analyzer can be applied to the study of mitochondrial sirtuins. The described methods focus on setting up the seahorse assays to measure the oxidation of specific substrates in a human hepatoma cell line, Huh7. We have optimized and validated the assay conditions. We also generated a stable SIRT4 knocked-down (KD) Huh7 cell line and compared the usage of substrates such as glucose, glutamine, and fatty acids in the SIRT4 KD and control cells.

Additionally, we optimized the conditions for the use of well-established mitochondrial drugs (oligomycin, FCCP, antimycin) in these cell lines and evaluated the mitochondrial function in SIRT4 KD, control, and uninfected Huh7 cells.

2 Materials

All media and solutions are prepared using cell culture grade distilled water.

2.1 Huh7 and SIRT4 KD Cells in Culture

1. Growth Media: add 10 % FBS to high-glucose DMEM, store it at 4 °C. Warm up the media to 37 °C every time before using in cells.
2. 0.25 % Trypsin-EDTA (1×).
3. Puromycin dihydrochloride.
4. Lipofectamine 2000.
5. Optimem.
6. The pSicoRMS2 lentiviral constructs encoding scramble and SIRT4 shRNA were kindly provided by Dr. Eric Verdin at the Gladstone Institutes, San Francisco, CA, USA.

2.2 Seahorse Assay

1. XF24 Extracellular Flux Analyzer (SeahorseBioscience, MA, USA).
2. Prep Station (SeahorseBioscience).
3. XF24 FluxPak (SeahorseBioscience) (*see* **Note 1**).
4. XF Cell Mito Stress Test Kit (SeahorseBioscience) contains four pre-weighed mitochondrial drugs: Oligomycin A, carbonyl cyanide 4-(trifluoromethoxy) phenylhydrazone (FCCP), Antimycin A and Rotenone (*see* **Note 2**). Dissolve each of the compounds in DMSO to make 2.5 mM stock solution. Aliquot and store at −20 °C.

5. Unbuffered base Dulbecco's Modified Eagle's Media (DMEM) (*see* **Note 3**): DMEM; 100× GlutaMax; NaCl; Phenol Red sodium salt. Dissolve 2 g of NaCl, 8.3 g of DMEM and 15 mg of Phenol Red in 900 mL of H_2O; and add 10 mL of 100× GlutaMax (*see* **Note 4**). Mix well and adjust pH with 1 N NaOH to pH 7.3. Make up to 1 L with water. Filter the media to sterilize and store at 4 °C.

2.3 Basal OCR and ECAR Measurements and Mitochondrial Function Tests in Huh7 Cells

1. Assay media used for basal OCR and ECAR measurements and mitochondrial function tests in Huh7 cells: take 148.5 mL unbuffered base DMEM (*see* **Note 5**), and add 1.5 mL of 45 % glucose solution to make a final concentration of 25 mM (*see* **Note 6**). Warm in 37 °C water bath and adjust pH to 7.4 with 1 N NaOH before use (*see* **Note 7**).
2. Remove one aliquot of Oligomycin, FCCP, and Antimycin stock solution (2.5 mM) from −20 °C freezer and thaw at room temperature for 30 min on the seahorse assay day. Dilute each of the compounds in above prepared assay media to a concentration at 10 times of the final working concentration.

2.4 Glutamine Oxidation in Huh7 and SIRT4 KD Cells

1. Take 148.5 mL unbuffered base DMEM, and mix with 1.5 mL 100× L-Glutamine (*see* **Note 8**).

2.5 Glucose Oxidation in Huh7 and SIRT4 KD Cells

1. Prepare new unbuffered base DMEM without adding glutamax, mix 148.5 mL of this media with only 1.5 mL of 45 % glucose solution.

2.6 Fatty Acid Oxidation (FAO) Measurement by Seahorse Assay

1. 10× KHB buffer stock: mix appropriate amount of buffer components to the following concentration: 1.1 M NaCl, 47 mM KCl, 20 mM $MgSO_4$, 12 mM Na_2HPO_4. Add 900 mL of H_2O and stir to allow each component to fully dissolve. Adjust pH to 7.4 and then adjust volume to 1,000 mL with water. Sterilize by filtering and store at 4 °C.
2. 50 mM carnitine stock: dissolve 81 mg carnitine in 10 mL H_2O. Use a syringe filter to sterilize the solution. Prepare 1 mL aliquots and store at −20 °C (*see* **Note 9**).
3. 1× KHB buffer: Mix 12 mL of 10× KHB, 1.2 mL of 50 mM carnitine with 106.8 mL of H_2O to make 120 mL of 1× KHB containing 0.5 mM carnitine (*see* **Note 10**).
4. Preparing 20 mL stock of BSA-conjugated palmitate solution (2 mM sodium palmitate/0.34 mM BSA) and control BSA solution (0.34 mM BSA) (*see* **Note 11**): weigh out 906.8 mg of ultra fatty acid-free bovine serum albumin (BSA) (*see* **Note 12**) and add to the glass beaker containing the pre-warmed (37 °C)

20 mL of 150 mM NaCl solution while maintaining the solution warmed at 37 °C (*see* **Note 13**). When BSA is fully dissolved, use the syringe filter to filter the entire 20 mL BSA solution. Take 10 mL of the BSA solution and quickly mix with 10 mL of 150 mM NaCl solution to make 0.34 mM stock, which will be used as control substrate in FAO measurement. Make 1 and 2 mL aliquots in glass vials and store in −20 °C. Keep the remaining 10 mL of non-diluted BSA solution in 37 °C water bath while preparing the palmitate solution. Weigh out 12.24 mg of the sodium palmitate, and add it to 8.8 mL of 150 mM NaCl solution in a small glass beaker with a stir bar. Place it into a beaker/water bath and heat to 70 °C while stirring. When the palmitate solution becomes clear, quickly transfer the 70 °C palmitate solution to the 10 mL of non-diluted BSA solution while stirring at 37 °C. Stir the mixture at 37 °C for 1 h (*see* **Note 13**). Adjust the pH to 7.4 with 1 N NaOH, and then adjust the final volume to 20 mL with 150 mM NaCl. Prepare 1 and 2 mL aliquot in glass vials and store in −20 °C.

3 Methods

3.1 Generation of SIRT4 Knocked-Down Huh7 Cells

1. The lentivirus was generated by standard molecular biology protocols using pSicoRMS2 constructs containing scramble or SIRT4 shRNA, as well as puromycin and mCherry selection markers [15]. Briefly, a four plasmid transfection mix was made containing 5 μg pSicoRMS2, 2.5 μg pMDLg/pRRE, 1.25 μg pRSV-Rev, and 1.5 μg pVSVg and 20 μL Lipofectamine 2000 in Optimem. The mixture was used to transfect a T75 plate of 25 % confluent HEK293T cells.
2. Change culture media 24 h after infection.
3. Collect media from cells 36 h after infection.
4. To infect Huh7 cells, mix viral supernatant 1:1 with fresh media and apply directly to cells.
5. Change culture media 24 h after infection. Use puromycin in fresh culture media and select for infected cells by culturing in the presence of puromycin for 72 h (*see* **Note 14**). The optimal concentration of puromycin was found to be 1 g/μL for Huh7 cells, as this caused complete cell death of uninfected cells after 24 h.
6. Visualize selected cells on a fluorescent microscope to confirm all cells express mCherry.
7. Isolate RNA using the Qiagen RNeasy isolation kit and prepare cDNA from 1 μg of RNA using the Bio-Rad iScript cDNA synthesis kit. Determine the efficiency of knockdown using

real-time qPCR with primers to SIRT4 and an endogenous control. Primers designed to mCherry can also be used to determine infection efficiency between scramble and SIRT4 knockdown cell lines.

3.2 Huh7 and SIRT4 KD Cells Maintenance

1. When the cells are confluent, remove the growth media and wash with 10 mL of PBS once. Aspirate PBS off and add 1 mL of trypsin to cover the whole surface of the cells. Incubate at 37 °C for 3 min before aspirating off the trypsin (*see* **Note 15**). Keep the cells for an additional 1–2 min in the incubator, and then tap the flask to detach the cells from the flask. Add 10 mL fresh growth media to flush all the cells off and collect them in a 50 mL conical tube. Mix 1.25 mL of these cells (1/8) with 14 mL of fresh growth media and plate them in a new T75 flask.
2. Maintain the Huh7 and KD cells in T75 flask at 37 °C with 5 % CO_2. The KD cells need to be cultured for one or two more passages in puromycin-free media before being used for seahorse measurement (*see* **Note 14**).

3.3 Basal OCR and ECAR Measurement and Mitochondrial Function Tests in Huh7 Cells (See Note 16)

Prepare materials as in below on day 1 (1 day before the seahorse assay):

1. Pre-incubate one sensor cartridge with 1 mL/well XF24 calibrant solution on one utility plate in a 37 °C non-CO_2 incubator overnight.
2. Turn on the XF24 analyzer with XF24 software running to allow instrument to stabilize at 37 °C before the assay.
3. Prepare cells as following. Collect the cells as described in Subheading 3.2 and dilute cells so that every 100 μL of the cell/media mixture should contain the desired number of cells for one well of the XF24 cell culture plate. Based on the results from the optimization experiments (*see* **Note 16**), plate Huh7 cells at 60,000 cells/well in the XF24 culture plate. Load 100 μL of the cell/media mixture into each well of the XF24 culture plate, and allow the cells to adhere for 2 h in the incubator before adding additional 150 μL of growth media for a total volume of 250 μL per well (*see* **Note 17**). Plate the cells in all wells except for A1, B4, C3, and D6, which will be filled with growth media only and used as background correction wells during the seahorse measurement. Place the same cell line at least in three replicates on each plate.
4. Leave the cells in the incubator at 37 °C with 5 % CO_2 overnight and proceed with the seahorse assay the next day.
5. Seahorse measurement takes place on day 2 following the steps described in below:

6. Inspect cells in the XF24 culture plate under microscope to assure confluence and even seeding.
7. Make fresh assay media as described in Subheading 2. Warm it up to 37 °C and adjust pH to 7.4. Transfer the assay media to a 500 mL square reagent bottle that can fit into the seahorse prep station. Leave the bottle in the prep station, which can maintain the temperature at 37 °C. The media change can be started any time from now.
8. At this point, load the designed seahorse assay program into XF24 software and have the program ready to go. The mitochondrial drugs also need to be taken out for thawing.
9. Perform media change (*see* **Note 18**) and incubate the cells with the assay media in a 37 °C non-CO_2 incubator for 60 min (±5 min in all of our assays) for equilibration. Start the seahorse assay program no more than 30 min after media change in the cells (*see* **Note 19**).
10. Dilute thawed mitochondrial drugs in assay media to 10 times of the final working concentration (*see* **Note 20**). Load 50–100 μL through each port as needed (*see* **Note 21**). Keep the cartridge loaded with compound in the 37 °C non-CO_2 incubator for 10 min or more to allow it to warm up before starting the calibration (*see* **Note 19**).
11. At 30 min post media change in the cells, load the sensor cartridge from **step 10** (containing compounds in ports) into XF24 flux analyzer instrument tray, and start calibration.
12. After calibration is complete, replace the utility plate with the pre-incubated (60 min) cell plate and continue the program. The detailed program is as seen in Table 1.
13. At the end of the assay, take the cell plate out and examine the cells under the microscope to ensure the cells are still attached and look healthy after the measurement. It is optional to lyze the cells in the plate for protein quantitation. Download and analyze the seahorse measurement results (Fig. 1).

3.4 Compare the Substrate Utilization and Mitochondrial Function in Uninfected, Scramble (SCR), and SIRT4 KD Huh7 Cells

1. 60,000/well of Huh7, Huh7-SIR4KD, and Huh7-SCR were placed in multiple replicates in one XF24 cell culture plate on day 1.
2. The rest of the procedures follow what has been described in Subheading 3.3.
3. Basal levels of OCR and ECAR, proton lead associated oxygen consumption, maximal mitochondrial respiration and non-mitochondrial respiration were analyzed in these cells (Fig. 2; *see* **Note 22**).

Table 1
Program used for basal OCR, ECAR measurements and mitochondrial function test

Command	Time (min)
Calibrate	Fixed
Equilibrate	Fixed
Loop Start	3
Mix	3
Wait	2
Measure	3
Loop End	
Inject from port A	
Loop Start	3
Mix	3
Wait	2
Measure	3
Loop End	
Inject from port B	
Loop Start	3
Mix	3
Wait	2
Measure	3
Loop End	
Inject from port C	
Loop Start	3
Mix	3
Wait	2
Measure	3
Loop End	
Program End	

3.5 Study Glutamine Oxidation in Uninfected, SCR, and SIRT4 KD Huh7 Cells

1. Use the assay media that contains only glutamine in the base DMEM for this experiment. All the rest of the procedures are the same as described in Subheadings 3.3 and 3.4 (Fig. 3; *see* **Note 23**).

3.6 Study Glucose Oxidation in Uninfected, SCR, and SIRT4 KD Huh7 Cells

1. Use the assay media that contains only glucose in the base DMEM without glutamax for this experiment. All the rest of the procedures are the same as described in Subheadings 3.3 and 3.4 (Fig. 4; *see* **Note 24**).

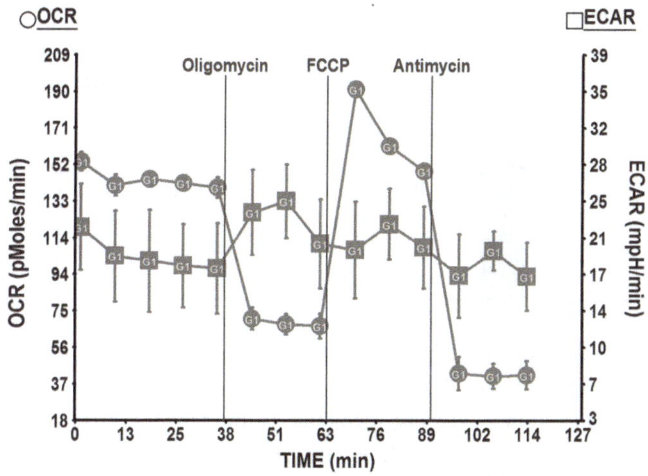

Fig. 1 OCR and ECAR measurement in Huh7 cells. In the assay media containing 25 mM glucose, Huh7 showed moderate level of basal OCR and relatively high ECAR indicating its preference using glucose for glycolysis. With the optimized concentration of mitochondrial drugs, we saw a ~60 % reduction in OCR following the injection of oligomycin. OCR jumped up by 40 % upon the injection of FCCP indicating the release of maximal respiration capacity. Finally, injection of antimycin blocked all the mitochondrial respiration and revealed the remaining OCR in Huh7 that was from the non-mitochondrial respiration

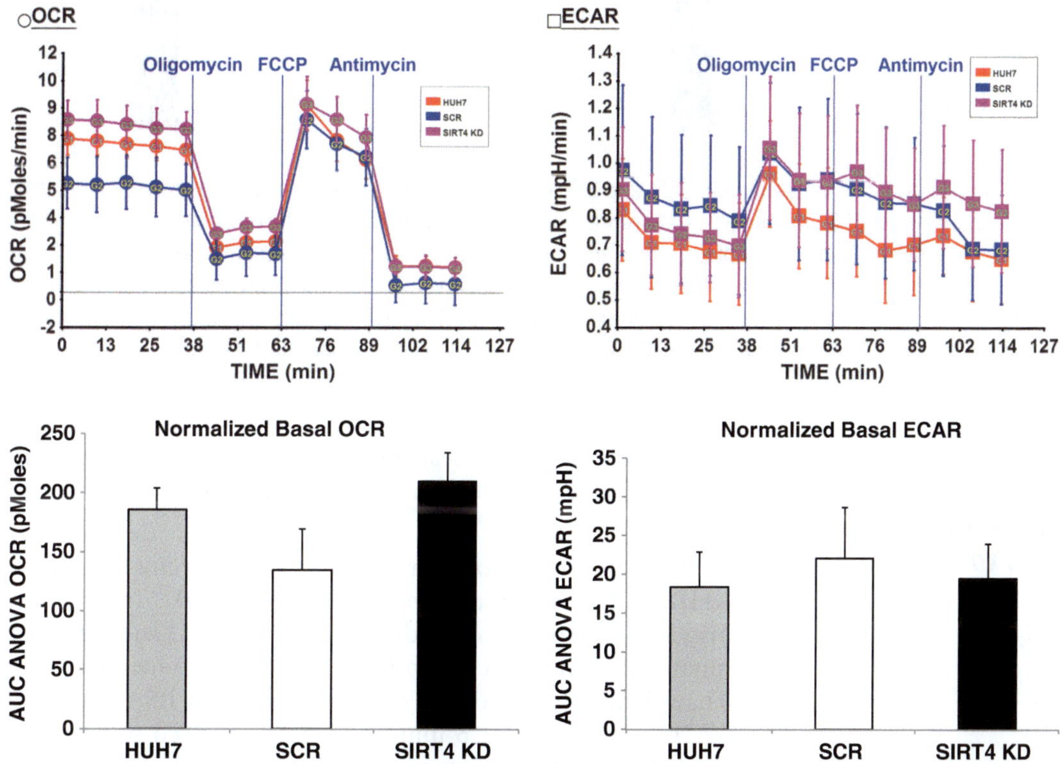

Fig. 2 OCR and ECAR comparison in uninfected, SCR, and SIRT4 KD Huh7 cells. Seahorse assay was run with the assay media containing 2 mM Glutamax and 25 mM Glucose. Data for each cell line are the average of six or seven replicates and have been normalized to the total protein amount. Basal OCR was higher in SIRT4 KD cells, while basal ECAR remained at similar level in all cell lines. The difference in basal OCR was more significant when comparing the SIRT4 KD cells line to the SCR control ($p = 0.014$) than to the uninfected Huh7 cells ($p = 0.071$). ECAR and the disturbed bioenergetics profile post-mitochondrial compounds injection did not show significant difference among the three analyzed cell lines

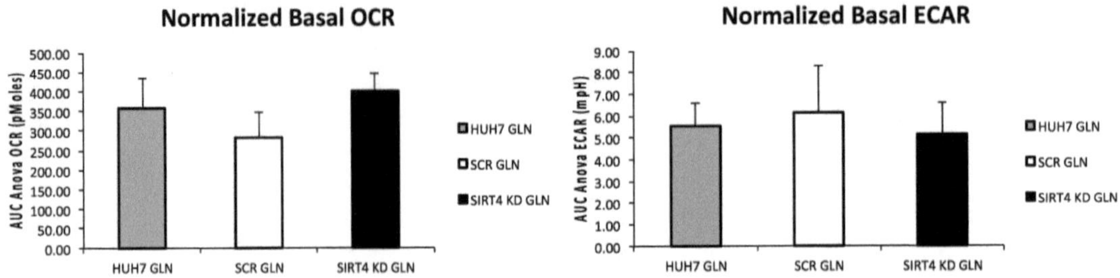

Fig. 3 Glutamine oxidation in uninfected, SCR, and SIRT4 KD Huh7 cells. Seahorse assay was run with the assay media containing 4 mM Glutamine. Data represents the average of five replicates for each cell line and has been normalized to the total protein amount. Basal OCR were at the maximal level, while basal ECRA were at minimal level in this media condition for all three analyzed cell lines. The glutamine oxidation in SIRT4 KD cells was the highest comparing to SCR and uninfected Huh7 cells. Again, the OCR difference was more significant when comparing to SCR ($p = 0.0089$) than comparing to the uninfected Huh7 cells ($p = 0.32$)

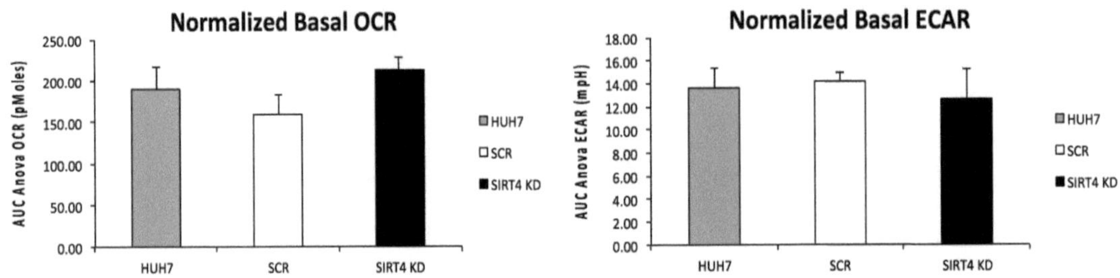

Fig. 4 Glucose oxidation in uninfected, SCR, and SIRT4 KD Huh7 cells. Seahorse assay was run with the assay media containing only 25 mM glucose. Data represents the average of six or seven replicates for each cell line and has been normalized to the total protein amount. With the presence of glucose in assay media, ECAR remained at regular level and no significant difference was observed in the three analyzed cell lines. Basal OCR was again increased in SIRT4 KD cells. The increase in basal OCR in SIRT4 KD cells was more significant when comparing to the SCR control ($p = 0.00037$) than to the uninfected Huh7 cells ($p = 0.08$)

3.7 FAO Measurement in Uninfected, SCR, and SIRT4 KD Huh7 Cells

1. On day 1, prepare the seahorse assay and the cells in the same way as described in Subheadings 3.3 and 3.4.
2. On day 2, follow the same workflow as described in Subheading 3.3. However, for FAO measurement, use 1× KHB buffer as the assay media (*see* **Note 10**). In addition, instead of using mitochondrial drugs, the FAO substrate palmitate-BSA and BSA control are used for FAO measurement. Thaw 1 mL aliquot of the palmitate-BSA and BSA control in 37 °C water bath for 5 min when both are used to test FAO procedure. When comparing the FAO between cell lines (Huh7, Huh7-SCR, and Huh7-SIRT4KD, etc.), thaw only one 2 mL aliquot of palmitate-BSA. The thawed palmitate-BSA or BSA control is loaded directly into port A at 75 μL/well without pre-dilution.

Table 2
Sample program for fatty acid oxidation measurements

Command	Time (min)
Calibrate	Fixed
Equilibrate	Fixed
Loop Start	5
Mix	3
Wait	2
Measure	3
Loop End	
Inject from port A	
Loop Start	3
Mix	3
Wait	2
Measure	3
Loop End	
Program End	

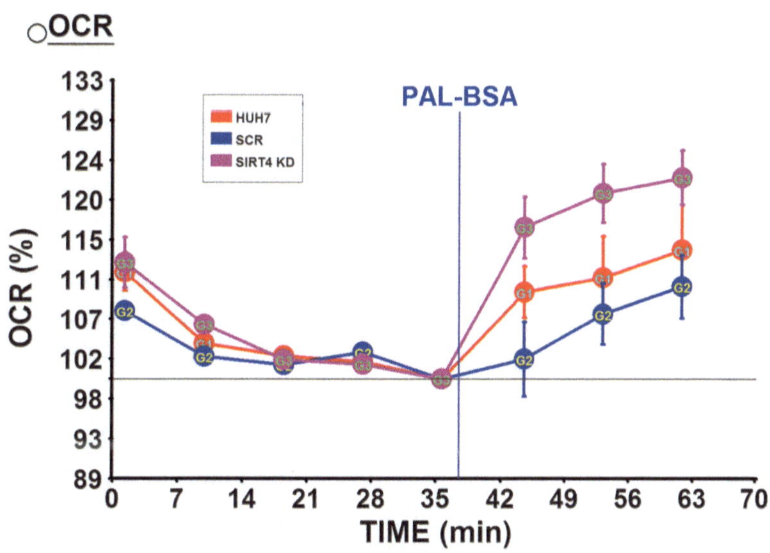

Fig. 5 Disruption of SIRT4 resulted in increased FAO in Huh7 cells. FAO was measured in a seahorse assay by looking at the percentage increase of OCR in response to the injection of palmitate. Data represents the average of five replicates for each cell line

3. Follow the rest of the procedure as described in Subheading 3.3 to run the seahorse assay program. The program used for FAO measurement is as seen in Table 2.
4. Collect and analyze the data (Fig. 5; *see* **Note 25**).

4 Notes

1. The XF24 FluxPak contains 20 XF24 cell culture microplates, 18 XF24 extracellular flux assay kit (each package of the assay kit includes one sensor cartridge, one utility plate, and one lid), and one bottle of 500 mL of XF24 calibrant solution, pH 7.4. The cell culture microplates and calibrant solution (SeahorseBioscience) can be purchased separately. However, the sensor cartridges can only be purchased with the FluxPak.
2. XF Cell Mito Stress Test Kit contains four mitochondrial drugs (Oligomycin, FCCP, Antimycin, and Rotenone) that can be subsequently added to the cells during a seahorse assay to examine the mitochondrial function. Normally, intracellular substrate oxidation produces ATP and causes oxygen consumption, which is predominantly controlled by the parallel re-entry pathways through the ATP synthase and proton leak. Addition of Oligomycin blocks the ATP synthase and the residual respiration is due to the proton leak. The decrease upon adding Oligomycin approximates to the proton current flowing through the ATP synthase before the inhibitor was added. The decrease compared to basal provides the coupling efficiency. Addition of a carefully calibrated concentration of the protonophore FCCP introduces a high artificial proton conductance into the membrane. This maximal respiration is now controlled by electron transport chain activity and/or substrate delivery. The increased respiratory capacity above basal respiration provides the spare respiratory capacity. Finally, electron transport chain inhibitors (Antimycin or Rotenone) are added; any residual respiration is non-mitochondrial. We have found in our seahorse assays, Antimycin and Rotenone can be equally effective as inhibitors of mitochondrial respiration. Only Antimycin was used in the seahorse assays described in this method. Antimycin A, a complex III inhibitor, binds to the Qi site of cytochrome C reductase, thereby inhibiting the oxidation of ubiquinol in the electron transport chain of oxidative phosphorylation. The inhibition of this reaction disrupts the formation of the proton gradient across the inner membrane. Therefore, the production of ATP is subsequently inhibited, as protons are unable to flow through the ATP synthase complex in the absence of a proton gradient.
3. Unbuffered Base DMEM can be purchased directly from SeahorseBioscience, which contains 2 mM GlutaMax. The powdered DMEM from sigma contains only the essential components of DME and gives the flexibility to remove L-glutamine from the media, which will potentially facilitate focusing the tests on the oxidation of substrates other than

L-glutamine. Using the self-prepared media will also dramatically reduce the cost.

When preparing self-made, unbuffered base DMEM, it is helpful to add the Phenol Red after all other components are well dissolved in water because it can be seen better this way.

4. It is optional to add or not to add GlutaMax in the base DMEM medium depending on the assay needs. We have used both media as described in this method.
5. The volume of the assay media is based on using the prep station to change the media in one full (24 wells) XF24 culture plate, because additional volume is required to prime the prep station. If the prep station is not available, prepare 50 mL of the assay media for each full XF24 culture plate and change media manually.
6. 25 mM glucose are used in the assay media to study the basal OCR and ECAR, as well as the mitochondrial function tests because Huh7 cells are normally cultured in media containing high glucose (25 mM). For additional consideration of the assay media, also *see* **Notes 10** and **16**.
7. The pH is temperature-dependent. Because the seahorse measurement is conducted at 37 °C, the assay media needs to be warmed up to 37 °C before adjusting the pH. Filtration of the assay media after the pH adjustment is usually not necessary, as the cells would not typically be put back in culture after the seahorse measurement.
8. Since we are going to test glutamine oxidation, glucose is completely removed from the assay media so that we can force the cells to use glutamine as their substrate. We add L-glutamine in addition to the glutamax in the media to ensure that the cells have enough fuel.
9. Frozen carnitine stock is good for up to 1 month and may be used within 3 days of thawing when kept at 4 °C [16]. We always use freshly made carnitine that has been kept frozen for no more than 2 weeks.
10. In our assays with Huh7 cells, we have found FAO worked the best when no glucose was added in the 1× KHB media. FAO was still measurable when 2.5 mM glucose was present in the assay media, but did not work so well when higher concentration of glucose was used. We think because Huh7 cells have the tendency to use glucose for glycolysis, the presence of glucose will prevent the cells from using other available substrates. Thus the glucose can be completely removed for Huh7 cells when measuring FAO or other substrate utilization to avoid the cells using glucose as the fuel. However, some other cell lines may require the presence of glucose for survival. Therefore, optimization experiments are required for each cell

line to determine the ideal glucose concentration that can be used in the assay media.

11. When the molar ratio of palmitate and BSA is kept at 6:1, we can easily get the clear well-dissolved palmitate/BSA solution. This ratio might be increased a little; however, plamitate will not be fully dissolved in the solution at the ratio of 12:1.
12. For FAO measurement, make sure to use the fatty acid-free BSA here.
13. This can be achieved by placing a larger beaker with water on a heated stir plate and maintaining the "water bath" temperature at 37 °C. The glass beaker containing BSA solution can be placed inside in the larger beaker water bath.
14. In our studies we found that continuous culture of Huh7 cells in puromycin caused a stress on the cells which lead to a decrease in oxidative rate. We recommend using puromycin for selection for 3–5 days, and then remove puromycin during continuous culture of the cells. Since cells are not constantly selected we advise monitoring the levels of infection by mCherry fluorescence, as well as SIRT4 and mCherry mRNA levels.
15. Removing trypsin may not be necessary, but we prefer to do so, especially when the cells are going to be used for the seahorse assay. Because Huh7 cells are strongly adherent cells, the initial incubation with trypsin usually will not disassociate them from the flask. Therefore, trypsin can be easily aspirated off without losing the cells.
16. Cells differ in their growth rate, energy demand, preferred nutrients, and tolerance to stress conditions. Optimization of ideal assay conditions for each cell line is the key ensuring the successful seahorse measurement and obtaining meaningful and reproducible results downstream. Typically, one would optimize the following parameters: assay media components; cell seeding density; the concentration of injected compounds; and mix/wait/measure cycle timing [16]. For seahorse assay media, we recommend starting with the media that has the same components as the growth media for that cell line, except that no FBS is added. The components in FBS can be complex and vary quite a bit from lot to lot. Additionally, FBS has buffering capacity and will interfere with the ECAR reading in the seahorse assay. For cells that require FBS for survival during the seahorse measurement, a small amount (<1 %) of FBS can be considered to use in the assay media. We did not use any FBS in our assays as Huh7 cells remain healthy in the media without it.

    Besides testing the assay media, a couple of optimization experiments were also done with the focus on determining the ideal seeding density using Huh7 cells in seahorse assays and the working concentration of the mitochondrial drugs that is not toxic to the cells but efficient to stimulate cellular response and allows for looking at the mitochondrial function. In the meantime, we made sure that DMSO (used as solvent for mitochondrial drugs) had no effect on our cells at the concentration we aimed to use. We did not include many replicates in the early optimization experiments. Instead, we set up those optimization experiments in a way that allow us to look for trends and gather information and indication for the later real experiment. Limited to the length of this article, we do not describe details of the setup of the optimization experiments.

17. The cells were prepared this way for all described seahorse experiments in this article. Always follow the two-step seeding process as recommended in the seahorse assay manual to ensure a momogenous cell layer. Since Huh7 cells are strongly adherent, they become well attached after 1 h. For cells that are not strongly adherent, plan 3–5 h of time to allow the cells fully attached before adding additional media.

18. Use the same program in the prep station to change media in each experiment for consistency. Basically, growth media were aspirated off first. 675 μL/well of assay media were then dispensed to wash the cells once. After the assay media was aspirated, a final of 675 μL of assay media were added in each well. The media change can be done manually by following the same steps. When manually aspirating the media off, make sure to leave a small amount (50 μL) of media behind in each well to avoid detaching the cells when media is added back. The volume of assay media per well can vary from 500 μL to 1 mL. We use 675 μL/well in our assays described here.

19. Calibration is the first step in the seahorse assay program checking on the sensors in calibrant buffer. This step takes about 30 min to complete. The measurement on the cells plate can only be started after the calibration is finished. Therefore, the seahorse assay program should be started no more than 30 min after media change in the cells to avoid extended incubation of the cells in the assay media. As mentioned earlier, seahorse assay media is unbuffered media usually containing a limited amount of nutrients and no FBS, therefore extended incubation of the cells in the assay media may result in depletion of nutrients or a dramatic change in media pH, which are unfavorable conditions for the cells to function normally. Therefore, the subsequent seahorse measurements will not be reliable, reproducible, and informative.

    On the other hand, the calibration should not be done too early either because the data accuracy may be compromised. Ideally, each experiment should follow the same time frame (±5 min), which will generate consistent results and allow for data comparison across the experiments. Good planning prior to a seahorse assay can usually ensure a smooth workflow. Also, having the seahorse assay program preloaded onto the computer before starting the media change in the cells is usually helpful to keep up with the time.

20. From the optimization experiments, we learned Oligomycin, FCCP, and Antimycin worked at 1, 0.4, and 1 μM, respectively when seeding Huh7 cells at 60,000/well. We therefore diluted Oligomycin, FCCP, and Antimycin in the assay media to 10, 4, and 10 μM, respectively, which were ready to be loaded in the ports of the sensor cartridge. Because the stock concentration of these drugs is 2.5 mM, dilution of each drug in the assay media is more than 2,500 fold in our assay. At this dilution, DMSO has been shown no effect on the performance of Huh7 cell. Additionally, when the dilution fold is this high, we do not have to worry about the pH change in the assay media that is possibly introduced by the drug solution. Therefore, it is important to make a stock compound or test agent at a high concentration when it is possible.
21. The volume load from each port should be 1/10 of the final volume in each well in the XF culture plate. That is why we have our compounds diluted to 10 times of their final working concentration. In our assay, we load 675 μL/well assay media in the XF culture plate, we therefore load 75 μL of 10 μM Oligomycin in port A, 83 μL of 4 μM FCCP in port B, and 93 μL of 10 μM Antimycin in port C, respectively. All wells are loaded with prepared compound, including the background correction wells.
22. At the end of the assay, assay media were aspirated off and 200 μL of protein lysis buffer were added into each well to lyse the cells. Collect the supernatant of the cell lysis and perform BCA assay to quantitate protein. The results shown in Fig. 2 through Fig. 4 were normalized to the protein amount. Figure 2 indicated that the disruption of SIRT4 increased the substrate oxidation (OCR) but not glycolysis (ECAR) at basal level in Huh7 cells. The perturbed OCR and ECAR in response to mitochondrial compounds did not vary in SIRT4 KD cells and the control cell lines. In our experiments, we have also noticed that the increased basal OCR in SIRT4 KD cells was consistently significant when comparing to the scramble control, but was not always significant when comparing to the uninfected Huh7 cells. We attributed the discrepancy in basal OCR between the scramble control and uninfected cell control to the viral infection and puromycin selection in scramble cells. The OCR difference between the SCR and uninfected cells was never significant. We think the scramble cells are more reasonable than the uninfected cells to serve as

a control in our experiments. However, we included uninfected Huh7 cells in all our assays for complete information.

23. Since we saw elevated substrate oxidation in SIRT4 KD cells from Fig. 2, we designed seahorse assays to study specific substrate oxidation. Figure 3 showed results from an experiment that only glutamine was used in the base DMEM assay media. The results indicated that glutamine oxidation was slightly increased in SIRT4 KD cells. Additionally, in the media condition without presence of glucose, glycolysis (ECAR) level is at minimal level in all cell lines.
24. Figure 4 showed that in the assay media containing only glucose, glycolysis remained at similar level between the SIRT4 KD cells and the control cells; however, basal OCR was again higher in SIRT4 KD cells. Together with the results in Figs. 2 and 3, SIRT4 KD Huh7 cells showed increased mitochondrial respiration. SIRT4 gene knockdown did not seem to have any impact on glycolysis level in Huh7 cells.
25. Because we inject the substrate of FAO through the port after the basal measurement, we look at the OCR increase percentage above the basal to compare the FAO level in each cell line. This can be achieved by normalizing the data to the last measurement at basal level as shown in Fig. 5. Previously, a report showed that shRNA-SIRT4 adenovirus-mediated Sirt4 KD in primary mouse myotubes and hepatocytes increased FAO, which was measured with [3H] palmitate [14]. Our data reproduced this finding in human Huh7 cells with the label-free seahorse assay.

Acknowledgment

We would like to acknowledge the American Heart Association grants 12SDG8840004 and 12IRG9010008 for funding support (MDH).

References

1. Clark LC Jr, Kaplan S, Matthews EC, Edwards FK, Helmsworth JA (1958) Monitor and control of blood oxygen tension and pH during total body perfusion. J Thorac Surg 36(4): 488–496
2. Ferrick DA, Neilson A, Beeson C (2008) Advances in measuring cellular bioenergetics using extracellular flux. Drug Discov Today 13(5–6):268–274. doi:10.1016/j.drudis. 2007.12.008
3. Hirschey Matthew D (2011) Old enzymes, new tricks: sirtuins are NAD+-dependent de-acylases. Cell Metab 14(6):718–719. doi:10.1016/j.cmet.2011.10.006
4. Anderson KA, Hirschey MD (2012) Mitochondrial protein acetylation regulates metabolism. Essays Biochem 52:23–35. doi:10.1042/bse0520023
5. Hirschey MD, Shimazu T, Goetzman E, Jing E, Schwer B, Lombard DB, Grueter CA, Harris C, Biddinger S, Ilkayeva OR, Stevens RD, Li Y, Saha AK, Ruderman NB, Bain JR, Newgard CB, Farese RV Jr, Alt FW, Kahn CR, Verdin E (2010) SIRT3 regulates mitochon-

drial fatty-acid oxidation by reversible enzyme deacetylation. Nature 464(7285):121–125. doi:10.1038/nature08778
6. Hirschey MD, Shimazu T, Jing E, Grueter CA, Collins AM, Aouizerat B, Stancakova A, Goetzman E, Lam MM, Schwer B, Stevens RD, Muehlbauer MJ, Kakar S, Bass NM, Kuusisto J, Laakso M, Alt FW, Newgard CB, Farese RV Jr, Kahn CR, Verdin E (2011) SIRT3 deficiency and mitochondrial protein hyperacetylation accelerate the development of the metabolic syndrome. Mol Cell 44(2):177–190. doi:10.1016/j.molcel.2011.07.019
7. Picklo MJ Sr (2008) Ethanol intoxication increases hepatic N-lysyl protein acetylation. Biochem Biophys Res Commun 376(3):615–619. doi:10.1016/j.bbrc.2008.09.039
8. Qiu X, Brown K, Hirschey MD, Verdin E, Chen D (2010) Calorie restriction reduces oxidative stress by SIRT3-mediated SOD2 activation. Cell Metab 12(6):662–667. doi:10.1016/j.cmet.2010.11.015
9. Shimazu T, Hirschey MD, Hua L, Dittenhafer-Reed KE, Schwer B, Lombard DB, Li Y, Bunkenborg J, Alt FW, Denu JM, Jacobson MP, Verdin E (2010) SIRT3 deacetylates mitochondrial 3-hydroxy-3-methylglutaryl CoA synthase 2 and regulates ketone body production. Cell Metab 12(6):654–661. doi:10.1016/j.cmet.2010.11.003
10. Du J, Zhou Y, Su X, Yu JJ, Khan S, Jiang H, Kim J, Woo J, Kim JH, Choi BH, He B, Chen W, Zhang S, Cerione RA, Auwerx J, Hao Q, Lin H (2011) Sirt5 is a NAD-dependent protein lysine demalonylase and desuccinylase. Science 334(6057):806–809. doi:10.1126/science.1207861
11. Peng C, Lu Z, Xie Z, Cheng Z, Chen Y, Tan M, Luo H, Zhang Y, He W, Yang K, Zwaans BMM, Tishkoff D, Ho L, Lombard D, He T-C, Dai J, Verdin E, Ye Y, Zhao Y (2011) The first identification of lysine malonylation substrates and its regulatory enzyme. Mol Cell Proteomics 10(12). doi:10.1074/mcp.M111.012658
12. Nakagawa T, Lomb DJ, Haigis MC, Guarente L (2009) SIRT5 deacetylates carbamoyl phosphate synthetase 1 and regulates the urea cycle. Cell 137(3):560–570. doi:10.1016/j.cell.2009.02.026
13. Haigis MC, Mostoslavsky R, Haigis KM, Fahie K, Christodoulou DC, Murphy AJ, Valenzuela DM, Yancopoulos GD, Karow M, Blander G, Wolberger C, Prolla TA, Weindruch R, Alt FW, Guarente L (2006) SIRT4 inhibits glutamate dehydrogenase and opposes the effects of calorie restriction in pancreatic beta cells. Cell 126(5):941–954
14. Nasrin N, Wu X, Fortier E, Feng Y, Bare OC, Chen S, Ren X, Wu Z, Streeper RS, Bordone L (2010) SIRT4 regulates fatty acid oxidation and mitochondrial gene expression in liver and muscle cells. J Biol Chem 285(42):31995–32002. doi:10.1074/jbc.M110.124164
15. Ventura A, Meissner A, Dillon CP, McManus M, Sharp PA, Van Parijs L, Jaenisch R, Jacks T (2004) Cre-lox-regulated conditional RNA interference from transgenes. Proc Natl Acad Sci U S A 101(28):10380–10385. doi:10.1073/pnas.0403954101
16. SeahorseBioscience (2009) XF24 extracellular flux analyzer and prep station installation and operation manual. Seahorse Bioscience Inc 1–159

Chapter 17

The Emerging Links Between Sirtuins and Autophagy

In Hye Lee, Jeanho Yun, and Toren Finkel

Abstract

Evidence suggests a role for acetylation and deacetylation in regulating autophagy. In this chapter, we describe the methods useful for understanding this important connection. In particular, we discuss methods for the measurements of sirtuin deacetylase activity, in vivo acetylation detection, and the common assays used to monitor both autophagy and the more selective process of mitophagy.

Key words Autophagy, p62, LC3, Autophagosome, Atg, Keima

1 Introduction

Macroautophagy, herein termed autophagy, is a highly regulated process that allows damaged proteins and organelles in the cytosol to be engulfed within specialized double membrane structures termed autophagosomes [1]. These autophagosomes are then delivered to lysosomes, where the contents of the autophagosomes can be degraded. This degradation provides several important functions. First, it removes damaged and potentially harmful proteins and organelles from the cell. Secondly, it provides biosynthetic materials for ongoing cellular needs. The latter function is presumably why autophagy is strongly induced under starved conditions. In addition, even under basal conditions, autophagy is needed to maintain cellular homeostasis. Indeed, in the absence of autophagy tissues accumulate a variety of damaged organelles including bizarre and abnormally functioning mitochondria that appear to result in conditions that mimic age-related pathologies [2, 3].

As mention, starvation is a condition known to stimulate autophagic flux. Similarly, starvation or caloric restriction is known to stimulate sirtuin activity [4, 5]. This link led us to hypothesize that there might be a connection between sirtuins and autophagy. Our data suggests that Sirt1, a mammalian NAD+-dependent deacetylase, interacts with autophagy-related proteins (Atgs) and regulates acetylation status of several essential Atg proteins

Matthew D. Hirschey (ed.), *Sirtuins: Methods and Protocols*, Methods in Molecular Biology, vol. 1077,
DOI 10.1007/978-1-62703-637-5_17, © Springer Science+Business Media, LLC 2013

including Atg5, Atg7, Atg8, and Atg12 [6]. This regulation was evident both in vitro and in vivo. Our data suggests that increasing Sirt1 activity stimulates autophagic flux, while removing Sirt1 inhibits autophagy. Subsequent studies revealed that the p300 acetyltransferase acted in counterbalancing fashion to regulate autophagy [7]. Other laboratories have substantially added to these observations suggesting a complex albeit incompletely understood role for acetylation and deacetylation in the regulation of autophagy [8, 9]. Nonetheless, given the increasing interest in both sirtuins and autophagy as therapeutic targets for aging and age-related pathologies, the link between these two biological pathways will undoubtedly receive additional focus in the future. Indeed, pharmaceutical approaches that target this connection are already emerging [10]. Here we provide a detailed guide to allow interested investigators the tools to assess the acetylation status of Atg proteins, measurements of autophagic flux, and newer strategies to measure the selective removal of mitochondria, a process known as mitophagy.

2 Materials

2.1 Assays to Assess In Vitro Deaceytylase Activity

1. Lysis Buffer: 50 mM Tris (pH 7.4), 1 % Triton X-100, 0.5 % Nonidet P-40, 150 mM NaCl, protease inhibitor cocktail, phosphatase inhibitor mixture, and 10 % (vol/vol) glycerol.
2. pCS-Flag-Sirt1 WT and HY plasmids from Addgene, pcDNA4-HisMax-Atg7, and Atg8 plasmids.
3. Effectene and Ni–nitrilotriacetic acid–agarose beads.
4. Anti-FLAG M2 affinity gel.
5. FLAG peptide, in Elution Buffer (50 mM Tris–HCl, 150 mM NaCl, pH 7.4).
6. Assay buffer: 50 mM Tris–HCl, 1 mM $MgCl_2$, and 150 mM NaCl (pH 8.0).
7. NAD.
8. Anti acetyl-lysine antibody from Cell Signaling (#9441; Cell Signaling Technology Inc., Danvers, MA).
9. Standard cell culture growth medium.

2.2 In Vivo Acetylation/Deacetylation of Atg Proteins

1. Sonicator XL 2020.
2. Lysis Buffer (see above).
3. Protein G-Sepharose.
4. Effectene.
5. Fugene 6.

6. Antibodies against acetyl-lysine, myc (SC40; SantaCrutz), atg7 (A2856; Sigma, XW-7984; ProSci Inc., Poway, CA, and #04-1055; Millipore, Billerica, MA).

2.3 Autophagy Assays

2.3.1 LC3 I/II Conversion

1. Pepstatin A and E64d from Sigma.
2. Lysis Buffer.
3. Antibody against LC3 (L8918; Sigma, NB100-2331; NOVUS Biologicals, Littleton, CO).

2.3.2 GFP-LC3 Assays

1. GFP-LC3 plasmid (Addgene Inc., Cambridge, CA).
2. Effectene (Qiagen).
3. Lab-Tek Chambered Coverglass w/cover.
4. Fixation Buffer: 4 % Paraformaldehyde in PBS.
5. Confocal laser-scanning microscope.

2.3.3 p62 Assay

1. Lysis Buffer.
2. Antibody for p62 (GP62; American Research Products, Inc., Belmont, MA).

2.3.4 Mitophagy Analysis Using Mitochondria-Targeted Keima (mtKeima)

1. Hela cells were obtained from the American Type Culture Collection (ATCC, Manassas, VA).
2. Lentivirus encoding mitochondria-targeted Keima (mtKeima).
3. Chambered Lab-Tek cover glass system (#1.0 German borosilicates; Nalge Nunc International).
4. Modular Incubator Chamber.
5. 500 mM 3-methyladenine, dissolved in water and stored at −20 °C.
6. 100 μM bafilomycin A1, dissolved in DMSO and stored at −20 °C.
7. Becton-Dickinson LSR II flow cytometer.
8. Zeiss LSM 510-UV laser-scanning confocal microscope.

3 Methods

3.1 In Vitro Deacetylase Assays

3.1.1 Transfection of HeLa Cells

1. Add 4 μg of DNAs (Sirt1 or His-Atgs) with 60 μL Effectene using the manufacturer's protocol of DNA transfection.
2. Wash cells in a 10 cm dish with 1× PBS and then add 7 mL of standard growth medium.
3. Add DNA–Effectene complexes diluted in 3 mL of complete media.
4. Incubate 37 °C for 24 h.

3.1.2 Cell Harvesting

1. Wash transfected plate once with 1× PBS.
2. Add 5 mL of PBS and scrape cells into a 15 mL conical tube.
3. Centrifuge cells at 160 × *g* for 5 min.
4. Remove PBS and add 1 mL of Lysis buffer.
5. Transfer resuspended cells to a 1.5 mL tube and incubate at 4 °C for 30 min (*see* **Notes 1** and **2**).
6. Briefly sonicate cell suspension (amplitude-#4, 20 % amplitude, 5–10 s).
7. Centrifuge lysates at 15,000 × *g* (4 °C for 10 min).
8. Collect supernatant.
9. BCA assay (at RT for 30 min) for measurement of protein concentration.

3.1.3 Purification of Sirt1 Enzymes

1. Bind 2 mg of Flag-tag Sirt1 transfected protein lysates with 40 μL of anti-Flag affinity gel (50 % v/v slurry of resin).
2. Incubate complexes at 4 °C for 1 h (*see* **Note 3**).
3. Wash beads containing Flag-tagged Sirt1 protein three times with Lysis Buffer.
4. Wash beads twice with Elution Buffer.
5. Elute Flag-Sirt1 protein using 100 μL of Elution Buffer including 2 μL of FLAG peptide (5 mg/mL of Stock) by gently shaking at 4 °C for 30 min (*see* **Notes 4–6**).
6. Spin down beads and collect supernatant.
7. BCA assay for eluted Flag-Sirt1 protein.

3.1.4 Purification of His-Tagged Atg Proteins

1. In a 1.5 mL tube, incubate 2 mg of protein cell lysate obtained from cells previously transfected with a His-Atg construct with 40 μL of Ni-resin (50 % v/v slurry) overnight at 4 °C.
2. Spin down briefly at 9,000 × *g* for 30 s.
3. Remove supernatant and wash resin three times with Lysis buffer (*see* **Note 7**).
4. Wash resin two times with Assay buffer.
5. Spin down and remove any residual supernatant.

3.1.5 Deacetylation Assay (See Note 8)

1. Prepare reactions by aliquoting purified Atg protein into a fresh 1.5 mL tube with or without approximately 300 ng of purified Sirt1 proteins with or without the addition of 5 μL of NAD (*see* **Note 9**) to obtain a final volume of 50 μL.
2. Incubate samples at 37 °C for 2 h (*see* **Note 10**).
3. Spin down and remove supernatant.

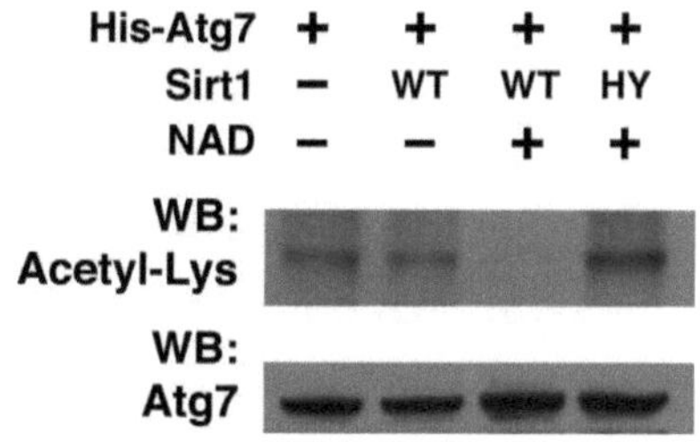

Fig. 1 In vitro Sirt1 deacetylation reactions with purified acetylated His-Atg7 as a substrate in the presence or absence of 10 mM NAD and purified Wild-type (WT) or deacetylase-inactive (HY) mutant Sirt1 protein (reprinted with permission from Lee et al. [6])

4. Add 35 μL of 2× protein sample buffer and heat at 95 °C for 5 min to stop the reaction.
5. Analyze samples using standard Western blotting procedures and detection with rabbit Acetyl-Lysine antibody (Fig. 1).

3.2 In Vivo Deacetylation/Acetylation of Atg Proteins (See Notes 11–16)

3.2.1 Transfection of Cells

1. For HeLa cells, dilute 4 μg of DNA (2 μg of myc-Atgs and 2 μg of empty vector or HA-Sirt1) with 40 μL Effectene reagent as previously described in 3.1 A.
2. For primary mouse embryonic fibroblast (MEF) transfection, dilute 18 μL of Fugene6 in 500 μL of serum-free medium (*see* **Note 17**) and incubate for 5 min at RT.
3. Add 6 μg of the desired myc-tagged Atg construct and incubate for 20 min at RT.
4. Wash MEF cells with 1× PBS and add 9 mL of complete growth media.
5. Add 1 mL of complete media into the tube containing the DNA-Fugene6 complex.
6. Add the above mixture to the MEF cell dish and incubate for 24 h at 37 °C.

3.2.2 Preparation of Mouse Tissues

1. Mince small pieces of fresh or frozen tissues (2–3 mm^3) with a sharp scalpel (*see* **Notes 18** and **19**).
2. To these finely minced tissues, add 1 mL of Lysis buffer and incubate at 4 °C for 30 min (*see* **Note 20**).
3. Sonicate in a 1.5 mL tube (amplitude-#4, 20 % amplitude, 10 s, 3–7 times) (*see* **Note 21**).
4. Centrifuge lysates at 13,000 rpm (4 °C for 20 min).
5. Collect supernatant and perform BCA assay for measurement of protein concentration.

3.2.3 Immunoprecipitation

1. Mix protein lysates (1–2 mg) with 10 μL of an acetyl-lysine antibody overnight at 4 °C.
2. Add 60 μL of protein G-Sepharose and incubate at 4 °C for an additional 2 h.
3. Spin down and remove supernatant.
4. Wash beads 3–5 times with Lysis buffer (*see* **Note 22**).
5. Add 35 μL of 2× sample buffer to the beads.
6. Boil samples at 95 °C for 5 min and load samples for Western blotting (Fig. 2) to detect protein of interest (*see* **Note 23**).

3.3 Autophagy Assay-Atgs

3.3.1 LC3 Assay (LC3 I/II Conversion)

1. Two hours prior to harvest of cells, add the lysosomal protease inhibitor pepstatin A and E-64d both at 10 μg/mL (*see* **Note 24**).
2. Harvest cells with previous protocol and resolve proteins on a 4–20 % gradient SDS-PAGE.
3. Transfer gel to solid membrane and detect LC3-I and LC3-II by western blotting (Fig. 3).

3.3.2 GFP-LC3 Assay: Transfection of GFP-LC3 in MEF Cells

1. Add 0.6 μg of the GFP-LC3 plasmid with 6 μL of Effectene in a total volume of 75 μL to each chamber of a two chamber Lab-Tek slide. Transfection of MEFs follows the manufacturer's protocol (*see* **Note 25**).
2. Incubate at 37 °C for 24 h.
3. Wash MEFs three times with PBS.
4. Add 500 μL of Fixation solution for 10 min.
5. Wash cells three times with PBS.
6. Observe cells using Confocal Microscopy (LSM510 Meta, Zeiss) (*see* **Notes 26–28**, Fig. 4).

3.3.3 p62 Assay (See Notes 29 and 30)

1. Prepare protein lysates from cells or tissues using Lysis buffer and sonication as described in Subheadings 3.1.2 and 3.2.2.
2. Analyze protein lysate by standard Western blotting using an antibody directed against p62. Clearance of p62 is dependent on autophagy and increased levels are consistent with impaired autophagic flux (*see* Fig. 5).

3.3.4 Mitophagy Analysis Using Mitochondria-Targeted Keima (mtKeima)

Mitophagy is the presumed selective removal of damaged or dysfunctional mitochondria from the cell by using the autophagic machinery [11]. The ability to monitor this process has recently been greatly aided by the development of a novel fluorescent reporter termed mitochondria-targeted Keima or mtKeima [12]. Keima is a coral fluorescent protein that is resistant to lysosomal degradation. By using the large shift in the peak excitation spectrum of Keima protein from 440 to 586 nm corresponding to neutral and acidic pH, mitophagy can be specifically assessed.

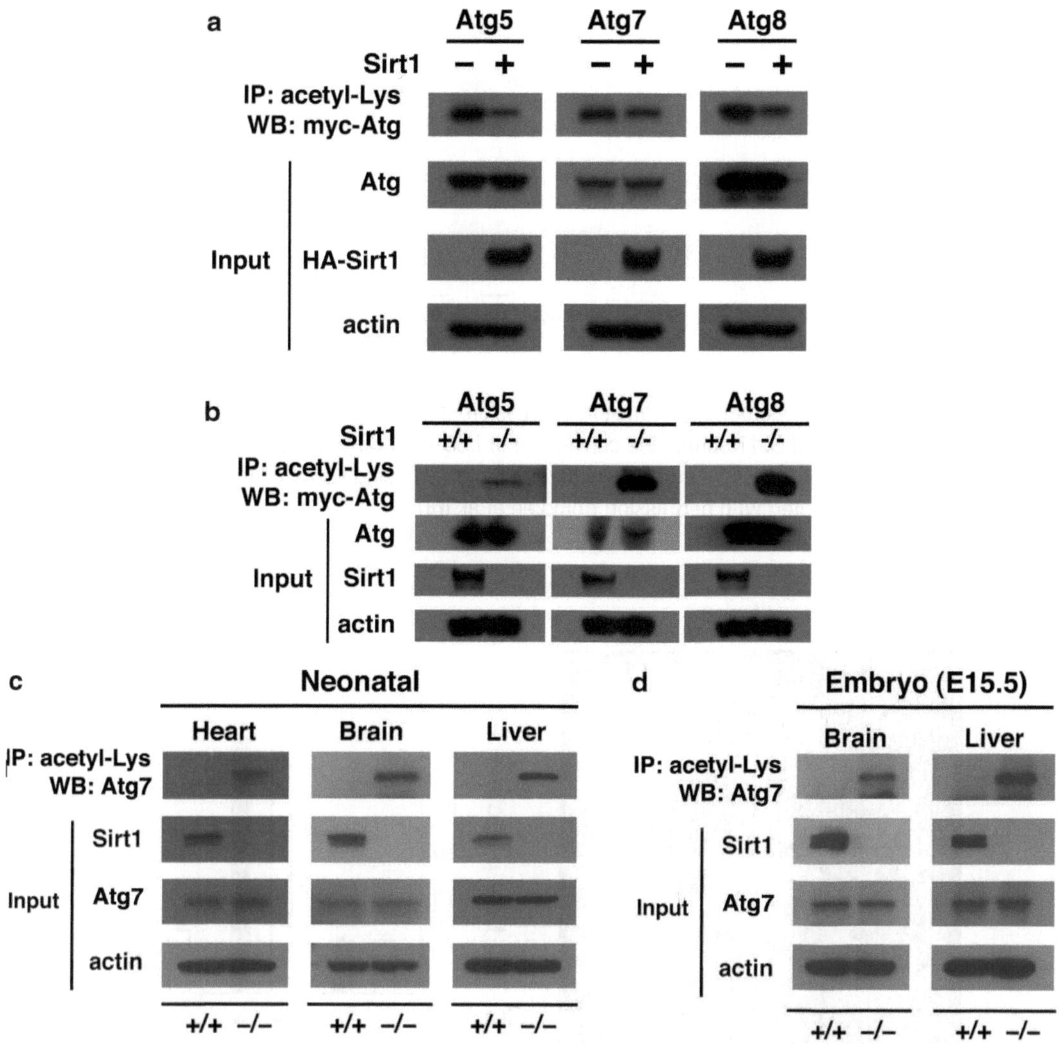

Fig. 2 Sirt1 regulates acetylation of autophagy gene products. (**a**) Transient-increased expression of Sirt1 reduces acetylation. HeLa cells were transfected with the indicated epitope-tagged Atg construct along with, where indicated, wild-type Sirt1 –/–. Measurements were done under fed conditions. *IP* immunoprecipitation, *WB* Western blotting. (**b**) Acetylation of autophagy proteins in wild-type or Sirt1–/– MEFs under fed conditions. Absence of Sirt1 dramatically increased acetylation. (**c**) Levels of endogenous Atg7 acetylation in various wild-type (+/+) and Sirt1–/– neonatal tissue. Tissues were harvested several hours after birth (reprinted with permission from Lee et al. [6]). (**d**) Similar analysis in E15.5 embryos

In particular, cells undergoing high levels of mitophagy have an increase in the ratio of the 586/440 excitation assessed by FACS analysis. Mitophagic cells expressing mtKeima also exhibit bright punctate structure with a high 586/440 ratio by confocal analysis. Hypoxia-induced mitophagy of Hela cells can be analyzed using the following procedure (*see* **Notes 31** and **32**).

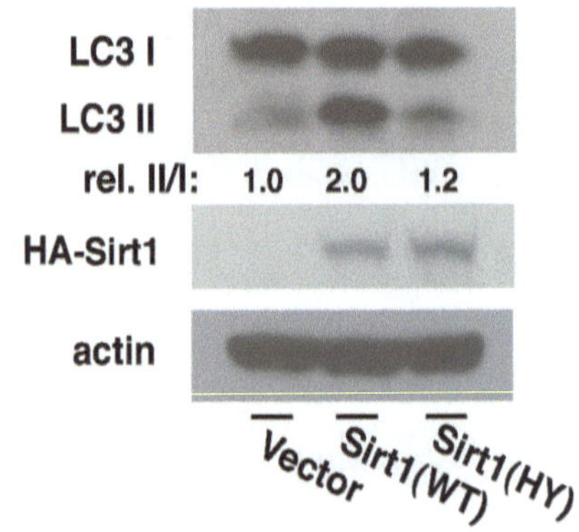

Fig. 3 Sirt1 is necessary for autophagy. Transient expression of wild-type (WT) Sirt1 but not a deacetylase-inactive (HY) point mutant of Sirt1 stimulates conversion of LC3-I to LC3-II in HCT116 cells. Quantification of the relative (rel.) levels of LC3-II/LC3-I is shown with the ratio of 1.0 being assigned to vector-transfected cells under fed conditions (reprinted with permission from Lee et al. [6])

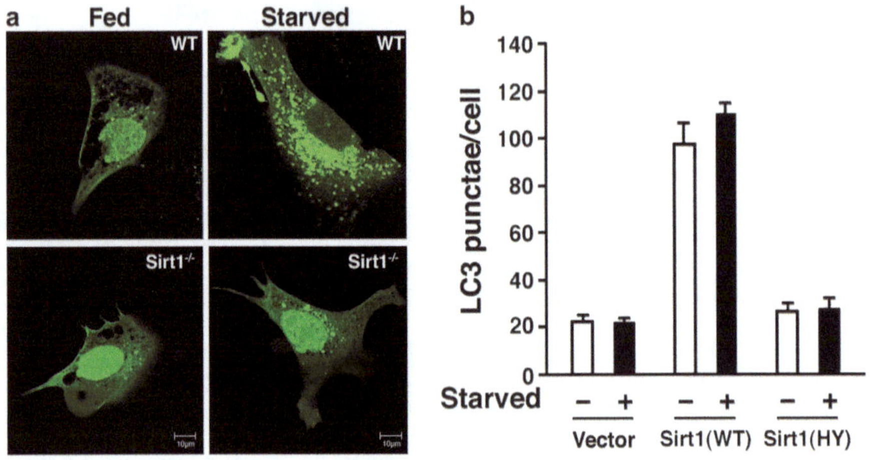

Fig. 4 Role of Sirt1 in autophagy. (**a**) Reduced autophagy in Sirt1−/− MEFs. High power view of wild-type or Sirt1−/− MEFs transfected with GFP-LC3 under fed or starved conditions. In wild-type (wt) cells starvation induces an increase in GFP-LC3 punctae. (**b**) Quantification of GFP-LC3 punctae in wild-type (+/+) and Sirt1−/− MEFs under fed or starved conditions (reprinted with permission from Lee et al. [6])

1. Plate Hela cells in a 6-well culture plate. After 1 day in culture, cells can be infected with a lentivirus encoding mtKeima for 24 h. Infected cells are selected with 2 μg/mL of puromycin for 2 days and surviving cells are used for further experiments.
2. Hela cells expressing mtKeima are plated in complete DMEM in a Lab-tek cover glass four well chamber slide. For FACS analysis, cells were plated in 60-mm culture dish.
3. For hypoxic treatment, Hela cells are transferred to a MIC-101 modular incubator chamber and were flushed with 1 % O_2, 5 % CO_2, 94 % N_2. Cells were placed in a 37 °C incubator

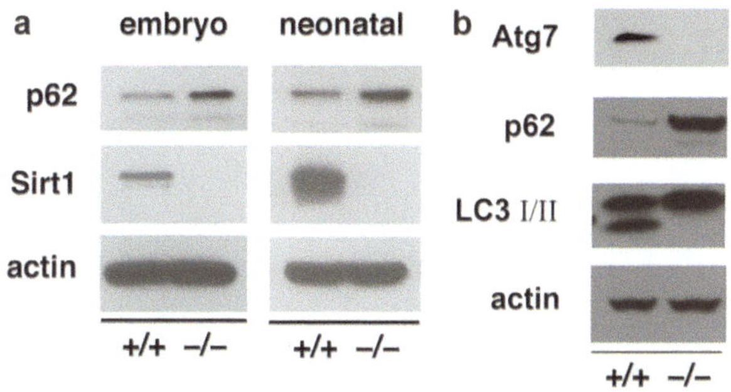

Fig. 5 Accumulation of p62 as a marker of autophagic flux. (**a**) Accumulation of the autophagy marker p62 in the myocardium of embryonic and neonatal Sirt1–/– tissue. Neonatal tissue was harvested in the first 3 h after birth (reprinted with permission from Lee et al. [6]). (**b**) Analysis of wild-type and Atg7-deficient MEFs. Levels of p62 are increased and levels of LC3-II are decreased in Atg7–/– MEFs. These two conditions allow one to conclude that there is a defect in autophagy

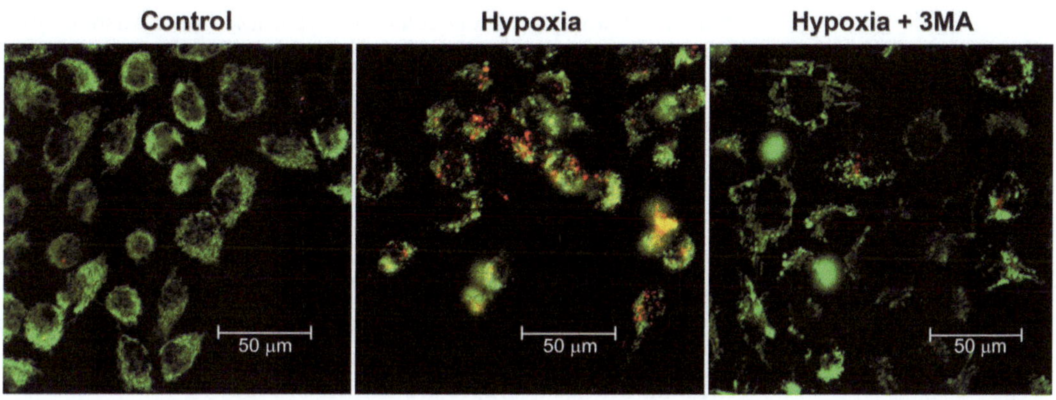

Fig. 6 Detection of mitophagy using mtKeima fluorescence. Hela cells expressing mtKeima were exposed to hypoxic conditions for 24 h with or without the addition of 5 mM of 3-methyladenine (3MA). In this confocal micrograph, mtKeima fluorescence signal from 561 nm (acidic) is *red* and the signal from 458 nm (neutral) is *green*. When mitophagy is induced, *red* punctate structures representing mitochondria contained within the acidic lysosome appear. The appearance of these *red* punctate structures can be inhibited by 3MA treatment

for the next 24 h. In order to pharmacologically inhibit mitophagy, 5 mM of 3-methyladenine or 100 nM of bafilomycin A1 can be added to the culture medium.

4. After a 24 h incubation, Hela cells in chambered Lab-tek cover glasses are examined by confocal microscopy. We use a Zeiss LSM 510-UV laser-scanning confocal microscope with 20× magnification and 458 and 561 nm excitation lasers to visualize the fluorescent signal from mtKeima in neutral and acidic pH, respectively. Images are acquired with Zeiss Zen 2009 software. Representative images of hypoxia-induced mitophagy are shown in Fig. 6. The excitation ratio at 561/458 nm of each images is calculated with Metamorph 7.6 software (Universal Imaging Corporation, West Chester, PA) using the region measurement function.

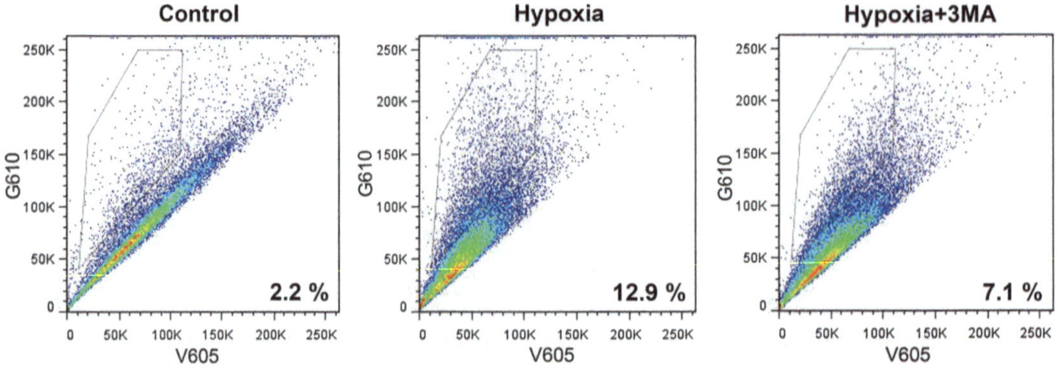

Fig. 7 Representative FACS analysis of hypoxia-induced mitophagy. Hela cells expressing mtKeima were exposed to hypoxic conditions for 24 h with or without the addition of 5 mM 3-methyladenine (3MA) and analyzed by FACS using V605 and G610 detectors. Cells with high G610/V605 ratio were selected (*boxed area*) and their proportion was calculated with FlowJo software

5. For FACS analysis, Hela cells are trypsinized and washed once with DPBS. Cells are resuspended into 1 mL of DPBS and analyzed with BD LSR II flow cytometer (Bioscience, San Jose, CA) equipped with 407, 488, 532, and 633 nm LASER lines using DIVA 6.1.2 software (BD). Cells are excited with a violet laser (407 nm) with emission detected by 605 ± 20 nm (V605) detector and with a green laser (532 nm) with emission detected by 610 ± 10 nm (G610) detector simultaneously. Ratio of emission at G610/V605 of whole sample or designated population is calculated using FlowJo software (FlowJo 7.6.5; Tree Star Inc., Ashland, OR). Representative data for hypoxia-induced sample are shown in Fig. 7.

4 Notes

1. Use a rotator in the cold room following cell lysis to prevent degradation of proteins.
2. Prior to adding Lysis buffer to cells, supplement the Lysis buffer with the deacetylase inhibitors TSA (1 μM) and nicotinamide (10 mM) along with 5 mM imidazole.
3. For maximum binding, incubate mixture at 4 °C overnight.
4. To prepare the stock solution, dissolve lyophilized Flag peptide in Elution Buffer to a final concentration of 5 mg/mL (50×).
5. The working concentration is 100–150 μg/mL peptide in order to elute FLAG fusion proteins from the ANTI-FLAG-M2 affinity resins.
6. Five to one column volumes of the working solutions are sufficient to elute most FLAG fusion proteins.
7. When you wash beads, add 20 mM imidazole.

8. Use fresh assay buffer, enzymes, and substrates.
9. Make 100 mM of stock of NAD with distilled water and filter it with 0.22 μm filter (Millipore).
10. Equal amounts of acetylated substrate were added to each tube and samples were subsequently shaken gently during the reaction.
11. To check acetylation status of individual Atg proteins we routinely used 1–2 mg of proteins. Poorly expressed Atg proteins can require up to 6 mg of initial lysate.
12. Rabbit antibody is more sensitive than the mouse antibody in detecting acetylated proteins.
13. Both expression and acetylation of tagged-Atgs can be variable.
14. The detection of acetylated protein often varies due to lot-to-lot variation of the polyclonal acetyl-lysine antibody.
15. Acetylation of proteins is sometime sensitive to experimental conditions including cell confluency, passage number, or overall health of the cells.
16. When harvesting cells or tissues, use Lysis buffer containing Triton X-100 and NP-40 followed by sonication.
17. Serum-free media should not have antibiotics, which may interfere with complex formation between Fugene 6 and DNA.
18. When you prepare tissue, begin with samples that are approximately 2–3 mm^3, roughly the size of ½ a grain of rice.
19. Before sonication of tissues, chop tissues into the smaller size possible.
20. While incubating in Lysis buffer, use a rotator in the cold room to prevent degradation of proteins.
21. Sonicate cell or tissue lysates until you cannot see aggregates. Over-sonication can cause degradation of proteins.
22. Please remove supernatant as much as you can in the last wash.
23. The information for one of the Atg7 antibodies and Western blotting are followed.
    A. Cat #: A2856.
    B. Blocking sol: 5 % Milk (Incubate membrane at RT for 30 min).
    C. First antibody: 5 % BSA (Dilution ratio: 1:1,000, Incubate membrane at 4 °C overnight).
    D. Second antibody: 5 % Milk (second rabbit antibody from Santa Cruz, dilution factor: 1:5,000–10,000, Incubate membrane with antibodies at RT for 1 h).
    E. Atg7 is a relatively abundant protein, so it can be detected using roughly 20 μg of initial protein lysate.

24. In some cases such as HeLa and HCT116 cells, a 30 min incubation of these lysosomal protease inhibitor including pepstatin and E64d is sufficient.
25. The general transfection efficiency of MEFs is less than 10 %. There is significant variation between different commercially available transfection reagents. We have been successful using Effectene (Qiagen), Fugene 6, Fugene 6 HD (Roche), TurboFect (Fermentas), or Lipofectamine 2000 (Invitrogen).
26. When you observe GFP-LC3 dots in cells using the confocal microscope, a 62× objective lens is recommended.
27. A minimum of 5–10 random fields should be taken for each sample.
28. A minimum of 300 cells per sample were counted using quadruplicate samples per construct per experiment. GFP-LC3 dot was considered to be a structure greater than approximately 1 μm in diameter. A positive cell was defined as having more than three GFP-LC3 dots.
29. When you detect p62 level in cell lysates usually around 25–50 μg is sufficient.
30. Block the membrane with 5 % milk at RT for 30 min, then incubate it with p62 antibody in 5 % BSA (Dilution ratio: 1:1,000–3,000) at 4 °C overnight. After washing membrane, incubate it with a secondary antibody in 5 % milk (secondary guinea pig antibody from Santa Crutz, dilution factor: 1:5,000) at RT for 1 h.
31. For granularity analysis, nuclear staining is advised prior to analysis by confocal microscopy. The cell nucleus can be stained with 1 μg/mL of Hoechst 33342 for 15 min.
32. Extent of mitophagy also can be expressed as percentage of cells exhibiting structures with a high 561/458 ratio. We considered cells to be positive for mitophagy if there was more than three punctate per cell. Analysis was performed using the Granularity function of Metamorph software.

Acknowledgment

We are grateful to A. Miyawaki for the gift of mtKeima. This work was supported by NIH Intramural funds.

References

1. Mizushima N, Yoshimori T, Ohsumi Y (2011) The role of Atg proteins in autophagosome formation. Annu Rev Cell Dev Biol 27:107–132. doi:10.1146/annurev-cellbio-092910-154005
2. Komatsu M, Waguri S, Chiba T, Murata S, Iwata J, Tanida I, Ueno T, Koike M, Uchiyama Y, Kominami E, Tanaka K (2006) Loss of autophagy in the central nervous system

causes neurodegeneration in mice. Nature 441(7095):880–884. doi:nature04723 [pii] 10.1038/nature04723

3. Hara T, Nakamura K, Matsui M, Yamamoto A, Nakahara Y, Suzuki-Migishima R, Yokoyama M, Mishima K, Saito I, Okano H, Mizushima N (2006) Suppression of basal autophagy in neural cells causes neurodegenerative disease in mice. Nature 441(7095):885–889. doi:nature04724 [pii] 10.1038/nature04724
4. Nemoto S, Fergusson MM, Finkel T (2004) Nutrient availability regulates SIRT1 through a forkhead-dependent pathway. Science 306(5704):2105–2108. doi:306/5704/2105 [pii] 10.1126/science.1101731
5. Cohen HY, Miller C, Bitterman KJ, Wall NR, Hekking B, Kessler B, Howitz KT, Gorospe M, de Cabo R, Sinclair DA (2004) Calorie restriction promotes mammalian cell survival by inducing the SIRT1 deacetylase. Science 305(5682):390–392. doi:10.1126/science.1099196 1099196 [pii]
6. Lee IH, Cao L, Mostoslavsky R, Lombard DB, Liu J, Bruns NE, Tsokos M, Alt FW, Finkel T (2008) A role for the NAD-dependent deacetylase Sirt1 in the regulation of autophagy. Proc Natl Acad Sci U S A 105(9):3374–3379. doi:0712145105 [pii] 10.1073/pnas.0712145105
7. Lee IH, Finkel T (2009) Regulation of autophagy by the p300 acetyltransferase. J Biol Chem 284(10):6322–6328. doi:M807135200 [pii] 10.1074/jbc.M807135200
8. Hamai A, Codogno P (2012) New targets for acetylation in autophagy. Sci Signal 5(231):pe29. doi:scisignal.2003187 [pii] 10.1126/scisignal.2003187
9. Yi C, Yu L (2012) How does acetylation regulate autophagy? Autophagy 8(10):1529–1530. doi:21156 [pii]
10. Morselli E, Marino G, Bennetzen MV, Eisenberg T, Megalou E, Schroeder S, Cabrera S, Benit P, Rustin P, Criollo A, Kepp O, Galluzzi L, Shen S, Malik SA, Maiuri MC, Horio Y, Lopez-Otin C, Andersen JS, Tavernarakis N, Madeo F, Kroemer G (2011) Spermidine and resveratrol induce autophagy by distinct pathways converging on the acetylproteome. J Cell Biol 192(4):615–629. doi:jcb.201008167 [pii] 10.1083/jcb.201008167
11. Youle RJ, Narendra DP (2011) Mechanisms of mitophagy. Nat Rev Mol Cell Biol 12(1):9–14. doi:nrm3028 [pii] 10.1038/nrm3028
12. Katayama H, Kogure T, Mizushima N, Yoshimori T, Miyawaki A (2011) A sensitive and quantitative technique for detecting autophagic events based on lysosomal delivery. Chem Biol 18(8):1042–1052.doi:S1074-5521(11)00204-3 [pii] 10.1016/j.chembiol.2011.05.013

Chapter 18

Methods to Study the Role of Sirtuins in Genome Stability

Paloma Martínez-Redondo and Alejandro Vaquero

Abstract

One of the most important roles of Sirtuins is to ensure the maintenance of genome integrity under stress conditions. In this chapter, we provide a methodology to study this role of Sirtuins at two different levels: detection of genomic instability (with the Neutral Comet Assay) and study of Sirtuin dynamics in chromatin under stress conditions (by isolating insoluble chromatin fractions).

Key words Genome stability, Chromatin, DNA damage, Neutral comet, SIRT1-7, Benzonase

1 Introduction

Among the different roles that Sirtuins have acquired over evolution from early eukaryotes onwards is the preservation of genome stability [1, 2]: they help protect the integrity of the genome under genotoxic, metabolic, or oxidative stress conditions [3–6]. To date, all Sirtuins except SIRT4 and 5 have been linked directly or indirectly to control of genome stability. SIRT1 and SIRT6 appear to be the two Sirtuins most directly involved in these processes [7, 8]. The direct mechanisms mediated by Sirtuins in this context encompass regulation of expression of stress-related genes, control of the structural integrity of chromatin, and signaling of DNA damage and DNA repair [9, 10]—processes that often entail relocalization of the Sirtuins to chromatin upon activation by specific stress signals. Here we provide two different methods to study the mechanisms involved in this protection: *the neutral Comet Assay*, used to detect global DNA damage (double strand breaks, DSB), and *isolation of insoluble chromatin fractions*, to study Sirtuin dynamics in chromatin under stress conditions.

The Comet Assay, introduced in 1984 by Ostling and Johanson [11], is a single-cell electrophoresis technique for analyzing DNA damage. The Neutral Comet Assay is a simple quantitative technique to visualize the DNA damage (mainly DSB) in a single eukaryotic cell by analyzing the denatured DNA fragments that

Matthew D. Hirschey (ed.), *Sirtuins: Methods and Protocols*, Methods in Molecular Biology, vol. 1077,
DOI 10.1007/978-1-62703-637-5_18, © Springer Science+Business Media, LLC 2013

migrate out of the cell nucleus during electrophoresis. It is an unbiased and general method that provides rich information for different research topics, such as the modulation of Sirtuin activity by specific drugs under different forms of stress; the study of specific physiological conditions that involve Sirtuin function; and the knockdown of a particular Sirtuin. In fact, another version of the Comet Assay has been employed to ascertain how loss of SIRT1 or SIRT6 affects DNA damage [12, 13].

Isolating the insoluble chromatin fractions enables one to distinguish the protein fraction that is intimately bound to chromatin (present in the insoluble fraction of chromatin) from the soluble protein fraction that is not directly acting on the chromatin. This very powerful method can be used as an *analytical technique* or as a *preparative step* before performing multiple studies at the molecular level (e.g., immunoprecipitation, enzymatic reactions, or chromatographic studies). Compared to other methods, the preparative version is very mild: it is based on use of an enzyme (the nuclease benzonase; Sigma) that degrades all forms of DNA and RNA but lacks any proteolytic activity and it does not require large amounts of detergents or other reagents that could compromise protein activity or integrity [14]. This method enables one to determine the dynamic changes in the localization, levels, possible associated posttranslational modifications, and partners of each Sirtuin upon activation under stress. It is also very useful in overexpression experiments, where the excess of exogenous protein tends to accumulate in the soluble fraction of the cell. Thus, in contrast to using a global soluble fraction, isolating and subsequently purifying the insoluble chromatin fraction should enable one to study only the chromatin-associated *active* fraction of the protein.

2 Materials

All buffers are prepared using ultrapure water. Cellular extract fractionation buffers are all supplemented with protease inhibitors before use.

2.1 Single-Cell Electrophoresis

1. Cells: Knockout (for each Sirtuin studied) and wild-type MEFs obtained by standardized methods and cultured in DMEM (GIBCO) supplemented with 0.1 mM nonessential amino acids, 1 mM sodium pyruvate, GlutaMAX-1, penicillin–streptomycin, and 10 % fetal bovine serum.
2. Comet Assay slides. Superfrost slides and 1.5–2 % solution of normal melting point agarose in PBS. First, dip the non-frosted area of the slides into methanol, and pass them through a flame to clean off any dust from the slide. Then, dip the slides up to the frosted area 2–3 times in the agarose solution, wipe off the remaining agarose from the bottom of the slide (the side opposite to the labeled one), and then try to balance

out the remaining amount of liquid agarose on the top of the slide. Store the slides at room temperature in a horizontal position to allow drying (*see* **Note 1**).

3. Prepare a 0.5 % solution and a 1 % solution of low melting point (LMP) agarose in PBS. Store the solutions at 37 °C (*see* **Note 2**).
4. Neutral Lysis Solution: 0.5 M EDTA, 0.5 mg/mL Proteinase K, 2 % Sarkosyl, pH 8. Instructions for 50 mL: To 46.7 mL of pre-prepared EDTA 0.54 M (pH 8), add 0.025 g of Proteinase K and mix 3.3 mL of 30 % Sarkosyl (*see* **Note 3**). Keep at 4 °C.
5. Running buffer: 90 mM Tris–HCl, 90 mM boric acid, 2 mM EDTA, pH 8.5. Add 2.2 g of Tris–HCl, 11.13 g of boric acid, and 8 mL of EDTA 0.5 M (pH 8), and then add distilled water up to a final volume of 2 L. Keep at RT (*see* **Note 4**).
6. Staining solution: Hoechst 10 mg/mL in PBS. Prepare just before use.
7. Agarose gel electrophoresis chamber (*see* **Note 5**).

2.2 Comet Assay Data Analysis

1. Epifluorescence microscope: Use the ATL filter, the same filter as for DAPI (*see* **Note 6**).
2. Software analysis: CometScore (from TriTek), freeware available online (*see* **Note 7**).

2.3 DNA Damage Treatment

1. X-ray treatment: 5 Gy, 2 Gy/min.

2.4 Cellular Extract Fractionation

Soluble fraction (cytoplasmic and nuclear extract): The Dignam method [15] is used for the extraction. Insoluble fraction: Obtained from the pellet derived from the extraction of the soluble fraction.

1. Buffer A: 10 mM Tris–HCl pH 7.8, 10 mM KCl, 1.5 mM $MgCl_2$. Instructions for 1 L: mix 10 mL of 1 M Tris–HCl pH 7.8, 2.5 mL of 4 M KCl, and 3 mL of 0.5 M $MgCl_2$, and then add distilled water up to a volume of 1 L. Keep at 4 °C.
2. Buffer C: 10 mM Tris–HCl pH 7.8, 0.42 M NaCl, 1.5 mM $MgCl_2$, 0.2 mM EDTA, 25 % glycerol. Instructions for 1 L: Mix 10 mL of 1 M Tris–HCl pH 7.8, 105 mL of 4 M NaCl, 400 μL of 0.5 M EDTA, 3 mL of 0.5 M $MgCl_2$, and 250 mL of glycerol, and then add distilled water up to a volume of 1 L. Keep at 4 °C.
3. BC500 + 0.05 % NP40: Tris–HCl pH 7.8, 0.5 M NaCl, 0.5 mM EDTA, 0.1 mM PMSF, 0.1 mM DTT, 100 mM NaCl, 10 % glycerol, 0.05 % Igepal (NP40). Instructions for 1 L: Mix 10 mL of 1 M Tris–HCl pH 7.8, 1 mL of 0.5 M EDTA, 1 mL of 0.1 M PMSF, 1 mL of 1 M DTT, 25 mL of 4 M NaCl, 100 mL of Glycerol, and then add distilled water up to a volume of 1 L. Keep at 4 °C.

4. Loading buffer: 250 Mm Tris–HCl pH 6.8, 17 % glycerol, 2 % sodium dodecyl sulfate (SDS), 0.025 g bromophenol blue, and 10 % 2-mercaptoethanol. Keep at 4 °C.
5. RIPA buffer: 150 mM NaCl, 50 mM HEPES pH 7.4, 0.5 % deoxycholate, 0.1 % SDS. Keep at 4 °C.
6. Benzonase. Keep at −20 °C.

3 Methods

3.1 Neutral Comet Assay

There are several protocols for the Neutral Comet Assay; they differ mainly according to the lysis time and the lysis solution. Here we provide a method to be completed over 2 days with an overnight lysis, which enables greater reproducibility and sensitivity. On the first day, complete **steps 1–6**; on the second day complete **steps 7–20**. This method, performed under neutral conditions, can detect 50–10,000 DSB per cell [16].

1. To prepare single-cell slides, dip fresh slides into molten agarose (1.5–2 %) and let them dry. *See* Subheading 2.1, **item 2** (*see* Fig. 1a, b).

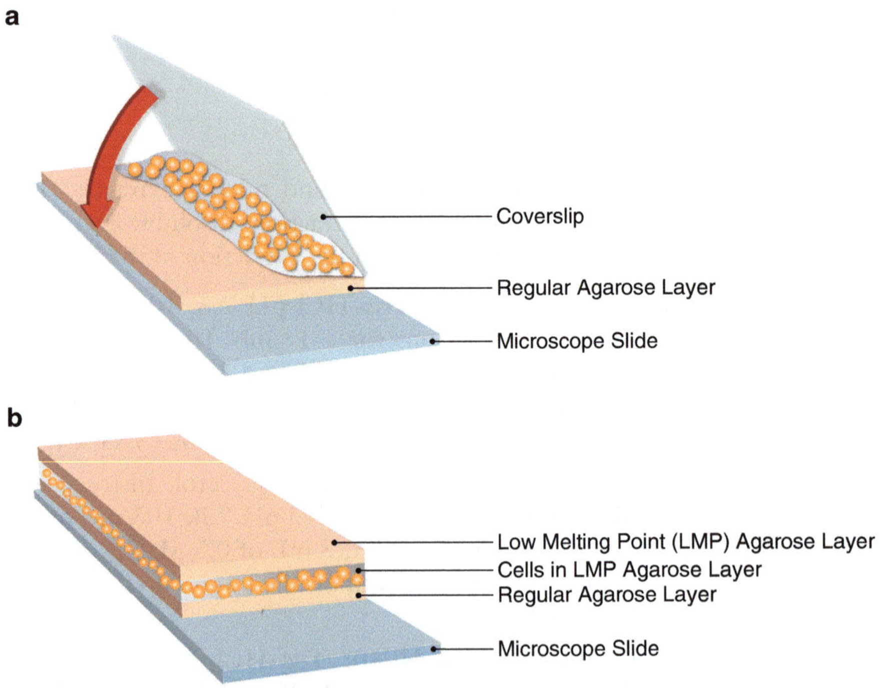

Fig. 1 Comet Assay slide. (**a**) Use a coverslip to equally distribute the low melting point (LMP) agarose over the previous agarose layer. Drop the coverslip onto the side touching the melted agarose. (**b**) The final Comet Assay slide: Suspend the cells in LMP agarose and then place them on a slide pre-coated with regular agarose. Add another layer of LMP agarose to fill any holes in the previous layer and to increase the distance between the cells and the surface

2. Trypsinize the cells with a 1:10 solution of trypsin in PBS (*see* **Note 8**), wash the pellet once with cold PBS, and then resuspend the washed pellet at 10^6 cells/mL in cold PBS.
3. Mix 10 μL of this solution with 75 μL of the 0.5 % LMP agarose, mix by pipetting, and distribute 80 μL equally on the agarose-coated slide, covering it with the coverslip (24×50 mm) (*see* Fig. 1a). Let the agarose solidify at 4 °C for at least 15 min (*see* **Note 9**).
4. Once the agarose has dried, carefully remove each coverslip (*see* **Note 10**). To each slide, add another layer of 1.0 % LMP agarose (90 μL). Cover each slide with another glass coverslip, let dry at 4 °C for at least 15 min, and then carefully remove the coverslip (*see* **Notes 9** and **10**).
5. Once the agarose has gelled, gently submerge the slides into a covered tray containing neutral lysis solution at 4 °C.
6. Cover the tray to protect it from light and place it in an incubator at 37 °C overnight (18–20 h) (*see* **Note 11**).
7. After overnight lysis, remove the lysis buffer, and then carefully add the electrophoresis buffer at RT for 25 min (*see* **Note 12**).
8. Repeat the above step twice.
9. Carefully remove the slides from the tray and place them in an electrophoresis chamber. Fill the camber with enough buffer to cover approximately 1–2 mm above the top surface of the agarose (*see* **Note 13**).
10. Perform electrophoresis in the electrophoresis buffer (fixed voltage: 0.6 V/cm; time: 25 min). Note: the current typically reaches 3 mA (*see* **Note 14** and Fig. 2a).
11. Once the electrophoresis has finished, remove the slides from the chamber and drain the excess electrophoresis buffer.
12. Immerse the slides in at least 200 mL of distilled water using the same tray used for the lysis and the washes.
13. Remove the water, add fresh water, and let the tray stand for 20 min.
14. Remove the slides from the tray, remove the excess water, and leave the slides resting horizontally.
15. Add 300 mL of Hoechst solution onto the agarose surface and incubate for 20 min.
16. Eliminate the excess Hoechst solution by rinsing the slides with PBS using a Pasteur pipette.
17. Cover the agarose layer with a glass coverslip and keep the slides at 4° (*see* **Note 15**).
18. For comet tail analysis, at least 50 Comet images should be analyzed, excluding doublets.

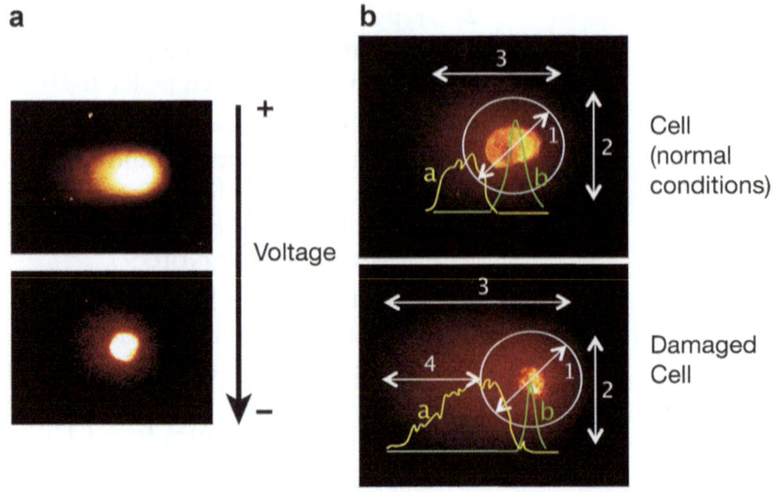

Fig. 2 (**a**) In the Comet Assay, the voltage used for the electrophoresis step is critical. The voltage value should be chosen according to the control cells; otherwise, all the cells will have a long comet tail. (**b**) Comet images of normal and damaged wild-type fibroblasts. Quantification of a single-cell comet by comet scoring programs is based on different measurements: *1*: Head diameter; *2*: Total height; *3:* Total length; *4*: Tail length; *a*: Tail area; and *b*: Head area

19. The image analysis software analyzes individual Comets for three parameters: *Tail length*, *%DNA in tail*, and *Tail moment* (*see* Fig. 2b).
20. Appropriate statistical analysis should be applied using means or medians, depending on population distributions. Error bars are typically used to represent the between-experiment variability (*see* Fig. 3).

3.2 Study of Sirtuin Dynamics Under Stress

1. For transfection and DNA damage treatment, prepare the adherent cells 1 day before transfection, to reach 60–70 % confluence by the day of transfection.
2. Transfect the cells with the Sirtuin-tagged plasmids using PEI (*see* **Note 16**).
3. Treat the cells 48 h after transfection (*see* **Note 17**). For X-ray treatment, radiate the cells with 5 Gy (2 Gy/min dose), and then scrape them at the following time points after treatment: 0, 1, and 5 h (*see* Fig. 4a). For untreated cells: collect as above.
4. For cellular extract fractionation, wash the harvested cells once with PBS. The pellet can be stored at –80 °C or used directly (*see* **Note 18**).
5. Obtain the soluble fraction by using the Dignam method: use buffer A for the nuclear fraction and buffer C for the cytoplasmic fraction. Alternatively, whole-cell extract can be obtained by mixing both buffers at the end.

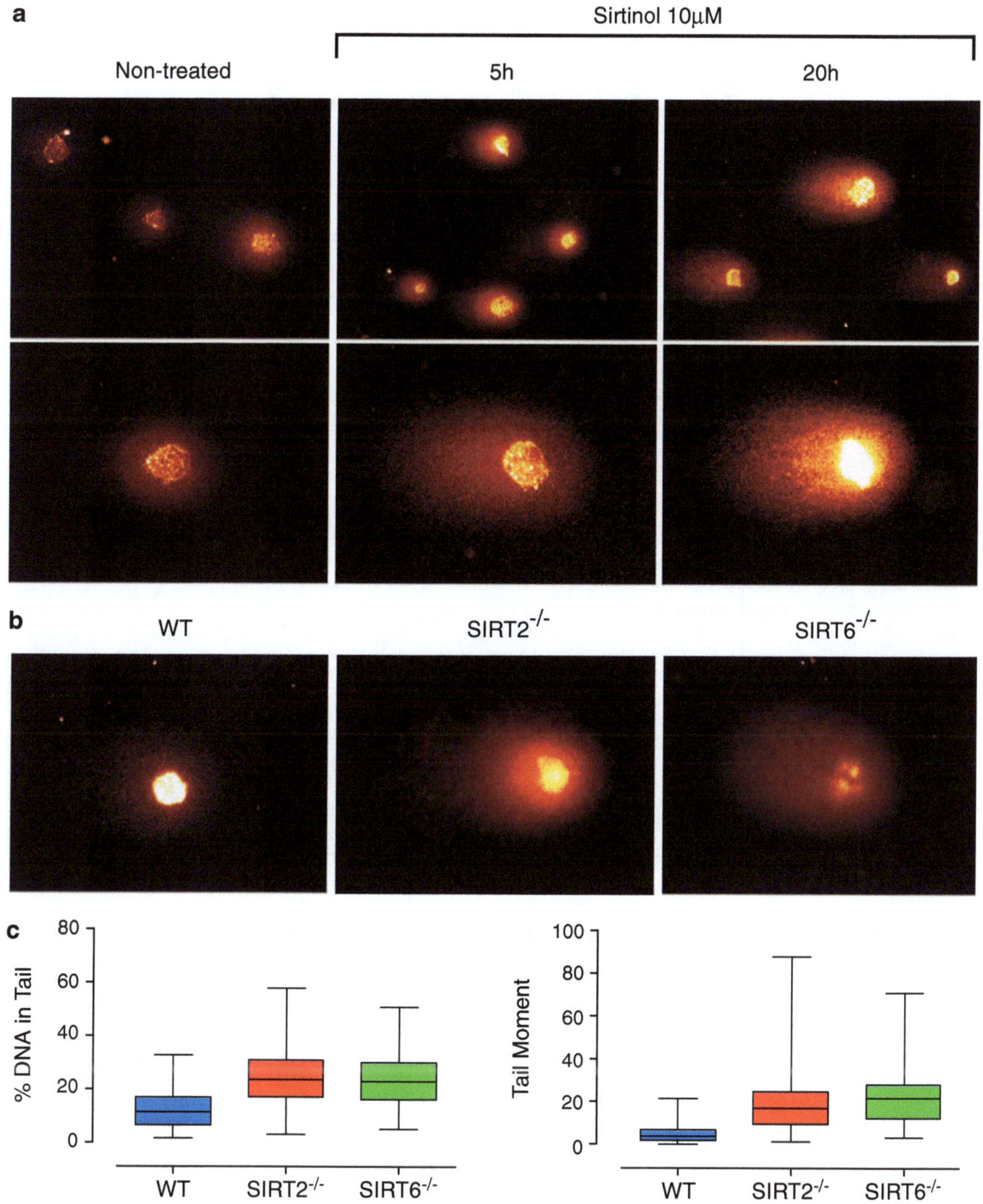

Fig. 3 (**a**) Comet Assay to evaluate Sirtinol treatment of DNA damage. The cells were incubated for 5 and 20 h with the SirT1/SirT2 inhibitor Sirtinol, harvested, and then used in the Comet Assay. Longer exposure to the drug led to higher levels of DNA damage. (**b**, **c**) Wild-type cells and Sirtuin (SirT2 or SirT6) knockout cells exhibited significant differences in their respective values for image analysis parameters typically used to measure DNA damage by the Comet Assay (%*DNA in tail* and *Tail moment*). The most informative parameter is %*DNA in tail*, as it enables characterization of the comet. The *Box* and *Whisker plots* show the median and distribution, respectively, of each population

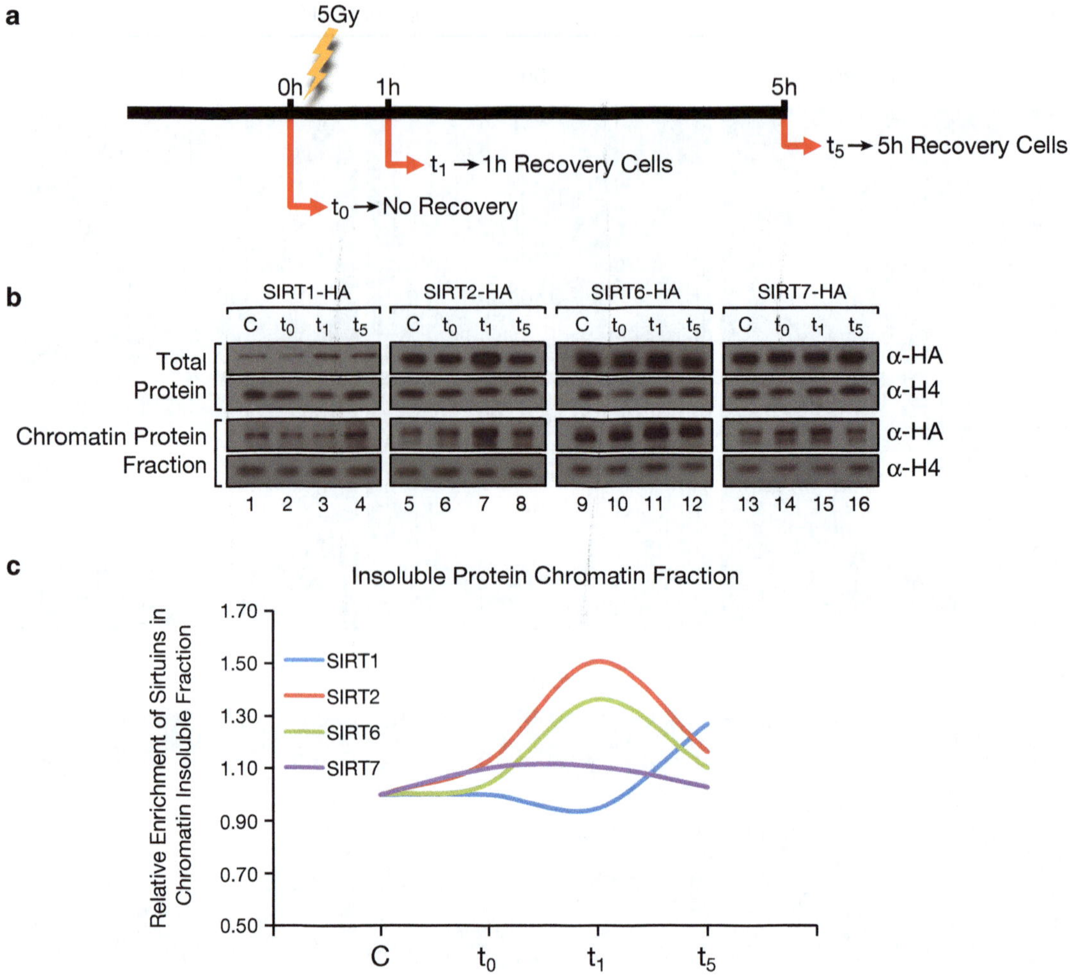

Fig. 4 (**a**) Transfected cells were either left untreated or irradiated with 5 Gy X-rays. The irradiated cells were harvested at different times after X-Ray exposure: 0, 1, and 5 h. (**b**, **c**) The different Sirtuins exhibited different patterns of chromatin binding after the DNA damage

6. Resuspend the pellet obtained from the Dignam extraction above in BC500 + 0.05 % NP40, and then centrifuge at 14,000 × *g* for 1 min (*see* **Note 19**).
7. The pellet is the insoluble fraction. At this point, one can choose between the *Analytical* or *Preparative* methods:
   (a) *Analytical* method: Loading buffer is added to the pellet and sonicated in a water bath sonicator (15 min; 30 s ON/30 s OFF). This enables dissolution of the insoluble fraction by breaking up the DNA (*see* **Note 20**). Another technique that provides the proteins bound in different insoluble chromatin levels is to dissolve the pellet in different NaCl concentrations (from 500 mM to 1 M) and then centrifuge at 14,000 × *g* for 1 min.

(b) *Preparative* method: RIPA buffer supplemented with benzonase nuclease (range: 25–50 U/mL per 5×10^6 cells) is added to the pellet (*see* **Note 20**). After vortexing, the sample is incubated at 4 °C with mild agitation. The incubation time varies from 4 h to overnight (*see* **Note 21**). After incubation, the sample is centrifuged at 7,000 × *g* for 10 min to eliminate any debris and the supernatant is kept at −80 °C to maintain protein integrity.

8. The *Analytical* method enables quantification of the Sirtuin and non-Sirtuin protein bound to the chromatin after different treatments or in different stages of the cell cycle. In this method, the result is analyzed by Western Blot and can be quantified (*see* **Note 22** and Fig. 4a, b).
9. The *Preparative* method is used when the protein integrity is important for subsequent use (e.g., immunoprecipitation, enzymatic activities). The samples can be used directly in enzymatic assays, chromatographic studies, etc.

4 Notes

1. Prepare the slides the same day, early in the morning, to avoid problems resulting from excessive drying. Ensure that you prepare enough glass slides (at least two slides per sample).
2. Caution: a few seconds in the microwave are enough! Cool down the agarose in 1.5 mL aliquots to 37 °C in a thermo block.
3. Using pre-prepared EDTA pH 8 is a fast and easy way to reach the required pH for the lysing solution. The pH adjustment is really important in this experiment to distinguish the *Neutral* Comet Assay (DSB only) from the *Alkaline* one (single or double strand breaks).
4. Be sure to adjust the pH to 8.5 by adding NaOH.
5. It is important to measure the length between both electrodes in order to calculate the voltage for the experiment.
6. Better images are obtained 1 day after the assay (store the slides in humidified boxes in a refrigerator overnight). Confocal microscopes do not provide adequate images: comet tail intensity and shape are poorly detected.
7. Note: the software requires ".bmp" files, so the microscopy images must first be converted into this format.
8. This is advantageous over cell scraping in that the cells are easier to count.
9. Ensure that the slides dry horizontally and are protected from light, in order to avoid any unspecific DNA damage to the cellular DNA. Do not cool the slides directly on ice or on

aluminum foil, as both attract humidity and therefore, the agarose would not solidify.

10. Remove the coverslip by pushing it off to the side: do not pull it off, which would tear the agarose.
11. Use a covered tray to keep the slides in a horizontal position. Avoid moving the slides unless absolutely necessary.
12. It is very important to protect the slides from light.
13. Be sure that the electrophoresis chamber is level so that the slide is fully covered with buffer (to a height of approximately 1–2 mm above the slide surface). This will prevent the appearance of different comet tails along the slide. Cover the chamber to protect it from light.
14. The voltage used varies depending on the cell type. It is important to choose an adequate voltage according to the control cells. Keep in mind that all cells harbor basal levels of DSBs, so it is important to set the Comet assay voltage for the control cells to behave as non-damaged cells (no comet) as an easier way to visualize and quantify the results.
15. Although the slides can be stored after the analysis for up to 3 or 4 days (at 4 °C and protected from light), the best results are obtained the day after finishing the protocol.
16. Non-transfected cells can be also used: plate them and then start the treatment 48 h after trypsinization.
17. Be sure that the cells have not reached more than 80–90 % confluence before the treatment.
18. Part of the cell pellet should be used for total protein extraction as follows: resuspend the part directly in loading buffer and then sonicate. This point is important to ensure the same transfection level for all treatments and to determine if and how the treatment affects the levels of the Sirtuin being studied.
19. It is important to resuspend the pellet in the buffer by pipetting carefully and then vortexing before the centrifugation step.
20. Do not use the pipette to resuspend the pellet, which is viscous and could stick to the tip. Vortex carefully instead.
21. The incubating time (from 4 h to overnight) varies according to the purpose of the experiment. Consider that shorter times of incubation require higher concentrations of enzyme (50 U/mL per 5×10^6cells) and that overnight incubation can affect the level of interaction between proteins. For immunoprecipitation assays, overnight incubation with benzonase and the antibody–beads incubation may affect the levels of interaction of the proteins associated to the insoluble fraction: these long incubation times can increase or decrease interaction independently of the cell-cycle stage or the treatment. In this case, we recommend a benzonase treatment time of 4 h.

22. The amount of protein should be normalized using specific antibodies for each fraction. We usually use H4 antibody to normalize the insoluble chromatin fraction. This enables studying changes in Sirtuin distribution under stress and in other conditions.

References

1. Lombard DB, Chua KF, Mostoslavsky R, Gostissa M, Franco S, Alt FW (2005) DNA repair, genome stability, and aging. Cell 120:497–512
2. Guarente L (1999) Diverse and dynamic functions of the Sir silencing complex. Nat Genet 23:281–285
3. Bosch-Presegué L, Vaquero A (2011) A dual role of sirtuins in cáncer. Genes Cancer 2: 648–662
4. Imai S, Armstrong CM, Kaeberlein M, Guarente L (2000) Transcriptional silencing and longevity protein Sir2 is an NAD-dependent histone deacetylase. Nature 403: 795–800
5. Landry J, Sutton A, Tafrov ST et al (2000) The silencing protein SIR2 and its homologs are NADdependent protein deacetylases. Proc Natl Acad Sci U S A 97:5807–5811
6. Tanny JC, Dowd GJ, Huang J, Hilz H, Moazed D (1999) An enzymatic activity in the yeast Sir2 protein that is essential for gene silencing. Cell 99:735–745
7. Wang RH, Sengupta K, Li C et al (2008) Impaired DNA damage response, genome instability, and tumorigenesis in SIRT1 mutant mice. Cancer Cell 14:312–323
8. Mostoslavsky R, Chua KF, Lombard DB et al (2006) Genomic instability and aging-like phenotype in the absence of mammalian SIRT6. Cell 124:315–329
9. Vaquero A (2009) The conserved role of sirtuins in chromatin regulation. Int J Dev Biol 53:303–322
10. Webster BR, Lu Z, Sack MN, Scott I (2012) The role of sirtuins in modulating redox stressors. Free Radic Biol Med 52(2):281–290
11. Ostling O, Johanson KJ (1984) Microelectrophoretic study of radiation-induced DNA damages in individual mammalian cells. Biochem Biophys Res Commun 123:291–298
12. Hajji N, Wallenborg K, Vlachos P, Füllgrabe J, Hermanson O, Joseph B (2010) Zpposing effects of hMOF and SIRT1 on H4K16 acetylation and the sensitivity to the topoisomerase II inhibitor etoposide. Oncogene 29:2192–2204
13. McCord RA, Michishita E, Hong T, Berber E, Boxer LD, Kusumoto R, Guan S, Shi X, Gozani O, Burlingame AL, Bohr VA, Chua KF (2009) SIRT6 stabilizes DNA-dependent protein kinase at chromatin for DNA double-strand break repair. Aging 1(1):109–121
14. Holden P, Horton WA (2009) Crude subcellular fractionation of cultured mammalian cell lines. BMC Res Notes 2:243
15. Dignam JD, Martin PL, Shastry BS, Roeder RG (1983) Eukaryotic gene transcription with purified components. Methods Enzymol 101:582–598
16. Olive PL, Banáth JP (2006) The comet assay: a method to measure DNA damage in individual cells. Nat Protoc 1:23–29

Chapter 19

Circadian Measurements of Sirtuin Biology

Kathryn Moynihan Ramsey, Alison H. Affinati, Clara B. Peek, Biliana Marcheva, Hee-Kyung Hong, and Joseph Bass

Abstract

Many of our behavioral and physiological processes display daily oscillations that are under the control of the circadian clock. The core molecular clock network is present in both the brain and peripheral tissues and is composed of a complex series of interlocking transcriptional/translational feedback loops that oscillate with a periodicity of ~24 h. Recent evidence has implicated NAD^+ biosynthesis and the sirtuin family of NAD^+-dependent protein deacetylases as part of a novel feedback loop within the core clock network, findings which underscore the importance of taking circadian timing into consideration when designing and interpreting metabolic studies, particularly in regard to sirtuin biology. Thus, this chapter introduces both in vivo and in vitro circadian methods to analyze various sirtuin-related endpoints across the light–dark cycle and discusses the transcriptional, biochemical, and physiological outputs of the clock.

Key words Circadian, Clock, Metabolism, Constant darkness, Synchronization, NAD, Sirtuins

1 Introduction

While the French astronomer JeanJacques d'Ortous de Mairan first noted that the diurnal leaf movements of the *Mimosa* plant persisted in constant darkness in the 1700s, it was not until more than 200 years later that it became established that circadian rhythms were driven by an internal endogenous circadian clock, rather than being just passive responses to a light–dark cycle. The highly conserved nature of the clock across diverse organisms, including cyanobacteria, fungi, plants, *Drosophila*, and mammals, suggests an evolutionary advantage to being able to synchronize behavioral and physiological processes to daily environmental cues in order to optimize energy homeostasis and enhance survival.

Many of our behavioral and physiological processes, including sleep–wake cycles, locomotor activity, hormone secretion, body temperature, and metabolism, display daily oscillations. Epidemiologic studies have revealed that many acute catastrophic events, including heart attacks, pulmonary edema, and hypertensive crises, tend

Matthew D. Hirschey (ed.), *Sirtuins: Methods and Protocols*, Methods in Molecular Biology, vol. 1077,
DOI 10.1007/978-1-62703-637-5_19, © Springer Science+Business Media, LLC 2013

to occur at specific times of day [1]. Furthermore, there is an increased risk of obesity and cardiovascular disease in shift workers, who have altered patterns of sleep/wake and fasting/feeding cycles, highlighting the close link between circadian rhythms and metabolism [2–4]. Other physiological processes, such as glucose metabolism, vary during the day. For example, humans are more glucose tolerant in the morning than in the evening, correlating with enhanced insulin sensitivity and insulin secretion in the early morning hours [5–10]. Of note, loss of rhythms in glucose metabolism is associated with the development of metabolic diseases such as type 2 diabetes [11, 12].

Although the main "driver" of circadian rhythms resides within the suprachiasmatic nucleus (SCN) of the hypothalamus in the brain, the circadian clock network is also present within most, if not all, peripheral tissues. The SCN neurons receive light input from the retinohypothalamic tract and synchronize peripheral tissue clocks through both neuronal and hormonal connections. The core molecular machinery underlying circadian rhythms within both the SCN and peripheral tissues has recently come to light through a combination of molecular and genetic studies. The core clock network consists of interwoven positive and negative transcription–translation feedback loops that function as a molecular oscillator with near 24 h precision (reviewed in ref. 13). The positive limb of the clock is composed of the basic helix-loop-helix transcription factors CLOCK and BMAL1, which heterodimerize and activate transcription of downstream targets, including the *Period* (*Per1*, *2*, *3*) and *Cryptochrome* (*Cry 1*, *2*) genes. PER and CRY proteins then heterodimerize, translocate back to the nucleus, and bind and inhibit the activity of CLOCK/BMAL1, thereby repressing their own transcription. Additional modifying components of the core clock network have been identified, including the casein kinases I epsilon and delta, which phosphorylate PER and CRY, ultimately leading to their degradation, and the orphan nuclear receptors REV-ERBα and ROR1α, both of which are transactivated by CLOCK/BMAL1, and in turn repress and activate *Bmal1* transcription, respectively (*see* Fig. 1).

Of particular interest, several recent studies have demonstrated that NAD^+ biosynthesis and the sirtuin family of NAD^+-dependent protein deacetylases constitute part of a novel feedback loop within the core clock network [14–17]. CLOCK/BMAL1 directly regulate the expression of the gene encoding the rate-limiting enzyme in NAD^+-biosynthesis, nicotinamide phosphoribosyltransferase (NAMPT), in peripheral tissues, and NAD^+ levels oscillate across the light–dark cycle. Not only is NAD^+ an essential cofactor for metabolic redox reactions, but it is also a substrate for SIRT1, which has been shown to deacetylate and regulate the activity of core clock proteins PER2 and BMAL1 [14, 15]. In addition, both genetic and pharmacologic modulation of NAD^+ levels has been

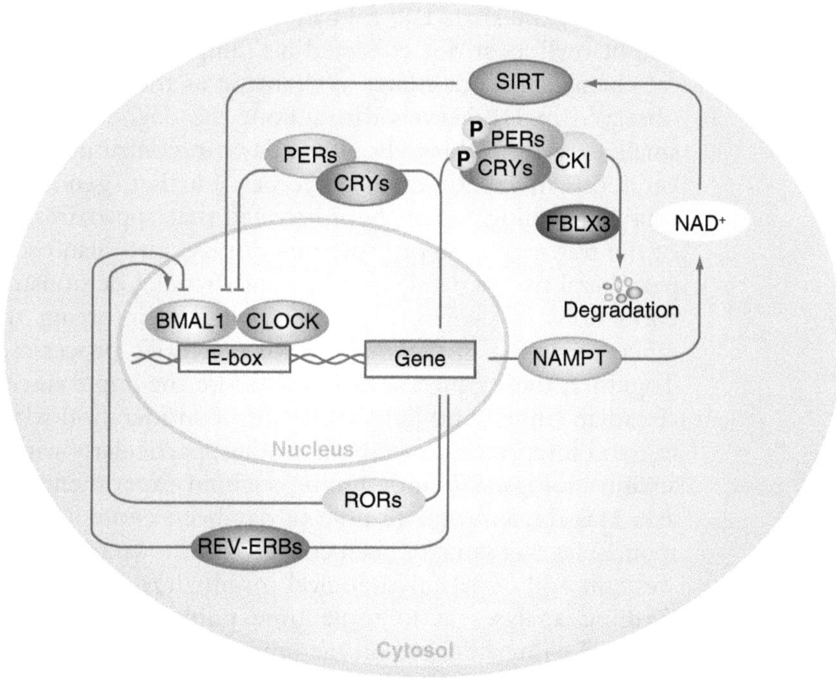

Fig. 1 Molecular overview of core clock network. The core molecular clock network consists of a series of positive and negative transcription/translation feedback loops. The positive limb consists of CLOCK/BMAL1, which heterodimerizes and transactivates expression of downstream target genes containing an E-box in their promoters, including the *Period*, *Cryptochrome*, *Ror*, *Rev-erb*, and *Nampt* genes. The PER and CRY proteins multimerize and inhibit the activity of CLOCK/BMAL1, until they are degraded following CKI- and FBLX3-mediated phosphorylation and ubiquitination, respectively. Rhythmic generation of *Nampt* results in oscillations in NAD^+ and SIRT activity, which feeds back to regulate the positive limb of the clock

shown to alter clock transcriptional activity and circadian behavior and metabolism [18–20], completing a novel metabolic feedback loop within the core clock machinery. The identification of NAMPT and the sirtuin deacetylases as nutrient sensors at the nexus of circadian and metabolic pathways will likely shed additional light on how peripheral tissues coordinate fuel utilization and metabolism with the fasting/feeding and sleep/wake cycles.

Given the relatively recent finding that NAD^+ biosynthesis and sirtuin biology are intimately connected to the circadian clock, it is important to keep the time of day in mind when designing sirtuin-related experiments. An interesting anecdote described by Ueli Schibler in his keynote lecture at the conference for the Society for Research of Biological Rhythms [21] demonstrates the significance of the time of day at which experiments are performed—a graduate student who had inherited a former postdoc's project involving DBP (albumin site D-binding protein) struggled for weeks to even detect its presence in liver, until he realized that DBP was essentially undetectable in the early morning hours (the time when he, as a "morning lark," collected his tissue samples), while it was induced more than 100-fold by the evening (the time when the "night-owl" post-doc collected his samples). While most biological changes are not nearly as dramatic as the 100-fold difference observed in DBP levels throughout the day, even differences as small as twofold throughout the day can contribute to significant variations in experimental outcomes. Further, genomic and proteomic profiling studies have revealed that approximately 5–20 % of the transcriptome or proteome displays circadian oscillations in peripheral tissues, while metabolomics studies demonstrate robust oscillations in many critical metabolites, reinforcing the importance of timing in the majority of metabolic processes [22–24]. Together, these observations underscore the importance of taking circadian timing and light cycles into consideration when designing and interpreting metabolic studies, particularly with regard to sirtuin biology. When designing such an experiment, one should ask: Has the appropriate time of day been chosen for the experiment? Have all samples been collected at the same time point? Can we gain additional physiological insight into the process by performing analyses at multiple time points throughout the day? Given the intense focus on mammalian sirtuins in the past several years, the goal of this chapter is to highlight both in vivo and in vitro methods to analyze various endpoints across the light–dark cycle, whether they are transcriptional, biochemical, physiological, or behavioral outputs of the clock. For example, given the role of the sirtuins in multiple metabolic processes, including gluconeogenesis, insulin secretion, and fatty acid oxidation, it will be important to understand the dynamics of how the sirtuins regulate these processes throughout the day at both physiological and molecular levels. Briefly, Subheading 3.1 outlines how to set up in vivo "circadian" studies in which the process of interest is analyzed over the course of 48 h in mice maintained in constant darkness, while Subheading 3.2 describes how to monitor in vivo circadian rhythms of locomotor activity as a behavioral output of the clock. Subheading 3.3 discusses how brief treatment of cultured cells with various compounds synchronizes rhythms of clock gene expression that persist for several days, thereby creating a powerful cell culture-based system to study sirtuin-related processes across the circadian cycle.

2 Materials

2.1 Circadian In Vivo Studies (48 h Constant Darkness Experiment)

1. C57BL/6J mice, aged 12 weeks (Jackson Labs).
2. Light-tight circadian cabinets with computer-controlled light–dark cycles.
3. Red light, red light-equipped room, and/or infrared night-vision goggles.

4. Standard housing mouse cages.
5. Dissection equipment (scissors, forceps, etc.).
6. Eppendorf tubes for tissue collection.
7. Liquid nitrogen.

2.2 Circadian In Vivo Studies (Locomotor Activity Rhythms)

1–3. Same as above.
4. Mouse cages equipped with running wheel and microswitch.
5. Computer equipped with data acquisition and analysis software.

2.3 Circadian In Vitro Studies

1. NIH-3T3 MEF cells (ATCC# CRL-1658).
2. NIH-3T3 MEF cells infected with PER2-LUC-expressing lentivirus [25, 26] or MEFs isolated from PER2-LUC mice [27].
3. MEF culture medium: Dulbecco's modified Eagle medium (DMEM) w/4.5 g/L glucose, 10 % FBS, 4 mM glutamine, 1 mM sodium pyruvate, 100 U/ml penicillin/streptomycin.
4. 3.5 and 10 cm tissue culture dishes.
5. 37 °C incubator (CO_2-free) with LumiCycle apparatus (Actimetrics).
6. Luciferin cell culture medium: DMEM (no phenol red) plus 10 % FBS, 10 mM HEPES buffer, 0.035 % (w/v) sodium bicarbonate, 4 mM glutamine, 1 mM sodium pyruvate, penicillin/streptomycin, 0.1 mM luciferin (added immediately before use).
7. 1× sterile PBS.
8. 4 cm glass cover slips.
9. Sterile vacuum grease.
10. Forskolin.
11. Cell scrapers.
12. 1.5 ml Eppendorf tubes for cell collection.

3 Methods

3.1 Measuring Circadian Rhythms In Vivo over the Course of 48 h in Constant Darkness

For a biological process to be classified as circadian, it must display an endogenous, entrainable oscillation of approximately 24 h that is invariant across a wide temperature range. Experimentally, the rhythm must (1) repeat its cycle every 24 h for at least two cycles and (2) persist in the absence of external cues (i.e., constant darkness). Therefore, to determine whether a process is under circadian control, such as whether NAD^+ levels are dynamically regulated by the clock, the procedure described below uses tissues that were collected for NAD^+ measurement every 4 h over the course of 48 h

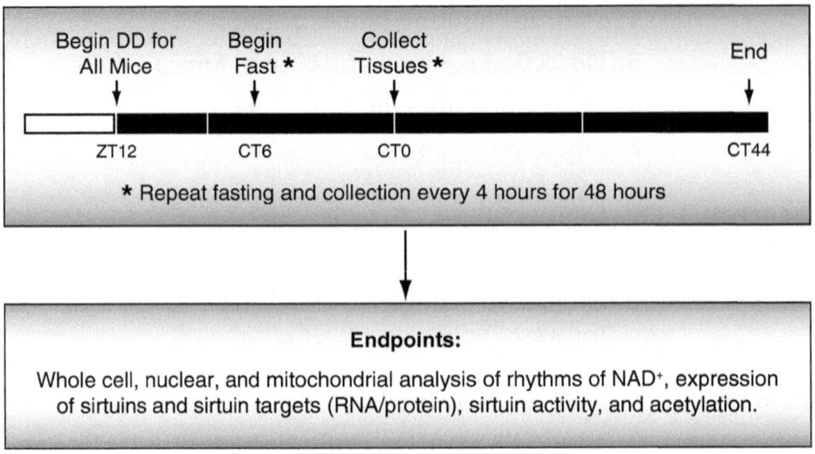

Fig. 2 In vivo experimental timeline for 48 h tissue collection in constant darkness. Mice are released into DD at ZT12 in order to coordinate with the start of the normal dark period. Food is removed from the first set of mice at CT6, which is 18 h prior to the first tissue collection (indicated by CT0). The fasting and tissue collection is then repeated every 4 h for 44 h in order to get a full set of samples to be analyzed for rhythms of the process of interest

in mice that have been maintained in constant darkness (12 total collection time points with 4 mice per time point = 48 mice). This approach can be applied to any process of interest, including gene or protein expression profiles, sirtuin activity, or sirtuin-related biochemistry (e.g., acetylation) or physiology (e.g., gluconeogenesis and fatty acid oxidation).

1. House C57BL/6J mice in light-tight boxes set to a 12:12 light–dark (LD) cycle in a temperature-controlled room (*see* **Notes 1** and **2**). Arrange 4 mice per cage in a total of 12 cages, each of which corresponds to 1 of the 12 collection time points. Maintain animals in these conditions for a minimum of 1 week prior to the start of the experiment to allow animals to acclimate to their new surroundings.
2. Change bedding, food, and water as needed, taking care not to disturb the light conditions, as even brief exposure to altered light conditions can affect the clock (*see* **Note 3**).
3. Switch the light cycle to constant darkness (DD) 36 h prior to the first tissue collection time point, which ensures that a full circadian period has passed without entraining cues (i.e., light), but not so long that all of the individual animals become desynchronized from one another (*see* **Note 4**, Fig. 2, and Table 1).
4. 18 h prior to the first collection, remove food from the first set of mice (circadian time 0 (CT0) time point). Other mice are

Table 1
Example experimental setup for 48 h tissue collection in constant darkness

Time Point	Remove Food		Tissue Collection	
CT0	Day 2	2pm	Day 3	6am
CT4		6pm		10am
CT8		10pm		2pm
CT12	Day 3	2am		6pm
CT16		6am		10pm
CT20		10am	Day 4	2am
CT24		2pm		6am
CT28		6pm		10am
CT32		10pm		2pm
CT36	Day 4	2am		6pm
CT40		6am		10pm
CT44		10am	Day 5	2am

then serially fasted for 18 h prior to each collection time (CT4-CT44) so that each animal experiences the same length of fasting. Since fasting and food consumption are centrally regulated circadian behaviors which in turn regulate sirtuin expression and activity, fasting each mouse minimizes the effect of feeding behavior on sirtuins across the circadian cycle (*see* **Note 5**, Fig. 2, and Table 1).

5. Continue to remove food from mice from subsequent time points every 4 h, taking care not to introduce light during the food removal (*see* **Note 6**, Fig. 2, and Table 1).
6. Thirty-six hours following the switch to DD, remove the first set of 4 mice from the light-tight box and euthanize using CO_2. Euthanasia should be carried out in complete darkness (with the use of infrared goggles) or in very dim red light. After the mice have been euthanized and the light-box sealed, the lights can be turned on for the dissection (*see* **Note** 7).
7. Dissect desired tissues and collect into the appropriately labeled tubes. Tissues used for NAD^+ measurement must be snap-frozen immediately in liquid nitrogen (details of methods to measure NAD^+ can be found in elsewhere in this volume). Tissues for RNA and protein analysis of clock, sirtuins, or sirtuin targets can also be collected. Nuclear and mitochondrial fractions can also be collected at this time for further subcellular analysis of sirtuin activity. Continue to collect tissues every 4 h for 48 h (*see* **Notes 8–10**).

3.2 Measuring Circadian Rhythms of Locomotor Activity *In Vivo*

The most robust and widely used method to examine the function of the endogenous circadian clock in rodents is monitoring behavioral rhythms of locomotor activity. Here, we describe how to monitor rhythms of activity by means of wheel running, a highly precise assessment of locomotor activity given that rodents actively run on the wheels for the majority of their wake time [28]. Infrared beam crossing, an alternative to wheel running, is discussed in Subheading 4. These experiments are useful to determine whether the sirtuin-related pathway of interest within brain or periphery might affect central behavioral activity rhythms.

1. Place mice in individual cages that contain a running wheel in light-tight boxes approximately 1 h prior to the start of the active (i.e., dark) period (*see* **Note 11**). Place just a handful of bedding in the cage so as to not prevent the running wheel from moving freely. A microswitch is attached to the side of the cage and transmits a signal for every revolution of the wheel. These signals are then passed to the collection computer (*see* **Note 12**).
2. Maintain mice in 12:12 LD conditions for a minimum of 1 week to allow acclimation to the new surroundings and to collect baseline activity data (*see* **Note 13**).
3. Switch the light cycle to constant darkness and continue to collect activity data for an additional 2–3 weeks (*see* **Note 14**).
4. Activity data is recorded continuously by data acquisition and analysis software such as ClockLab (http://www.actimetrics.com). Wheel running records (i.e., actograms) are then double plotted by vertically stacking the data in overlapping 48-h traces (*see* Fig. 3). The following fundamental circadian parameters can then be estimated using both visual inspection and available analyses programs (*see* **Note 15**):

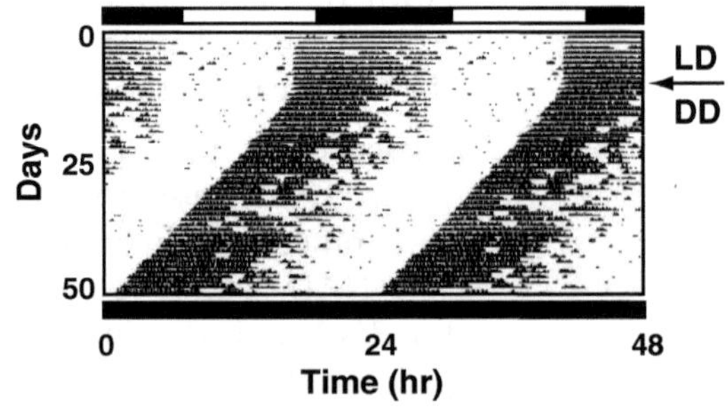

Fig. 3 Example of double-plotted actogram of locomotor activity. Wheel turns are recorded in 5-min blocks and plotted as histograms along a horizontal line representing the time of day. The actogram is double plotted, meaning that 2 consecutive days are positioned next to each other both horizontally and vertically. Lighting conditions are indicated to the right. The *arrow* denotes the switch from 12:12 LD to constant darkness (DD) on day 10

(a) Free-running period: The time interval between phase reference points (i.e., two peaks).

(b) Amplitude: The difference between peak and trough values of the rhythm.

(c) Phase: The timing of a reference point in the cycle (i.e., the peak) relative to a fixed event (i.e., the beginning of the night phase).

(d) Daytime activity profile: The total number of wheel running turns recorded during subjective day (or night) over the total activity during a circadian cycle or 24 h.

3.3 Measuring Circadian Rhythms In Vitro

A major breakthrough in our understanding of the molecular components of the clock came with the discovery that isolated cultured fibroblasts are able to display approximately 24 h oscillations of core clock genes following brief exposure to 50 % serum, demonstrating that cells have the ability to generate cell-autonomous rhythms independent of signals from the SCN [29]. In addition to serum shock, several other compounds that activate a variety of signaling pathways have also been shown to induce circadian rhythms of gene expression in cultured cells. These include activators of protein kinase A (forskolin and butyryl cAMP), the glucocorticoid receptor (dexamethasone) [30], G-protein-coupled receptors (endothelin), as well as temperature cycles and heat shock [31, 32]. The ability to study circadian rhythms in cultured cells is a valuable tool and has numerous advantages over more time-consuming in vivo studies, including the ease in which cells can be genetically manipulated to either overexpress or knockdown clock- or sirtuin-related genes. Here, we outline an experiment in which MEF cells are pharmacologically synchronized using forskolin to generate coordinated ~24 h oscillations of the molecular clock transcription factors and their downstream metabolic targets. Rather than synchronization of all of the cells at one time point, followed by a continuous collection of samples every 4 h for 44 h, we describe here a more convenient method whereby cells are synchronized at four time points over 36 h and then collected at three time points on the final day of the experiment (*see* Fig. 4 (Method 3) and Table 2). In addition, we describe a method to simultaneously monitor circadian rhythmicity during the time course experiment using a biological circadian clock reporter construct, Period2-luciferase (Fig. 5) [27]. Various outputs, including rhythms of gene or protein expression, NAD^+, or other biochemical or physiological endpoints of interest, can then be examined in the samples collected from the WT-MEFs.

1. Passage MEF cells in 10 cm dishes according to ATCC instructions (*see* **Note 16**). MEF cells that have been stably infected with PER2-LUC-expressing lentivirus can also be plated in

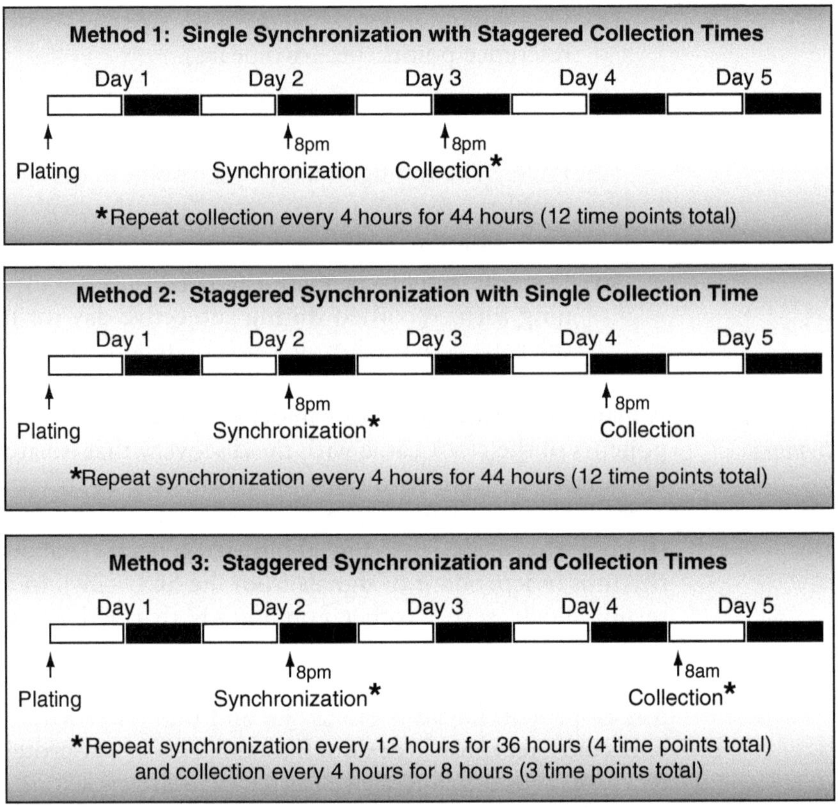

Fig. 4 In vitro experimental timelines for synchronization and 48 h collection time points in cells. Three different experimental setups for synchronizing cells in culture are presented. Method 1 entails synchronization of all cells at one time point, followed by collection of samples every 4 h over the course of 48 h. Method 2 entails synchronization of the cells every 4 h, enabling a single collection time point (which may be necessary if a particular assay requires all samples be run at the same time). Finally, Method 3 (described in text) outlines a convenient approach whereby cells are synchronized at four time points over 36 h and then collected at three time points on the final day of the experiment such that no overnight stays are required

parallel, allowing for real-time bioluminescence monitoring in order to assess synchronization (*see* **Note 17**).

2. One day prior to synchronization, passage 12 near-confluent (~90 %) 10 cm dishes of MEFs into twenty-four 10 cm dishes in normal MEF culture media (12 time points × 2 duplicates = 24 plates) and plate four 3.5 cm dishes of the PER2-LUC MEFs, such that all dishes are at least 70 % confluent at the start of the experiment the following day (~1.2×10^6 cells per 10 cm dish, ~2×10^5 cells per 3.5 cm dish).
3. For the 10 cm dishes, add forskolin directly to media in each dish at the designated time (*see* Fig. 4 and Table 2) to a final concentration of 10 μM. Incubate at 37 °C for 30 min, wash once with 1× PBS, replace with fresh MEF culture medium, and return to 37 °C until the time of collection (continue to **step 5**) (*see* **Notes 18** and **19**).

Table 2

Example experimental setup for synchronization and 48 h collection in cells

Time Point	Synchronization Time		Collection Time	
44	Day 2	8pm	Day 5	4pm
40				12pm
36				8am
32	Day 3	8am		4pm
28				12pm
24				8am
20		8pm		4pm
16				12pm
12				8am
8	Day 4	8am		4pm
4				12pm
0				8am

4. For the 3.5 cm dishes, add forskolin directly to media in each dish at the designated time (*see* Fig. 4 and Table 2) to a final concentration of 10 μM and incubate at 37 °C for 30 min. Wash once with 1× PBS and add 1.2 ml luciferin-containing medium. Coat the top rim of the plate with a thin layer of sterile vacuum grease and place cover glass on top to prevent evaporation. Gently press on cover glass to create a seal and place plates in the Lumicycler (maintained at 37 °C without CO_2) in order to monitor bioluminescence for the remainder of the experiment (*see* Fig. 5). Each plate should display ~24 h rhythms of luminescence beginning from 12 to 24 h after forskolin treatment (data analysis software (Clocklab) is available with the Actimetrics Lumicycle apparatus). In addition, plates synchronized at different times should display similar period lengths, amplitudes, and be in phase with one another (i.e., peaks and troughs should occur at similar times post-synchronization).
5. For the 10 cm dishes, aspirate media and wash cells once with PBS at the designated collection time points (*see* **Note 20**). Add 1.5 ml PBS, scrape cells and collect in microcentrifuge tubes, and centrifuge at 1,000 × *g* for 2 min. Aspirate supernatant and resuspend cells in 1 ml PBS, repeat spin, and freeze cell pellets at −80 °C. Depending on the goals of the experiment, various endpoints can then be examined, including RNA, protein, biochemical analyses, or NAD^+ quantification. For NAD^+ analysis, an aliquot of the cells can be set aside prior

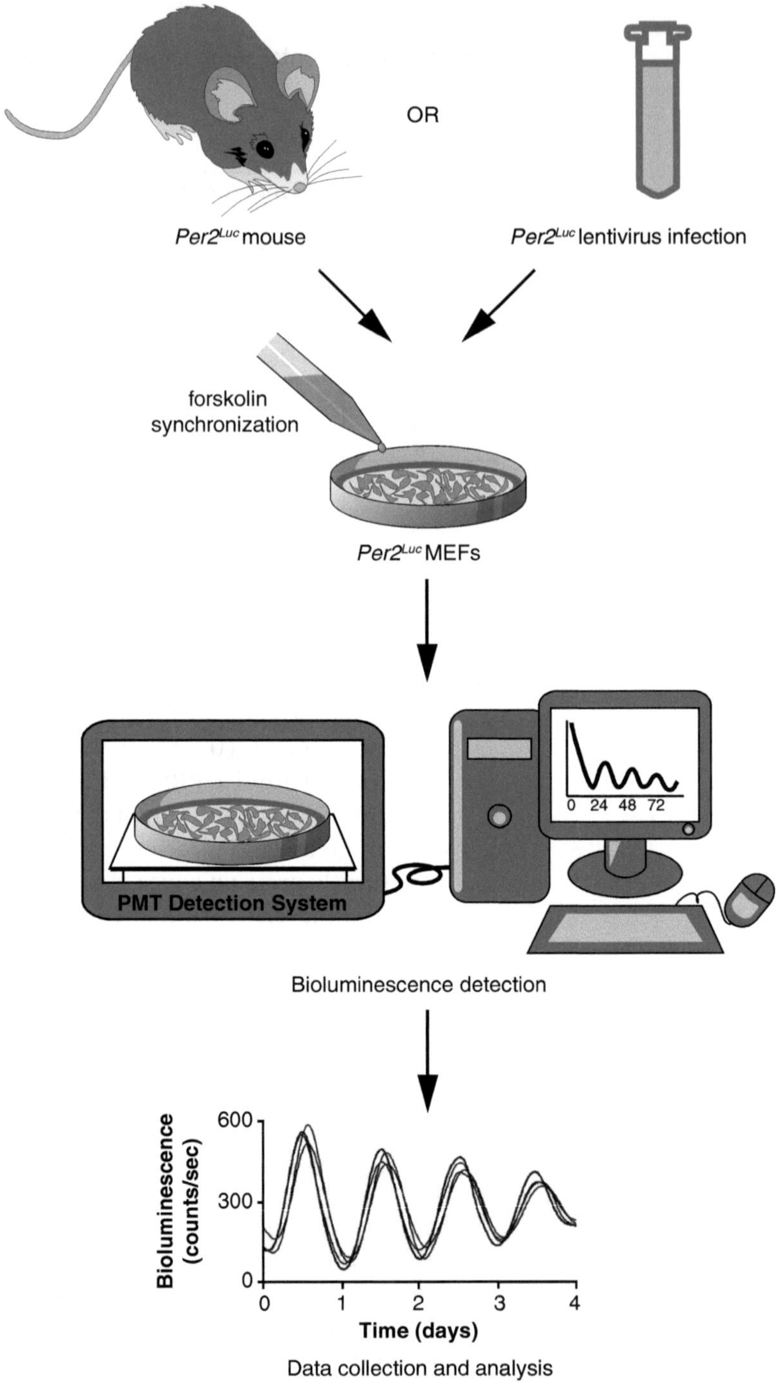

Fig. 5 Monitoring PER2-LUC oscillations in vitro. MEFs can be isolated from either PER2-LUC mice or can be infected with PER2-LUC lentivirus. Cells are then synchronized by the addition of forskolin directly to the media, and bioluminescence can be monitored for the remainder of the experiment in a Lumicycler machine. Each plate should display ~24 h rhythms of luminescence beginning from 12 to 24 h after forskolin treatment. An example trace is shown

to freezing for protein quantification in order to normalize for any difference in cell number between samples. NAD^+ measurement is described elsewhere in this volume.

6. If PER2-LUC bioluminescence monitoring is not performed, then it will be necessary to confirm proper synchronization by the analysis of transcript levels of canonical clock genes. RNA should be extracted from the frozen cell pellets, reverse transcribed, and the resultant cDNA should be amplified by RT-PCR with specific primers for canonical clock genes, including *Clock*, *Bmal1*, *Per1-3*, *Cry1-2*, *Rev-Erbα*, *Rorα*, and *Dbp*. Rhythms of expression of other genes of interest, including *Nampt*, the sirtuins, and downstream targets can also be analyzed in parallel.

4 Notes

1. Here, we describe the use of wild-type C57BL/6J mice for circadian analysis of sirtuin-related endpoints. However, this strategy can also be applied to transgenic or knockout mouse models of the sirtuin-related genes in order to gain insight into the role of the sirtuins or their targets in the circadian regulation of the process of interest.
2. Light-tight circadian cabinets with computer-controlled light–dark cycles can be purchased through Phenome Technologies. However, an alternative is to simply place the mouse cages in a temperature-controlled room with light-tight doors and a timed light cycle that can be manipulated (*see* **Note 3**).
3. Small amounts of light contamination in animal facilities during the night have been shown to not only shift the phase of the clock but also disrupt normal rhythms of endocrine metabolism and physiology, including plasma glucose, fatty acid, and corticosterone levels [33, 34]. It is therefore very important to make sure that doors are fitted with gaskets and sweeps and that any windows on the doors should be sealed in order to minimize light exposure to the mice. It is also critical to use either infrared goggles or a very dim red light (≤1 lx) anytime the light-tight box is opened during the dark phase. It can also be useful to have a dead bolt on the door on the inside of the room to prevent any accidental entries when performing experiments in the dark.
4. In normal light–dark cycles, light is a strong entraining cue that synchronizes the rhythms of all of the mice with the environment. In constant darkness, individual mice will continue to experience rhythms governed by the period of their endogenous clock (which is slightly less than 24 h in mice), a condition

called free-running (*see* Fig. 3). However, when the animals are free-running, each mouse will have a slightly different period length and eventually the colony will be desynchronized from each other.

5. Animals display a strong feeding rhythm that corresponds with their locomotor activity. Mice eat primarily during the dark period and fast during the light period, thus generating rhythms of feeding and fasting that affect the rhythm of sirtuin expression and function. For example, SIRT1 expression is induced during fasting and peaks after 18 h of fasting in liver tissue. Thus the fasting length in this experiment was set to 18 h in order to normalize fasting-induced rhythms of expression [35]. However, depending on the goals of the experiment, they can also be performed on mice fed ad libitum.
6. During alternating light–dark conditions, time is measured with respect to the time cues (zeitgebers), such that zeitgeber time (ZT) 0 corresponds with the start of the light period. In constant darkness, zeitgeber time is replaced with circadian time (CT) in which CT12 corresponds with the onset of the animal's normal active phase.
7. If brain tissue is collected, optic nerve transection at the chiasm should be performed, followed by excision of the eye, before turning on the light.
8. As an alternative to performing a continuous 48 h experiment, it is also possible to break the experiment into two separate 24 h experiments consisting of collections from CT0–CT20 and CT24–CT44. In this case, the second set of mice should be placed into constant darkness 60 h prior to the first collection (CT24) to achieve equivalent free-running conditions.
9. Sampling every 4 h is sufficient to detect six different phases of the circadian cycle; sampling less frequently than 4 h results in a dramatic decrease in the ability to detect 24 h rhythms [36].
10. If the circadian rhythms of the variable of interest have previously been determined and mice number are limiting, one can choose to perform experiments simply at the peak and trough of expression or activity.
11. Mice are transferred to wheel running cages 1–2 h prior to the start of the active (i.e., lights off) period in order to (1) avoid disturbing the mice during their normal rest period and (2) to minimize any novelty-induced phase shifts in circadian behavior that might arise due to introducing wheel cages during their normal sleep time [37, 38].
12. While wheel running is most standard method to monitor rhythms of locomotor activity, an alternative method involves the use of infrared beams to monitor the horizontal locomotor

activity of an individual mouse by recording the number of times the mouse breaks the IR beams. An advantage of this method is that it eliminates potential effects of exercise promoted by the wheels, an important consideration when dealing with metabolic genes or pathways.

13. To ensure proper functioning of the wheels, daily activity can be monitored using the data collection software or by visually inspecting the mice during either the light period or using infrared night-vision goggles during the dark period.
14. In order to simultaneously monitor rhythms of food intake, one can perform daily manual measurements of food intake (e.g., measurements of the weight of food at the start of both the light and dark periods over the course of several days). Alternatively, one can also take advantage of Feed-Scale electronic food monitoring chambers (Columbus Instruments) to measure the mass of food removed from a container that resides on a precision balance. Food measurements are taken every 6 min and recorded by the computer. Locomotor activity can be simultaneously measured using cages equipped with infrared beams.
15. More detailed descriptions of how to analyze and interpret actogram data using Clocklab software can be found in [28].
16. While the synchronization of immortalized NIH-3T3 MEF cell lines is described here, similar experiments can be performed in other cell lines, though optimization is required to determine the proper synchronization method for each cell line. In addition, synchronization experiments can be performed in isolated primary cells such as islets [39], organ explant culture such as liver, lung, or SCN [31], or in MEFs isolated from mice harboring mutations in clock genes and/or sirtuin-related factors. Typically, rapidly dividing cells, such as tumor cells, are harder to synchronize than non-transformed cells because rapid cell division leads to desynchronization [40].
17. As an alternative to monitoring PER2-LUC bioluminescence in lentivirally-infected MEFs, similar experiments can be performed using non-immortalized primary MEF cells isolated from mice expressing PER2-LUC [27]. If non-immortalized MEF cells are used, however, special considerations must be taken into account for cell death over time. It is best to use fresh cells (thawed 1–2 days before the start of experiment).
18. Several approaches can be taken to synchronize cells, as multiple cell lines have differential responses to the various synchronizing agents. Here, we outline experiments in which MEFs are synchronized with the adenylyl cyclase agonist forskolin [41]. Activators of the adenylyl cyclase pathway have been shown to induce circadian clock phase shifts and "reset"

circadian oscillations in cell culture systems potentially through the cAMP/PKA-dependent transcription factor CREB [42, 43]. Other stimuli that have been shown to control clock phase resetting in cell and tissue culture-based systems include: (1) horse/bovine serum shock (2 h treatment in 50 % serum, followed by culturing cells in 0.5 % serum-containing media) [29, 44], (2) glucocorticoid receptor agonist dexamethasone (2 h treatment with 0.1 μM dexamethasone) [30, 45], (3) heat shock (multiple methods, including a single shock treatment of ~38–39 °C for 10–60 min (1–6 h for tissue explants) or repeated daily temperature shifts (~35.5–38.5 °C) using a tightly temperature-controlled incubator [31, 32, 46, 47]).

19. In some cases, forskolin treatment can significantly slow cell growth (particularly when using non-immortalized MEFs). Therefore it is sometimes recommended to plate later time points at a higher cell density to ensure similar culture confluence at the end of the experiment for each plate. Large differences in confluence between plates can affect total NAD^+ concentrations.
20. This method allows for 24 h of recovery after shock prior to measurement. Thus, cell collections are done at 24–68 h after forskolin treatment. This delay in collection is due to the fact that cells display poor PER2-LUC rhythmicity until ~12–24 h after shock. Approximately 24 h after shock, robust ~24 h PER2-LUC oscillations begin and can be observed for 3–7 days.

References

1. Maron BJ, Kogan J, Proschan MA, Hecht GM, Roberts WC (1994) Circadian variability in the occurrence of sudden cardiac death in patients with hypertrophic cardiomyopathy. J Am Coll Cardiol 23(6):1405–1409, pii: 0735-1097 (94)90384-0
2. Di Lorenzo L, De Pergola G, Zocchetti C, L'Abbate N, Basso A, Pannacciulli N, Cignarelli M, Giorgino R, Soleo L (2003) Effect of shift work on body mass index: results of a study performed in 319 glucose-tolerant men working in a Southern Italian industry. Int J Obes Relat Metab Disord 27(11):1353–1358. doi:10.1038/sj.ijo.0802419, pii: 0802419
3. Karlsson B, Knutsson A, Lindahl B (2001) Is there an association between shift work and having a metabolic syndrome? Results from a population based study of 27,485 people. Occup Environ Med 58(11):747–752
4. Karlsson BH, Knutsson AK, Lindahl BO, Alfredsson LS (2003) Metabolic disturbances in male workers with rotating three-shift work. Results of the WOLF study. Int Arch Occup Environ Health 76(6):424–430. doi:10.1007/s00420-003-0440-y
5. Jarrett RJ, Keen H (1969) Diurnal variation of oral glucose tolerance: a possible pointer to the evolution of diabetes mellitus. Br Med J 2(5653):341–344
6. la Fleur SE, Kalsbeek A, Wortel J, Fekkes ML, Buijs RM (2001) A daily rhythm in glucose tolerance: a role for the suprachiasmatic nucleus. Diabetes 50(6):1237–1243
7. Lee A, Ader M, Bray GA, Bergman RN (1992) Diurnal variation in glucose tolerance. Cyclic suppression of insulin action and insulin secretion in normal-weight, but not obese, subjects. Diabetes 41(6):742–749
8. Roberts HJ (1964) Afternoon glucose tolerance testing: a key to the pathogenesis, early diagnosis and prognosis of diabetogenic hyperinsulinism. J Am Geriatr Soc 12:423–472
9. Boden G, Ruiz J, Urbain JL, Chen X (1996) Evidence for a circadian rhythm of insulin secretion. Am J Physiol 271(2 pt 1):E246–E252

10. Carroll KF, Nestel PJ (1973) Diurnal variation in glucose tolerance and in insulin secretion in man. Diabetes 22(5):333–348
11. Velasco A, Huerta I, Marin B (1988) Plasma corticosterone, motor activity and metabolic circadian patterns in streptozotocin-induced diabetic rats. Chronobiol Int 5(2):127–135
12. Van Cauter E, Polonsky KS, Scheen AJ (1997) Roles of circadian rhythmicity and sleep in human glucose regulation. Endocr Rev 18(5):716–738
13. Lowrey PL, Takahashi JS (2004) Mammalian circadian biology: elucidating genome-wide levels of temporal organization. Annu Rev Genomics Hum Genet 5:407–441. doi:10.1146/annurev.genom.5.061903. 175925
14. Asher G, Gatfield D, Stratmann M, Reinke H, Dibner C, Kreppel F, Mostoslavsky R, Alt FW, Schibler U (2008) SIRT1 regulates circadian clock gene expression through PER2 deacetylation. Cell 134(2):317–328. doi:10.1016/j.cell.2008.06.050, pii: S0092-8674(08)00837-4
15. Nakahata Y, Kaluzova M, Grimaldi B, Sahar S, Hirayama J, Chen D, Guarente LP, Sassone-Corsi P (2008) The NAD+-dependent deacetylase SIRT1 modulates CLOCK-mediated chromatin remodeling and circadian control. Cell 134(2):329–340. doi:10.1016/j.cell.2008.07.002
16. Ramsey KM, Yoshino J, Brace CS, Abrassart D, Kobayashi Y, Marcheva B, Hong HK, Chong JL, Buhr ED, Lee C, Takahashi JS, Imai SI, Bass J (2009) Circadian clock feedback cycle through NAMPT-mediated NAD+ biosynthesis. Science 324(5927):651–654. doi:10.1126/science.1171641, pii: 1171641
17. Nakahata Y, Sahar S, Astarita G, Kaluzova M, Sassone-Corsi P (2009) Circadian control of the NAD+ salvage pathway by CLOCK-SIRT1. Science 324(5927):654–657. doi:10.1126/science.1170803, pii: 1170803
18. Rutter J, Reick M, Wu LC, McKnight SL (2001) Regulation of clock and NPAS2 DNA binding by the redox state of NAD cofactors. Science 293(5529):510–514. doi:10.1126/science.1060698, pii: 1060698
19. Sahar S, Nin V, Barbosa MT, Chini EN, Sassone-Corsi P (2011) Altered behavioral and metabolic circadian rhythms in mice with disrupted NAD+ oscillation. Aging (Albany NY) 3(8):794–802, pii: 100368
20. Yoshino J, Mills KF, Yoon MJ, Imai S (2011) Nicotinamide mononucleotide, a key NAD(+) intermediate, treats the pathophysiology of diet- and age-induced diabetes in mice. Cell Metab 14(4):528–536. doi:10.1016/j.cmet.2011.08.014
21. Schibler U (2009) The 2008 Pittendrigh/Aschoff lecture: peripheral phase coordination in the mammalian circadian timing system. J Biol Rhythms 24(1):3–15. doi:10.1177/0748730408329383
22. Panda S, Antoch MP, Miller BH, Su AI, Schook AB, Straume M, Schultz PG, Kay SA, Takahashi JS, Hogenesch JB (2002) Coordinated transcription of key pathways in the mouse by the circadian clock. Cell 109(3):307–320, pii: S0092867402007225
23. Reddy AB, Karp NA, Maywood ES, Sage EA, Deery M, O'Neill JS, Wong GK, Chesham J, Odell M, Lilley KS, Kyriacou CP, Hastings MH (2006) Circadian orchestration of the hepatic proteome. Curr Biol 16(11):1107–1115. doi:10.1016/j.cub.2006.04.026, pii: S0960-9822(06)01487-4
24. Pronina EA, Shvidenko IG, Shub GM, Shapoval OG (2011) [Influence of electromagnetic emission at the frequencies of molecular absorption and emission spectra of oxygen and nitrogen oxide on the adhesion and formation of Pseudomonas aeruginosa biofilm]. Zhurnal mikrobiologii, epidemiologii, i immunobiologii (6):61–64
25. Liu AC, Tran HG, Zhang EE, Priest AA, Welsh DK, Kay SA (2008) Redundant function of REV-ERBalpha and beta and non-essential role for Bmal1 cycling in transcriptional regulation of intracellular circadian rhythms. PLoS Genet 4(2):e1000023. doi:10.1371/journal.pgen.1000023
26. Liu AC, Welsh DK, Ko CH, Tran HG, Zhang EE, Priest AA, Buhr ED, Singer O, Meeker K, Verma IM, Doyle FJ III, Takahashi JS, Kay SA (2007) Intercellular coupling confers robustness against mutations in the SCN circadian clock network. Cell 129(3):605–616. doi:10.1016/j.cell.2007.02.047
27. Yoo SH, Yamazaki S, Lowrey PL, Shimomura K, Ko CH, Buhr ED, Siepka SM, Hong HK, Oh WJ, Yoo OJ, Menaker M, Takahashi JS (2004) PERIOD2::LUCIFERASE real-time reporting of circadian dynamics reveals persistent circadian oscillations in mouse peripheral tissues. Proc Natl Acad Sci U S A 101(15):5339–5346
28. Siepka SM, Takahashi JS (2005) Methods to record circadian rhythm wheel running activity in mice. Methods Enzymol 393:230–239. doi:10.1016/S0076-6879(05)93008-5
29. Balsalobre A, Damiola F, Schibler U (1998) A serum shock induces circadian gene expression in mammalian tissue culture cells. Cell 93(6):929–937, pii: S0092-8674(00)81199-X
30. Balsalobre A, Brown SA, Marcacci L, Tronche F, Kellendonk C, Reichardt HM, Schutz G, Schibler U (2000) Resetting of circadian time

in peripheral tissues by glucocorticoid signaling. Science 289(5488):2344–2347, pii: 8856
31. Buhr ED, Yoo SH, Takahashi JS (2010) Temperature as a universal resetting cue for mammalian circadian oscillators. Science 330(6002):379–385. doi:10.1126/science. 1195262
32. Saini C, Morf J, Stratmann M, Gos P, Schibler U (2012) Simulated body temperature rhythms reveal the phase-shifting behavior and plasticity of mammalian circadian oscillators. Genes Dev 26(6):567–580. doi:10.1101/gad.183251.111
33. Dauchy RT, Dauchy EM, Tirrell RP, Hill CR, Davidson LK, Greene MW, Tirrell PC, Wu J, Sauer LA, Blask DE (2010) Dark-phase light contamination disrupts circadian rhythms in plasma measures of endocrine physiology and metabolism in rats. Comp Med 60(5): 348–356
34. Dauchy RT, Dupepe LM, Ooms TG, Dauchy EM, Hill CR, Mao L, Belancio VP, Slakey LM, Hill SM, Blask DE (2011) Eliminating animal facility light-at-night contamination and its effect on circadian regulation of rodent physiology, tumor growth, and metabolism: a challenge in the relocation of a cancer research laboratory. J Am Assoc Lab Anim Sci 50(3):326–336
35. Kim HS, Xiao C, Wang RH, Lahusen T, Xu X, Vassilopoulos A, Vazquez-Ortiz G, Jeong WI, Park O, Ki SH, Gao B, Deng CX (2010) Hepatic-specific disruption of SIRT6 in mice results in fatty liver formation due to enhanced glycolysis and triglyceride synthesis. Cell Metab 12(3):224–236. doi:10.1016/j.cmet.2010.06.009
36. Hughes ME, DiTacchio L, Hayes KR, Vollmers C, Pulivarthy S, Baggs JE, Panda S, Hogenesch JB (2009) Harmonics of circadian gene transcription in mammals. PLoS Genet 5(4):e1000442. doi:10.1371/journal.pgen.1000442
37. Janik D, Godfrey M, Mrosovsky N (1994) Phase angle changes of photically entrained circadian rhythms following a single nonphotic stimulus. Physiol Behav 55(1):103–107
38. Van Reeth O, Turek FW (1989) Stimulated activity mediates phase shifts in the hamster circadian clock induced by dark pulses or benzodiazepines. Nature 339(6219):49–51. doi:10.1038/339049a0
39. Marcheva B, Ramsey KM, Buhr ED, Kobayashi Y, Su H, Ko CH, Ivanova G, Omura C, Mo S, Vitaterna MH, Lopez JP, Philipson LH, Bradfield CA, Crosby SD, Jebailey L, Wang X, Takahashi JS, Bass J (2010) Disruption of the clock components CLOCK and BMAL1 leads to hypoinsulinaemia and diabetes. Nature 466(7306):571–572. doi:10.1038/nature09253, pii: nature09253
40. Nagoshi E, Saini C, Bauer C, Laroche T, Naef F, Schibler U (2004) Circadian gene expression in individual fibroblasts: cell-autonomous and self-sustained oscillators pass time to daughter cells. Cell 119(5):693–705. doi:10.1016/j.cell.2004.11.015
41. Seamon KB, Padgett W, Daly JW (1981) Forskolin: unique diterpene activator of adenylate cyclase in membranes and in intact cells. Proc Natl Acad Sci U S A 78(6):3363–3367
42. Eskin A, Takahashi JS (1983) Adenylate cyclase activation shifts the phase of a circadian pacemaker. Science 220(4592):82–84
43. Yagita K, Okamura H (2000) Forskolin induces circadian gene expression of rPer1, rPer2 and dbp in mammalian rat-1 fibroblasts. FEBS Lett 465(1):79–82
44. Welsh DK, Yoo SH, Liu AC, Takahashi JS, Kay SA (2004) Bioluminescence imaging of individual fibroblasts reveals persistent, independently phased circadian rhythms of clock gene expression. Curr Biol 14(24):2289–2295. doi:10.1016/j.cub.2004.11.057, pii: S0960982204009157
45. Meng QJ, McMaster A, Beesley S, Lu WQ, Gibbs J, Parks D, Collins J, Farrow S, Donn R, Ray D, Loudon A (2008) Ligand modulation of REV-ERBalpha function resets the peripheral circadian clock in a phasic manner. J Cell Sci 121(Pt 21):3629–3635. doi:10.1242/jcs.035048, pii: 121/21/3629
46. Brown SA, Zumbrunn G, Fleury-Olela F, Preitner N, Schibler U (2002) Rhythms of mammalian body temperature can sustain peripheral circadian clocks. Curr Biol 12(18):1574–1583
47. Prolo LM, Takahashi JS, Herzog ED (2005) Circadian rhythm generation and entrainment in astrocytes. J Neurosci 25(2):404–408

Chapter 20

Utilizing Calorie Restriction to Evaluate the Role of Sirtuins in Healthspan and Lifespan of Mice

Jessica Curtis and Rafael de Cabo

Abstract

Calorie restriction is the most powerful method currently known to delay aging-associated disease and extend lifespan. Use of this technique in combination with genetic models has led to identification of key metabolic regulators of lifespan. Limiting energy availability by restricting caloric intake leads to redistribution of energy expenditure and storage. The signaling required for these metabolic changes is mediated in part by the sirtuins at both the posttranslational and transcriptional levels, and consequently, sirtuins are recognized as instigating factors in the regulation of lifespan. This family of class III protein deacetylases is responsible for directing energy regulation based on NAD^+ availability. However, there are many effectors of NAD^+ availability, and hence sirtuin action, that should be considered when performing experiments using calorie restriction. The methods outlined in this chapter are intended to provide a guide to help the aging community to use and interpret experimental calorie restriction properly. The importance of healthspan and the use of repeated measures to assess metabolic health during lifespan experiments are strongly emphasized.

Key words Sirtuins, Calorie restriction, Lifespan, Healthspan

1 Introduction

In most animal models, limiting available energy by reducing caloric intake improves many metabolic health parameters, delaying onset of cancer, diabetes, cataracts, hypertension, and dyslipidemia [1–3]. It also enhances neuroprotection against age-related decline of cognitive function, stroke, and neurodegenerative diseases [4–6]. Consequently, calorie restriction (CR) serves as a strong positive effector on lifespan. As early as the 1930s, laboratory models of CR demonstrated extended lifespan in rodents exposed to a low calorie diet, predominantly due to reduced incidence of cancer [7]. Since then, efforts of CR studies have focused on elucidating the mechanism by which CR enhances lifespan and developing genetic models or pharmaceutical CR mimetics.

Several molecular candidates have been implicated as effectors of CR, such as insulin-like growth factor (IGF-1), mammalian

Matthew D. Hirschey (ed.), *Sirtuins: Methods and Protocols*, Methods in Molecular Biology, vol. 1077,
DOI 10.1007/978-1-62703-637-5_20,

target of rapamycin (mTOR), and silent mating-type information regulator (SIRT1) [8]. The commonality between these molecules is their regulation of bioenergetics. SIRT1 belongs to the sirtuin family of class III protein deacetylases and contributes to the coordination of energy metabolism by directing cellular processes in accordance with the availability of the metabolic currency, NAD^+. Low energy conditions like fasting or exercise increase the NAD^+/NADH ratio leading to sirtuin activation and deacetylation of protein targets such as AMPK, PGC-1α, p53, FOXO, and NFκB [9]. Sirtuin-activated signaling promotes energy redistribution in the form of lipolysis in the adipocyte, gluconeogenesis in the liver, and insulin secretion from the pancreas [10–12]. These efforts to maintain bioenergetic homeostasis inherently preserve organismal survival, illuminating an intimate relationship between energy sensing and longevity.

The essential role of SIRT1 in mediating the bioenergetic effects of CR is still controversial, which may be due to subtle experimental variations. In lower organisms, such as yeast and worms, the SIRT1 orthologue Sir2 is required for CR to extend lifespan [13, 14]. CR induces SIRT1 expression in a variety of rodent tissues as well as in human cells treated with serum from CR rats [15]. SIRT1 null mice do not have extended lifespan when maintained on a CR diet for 10 months [16]. However, they also present with developmental defects and only survive on an outbred genetic background [17, 18]. These developmental issues may confound the interpretation of lifespan extension. The predominant evolutionary theory for CR-mediated lifespan extension is that survival in times of nutritional stress depended on effective shifts in energetic usage [8]. Therefore, it is easy to envisage that SIRT1 null mice may possess the stress pathways relevant to confer CR benefits but developmental defects inundate these overlapping signals, preventing the manifestation of benefits. Conditional SIRT1 knockout models will provide better evidence for the effects of SIRT1 on CR-mediated longevity by bypassing problems associated with loss of SIRT1 expression during embryonic and adolescent development [17].

Gain-of-function data on the role of SIRT1 in healthspan is notably less deniable. Whole-body SIRT1 transgenic mice exhibit the beneficial qualities invoked by the CR regimen such as lean body composition, reduced blood insulin and glucose levels, reduced levels of DNA damage, and decreased incidence of cancer but not extended lifespan [19, 20]. Furthermore, with the added stressor of a high-fat diet, SIRT1 overexpression protects against diet-induced metabolic damage [21], and activation of SIRT1 by pharmacological treatments, such as SRT1720, leads to extended healthspan and survival under high-fat diet conditions [22].

The survival effects of CR are not universal. Recently our lab has demonstrated in a longitudinal study that CR in rhesus monkeys does not extend survival [23], contrasting results of a previously published monkey study showing that CR did extend lifespan [24]. This outcome is speculated to arise from differences in diet composition, genetic background, feeding patterns, and statistical analysis. These publications highlight the importance of complete methodological reporting in the literature. There are many factors to consider when interpreting results of a CR experiment including circadian rhythm, diet composition, animal enrichment, and background strain.

Many biomarkers of aging have been proposed in the field of aging, but lifespan analysis remains the gold standard. This is an easily measurable parameter and typically is demonstrated by a Kaplan–Meier curve. However, quality of life is as important as length of life, and in rodent models this is more difficult to characterize. Definitions of healthspan vary depending on the focus of each given research project, but generally, healthspan is considered to be the time that passes prior to the onset of age-related diseases. In humans, this can be monitored by parameters such as body-mass index, glucose tolerance, serum triglycerides, blood pressure, pulse wave velocity, bone density, waist:hip ratios, or even visual accommodation and a variety of mild physical stress tests, i.e., progressively inclined treadmill performance. These measures are tracked through the life of a patient and indicate when health issues arrive. Regardless of the parameters selected, in order to properly define healthspan in any particular model, it is crucial to have repeated measures of health parameters to identify the onset of impairments. Healthspan cannot be assessed by endpoint analysis. A longitudinal approach detailing metabolic alterations must be used to monitor metabolic health throughout the lifetime of an organism. No single panel of parameters characterizing healthspan has been defined, and measurements of healthspan are so inconsistent in the literature that meta-analyses and inter-study comparisons are difficult or impossible. Standardization of these measures is needed to ensure congruency between reports concerning healthspan. Our lab has developed a longitudinal approach for lifespan studies that integrates repeated measures to assess healthspan in addition to lifespan (Fig. 1). The methods outlined in this paper can be used to provide a more complete analysis of healthy aging, and specifically, how to properly use the CR paradigm in mice.

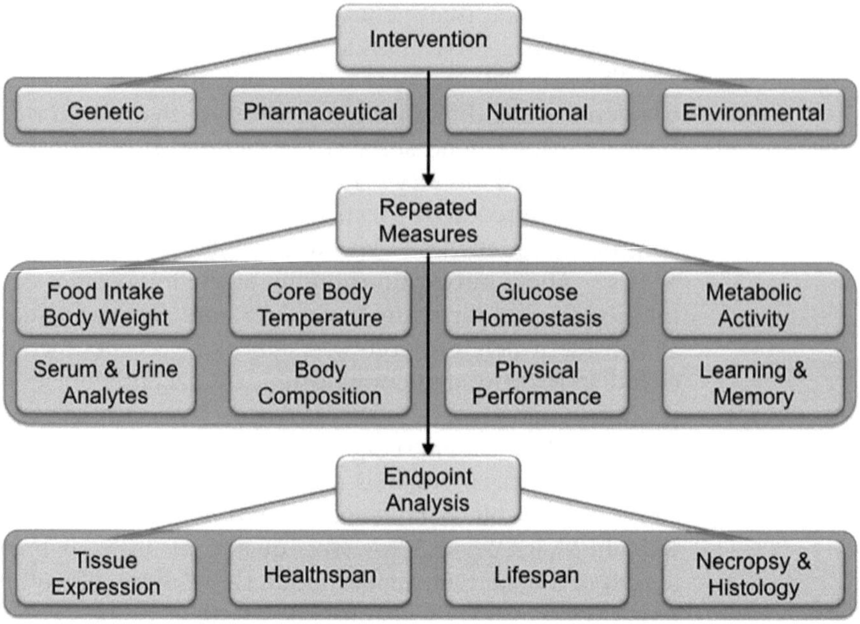

Fig. 1 Testing interventions for longitudinal assessment of healthspan and lifespan. In order to determine the effect of the intervention on healthspan and lifespan, data should be collected at regular intervals throughout the study

2 Materials

2.1 General Equipment

1. Scale to weigh food and animals.
2. Mouse housing cages (*see* **Note 1**).
3. Purified diet (*see* **Note 2**).

2.2 Animals

Animal care and handling should be performed strictly according to approved institutional animal care protocols. Twelve-month-old C57Bl/6 mice are often the type of animals used in CR studies, but various ages, strain backgrounds, and genetic modification have also been used in published studies (*see* **Notes 3** and **4**).

3 Methods

3.1 Monitoring Food Consumption in Ad Libitum Animals

1. Place pre-weighed food in the hopper (*see* **Notes 5** and **6**).
2. After 7 days, weigh the food remaining in the hopper and calculate the food consumed per animal per day.
3. Monitor food consumption in AL animals for duration of the study or until the body weights of the CR cohort plateaus for 2–3 weeks. As the AL animals become moribund or die, it is customary to fix the CR intake at the final, accurately measured

level (Basing the CR intake on that of a small group of moribund AL controls is inappropriate).

3.2 Calorie Restriction

1. Monitor food consumption and body weights weekly in control and experimental mice for at least 2 weeks before starting food restriction (*see* **Note 7**).
2. Once food consumption has stabilized, calculate food weights for CR cohort from the weekly food consumption of AL controls. Use a step-down approach for restriction so that animals can adjust to reduced calorie availability. If performing 40 % CR, start with 20 % CR for week one, 30 % CR during week two, and proceed to 40 % CR at week three. If animal health declines, extend step-down time or consider a lower percentage of caloric restriction.
3. Feed CR animals on the floor at the same time each day (*see* **Note 8**). Our mice are typically fed at the beginning of the light cycle.
4. Body weight of CR animals should plateau after approximately 10–15 weeks on the diet. Once this plateau is reached, continue feeding the CR animals a constant food allotment for the duration of the study.

3.3 Healthspan Analysis (Fig. 1)

1. Food intake and body weights should be collected biweekly for the duration of the study. The influence of a full stomach should be taken into account, since CR animals tend to gorge when presented with food.
2. Core body temperature can be monitored biweekly for the duration of the study with the use of implantable transponders (such as IPTT-300, BioMedic Data Systems).
3. Oral glucose and insulin tolerance tests should be performed every 6–12 months for the duration of the study [25].
4. Body composition by NMR or dual-energy X-ray absorptiometry and indirect calorimetry (metabolic cages) should be performed every 6–12 months for the duration of the study [26].
5. Physical performance assays (homecage activity, treadmill, strength tests) and behavioral testing (rotarod, open field, morris water maze) should be performed every 6–12 months for the duration of the study.
6. In a small subset of animals separate from the lifespan study, perform relevant immunological stress tests, such as the injection of LPS, tumor cells, or cold stress.

3.4 Lifespan Analysis (Fig. 2)

1. Viability should be monitored daily and as the animals advance in age the mice should be examined regularly (1–2 weeks). If an animal appears moribund, it should be euthanized per institutional protocols. Symptoms of moribund mice include lack

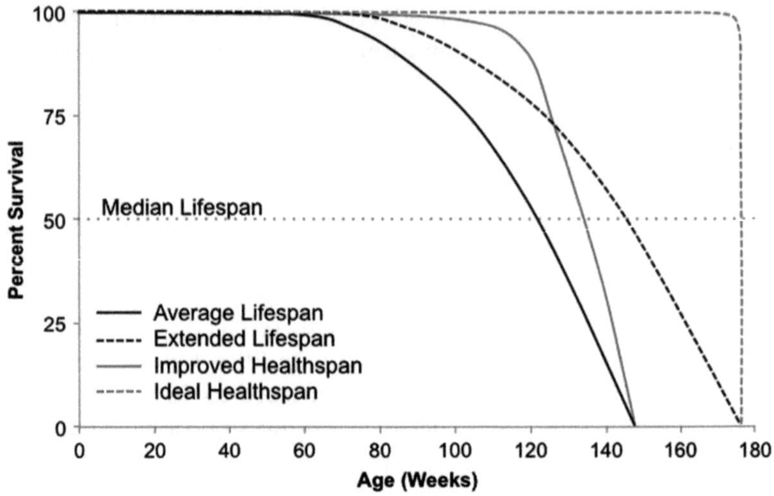

Fig. 2 Interpreting intervention effects with Kaplan–Meier survival analysis. Interventions can have several effects on the shape of the survival curves, plotted as percent survival versus age. Average lifespan (*solid black line*) and extended lifespan (*dashed black line*) are depicted, where the shape is similar but the maximal lifespan is increased. Extending the healthspan of a cohort may be independent of lifespan extension and manifest as less early deaths (*solid gray line*). Animal cohorts with ideal healthspan would experience very few deaths before the maximum survival period (*dashed gray line*)

of response to touch stimuli, hunched spine, sudden weight loss, matted fur, slowed breathing, and lack of appetite.

2. As animals expire throughout the study, complete a pathological necropsy as soon as possible, recording observations of the external appearance of the animals including hair color and evidence of kyphosis (hunched spine). Preserve the heart, lungs, liver, spleen, kidneys, seminal vesicles, and bladder in 4 % formalin. Take note of enlarged, discolored, or tumor-laden tissues.
3. Maximal lifespan is calculated as the mean lifespan of the top 10 % of survivors.
4. Median lifespan is calculated as the age at which 50 % of animals in a cohort survive.
5. Lifespan of control animals should be consistent between studies.

4 Notes

1. Population density should be considered when designing a CR study. The mouse is a social creature and prefers group-housing conditions, but this can lead to increased variation in

body weights and food consumption due to social relationships. It is certainly easier and cheaper for the researcher to use group housing, but it is important to be aware of the effects of housing density. Singly housed, C57Bl/6J male mice eat ~15 % more than group housed animals on average due to lack of social conflicts (Dawn Boyer and Rafael de Cabo, unpublished). These results are after 24 weeks on diet, and no significant changes in body weight have been observed yet. The number of mice per cage will depend on the size of the cage, but delivery of food to animals should consider the number of animals in a cage. For example, if there are four animals in the cage, at least four pellets of food should be delivered to help alleviate social behavioral effects of food consumption. Our lab uses Thoren caging model #15 for group housing (four male adult mice preferred) and model #5 for single housing.

2. Using a purified, chemically defined diet is essential for experimental standardization and reproducibility. These diets come as hard packed pellets and nutrient formulation is independent of material sourcing or batch, which can vary in naturally sourced "chow" diets. The American Institute of Nutrition (AIN) provides many open source formulations, such as the AIN-93 series, that are suitable for CR studies. These diets may not provide trace nutrients found in naturally sourced diets. Addition of pharmacological compounds to the food may decrease palatability and inadvertently induce voluntary CR. When testing food-delivered compounds, it is important to include a control-diet cohort to ensure that food consumption has not changed due to food taste. It is also important to realize that differing body composition between the control and CR animals influences micronutrient utilization and storage, particularly if a study will consider such things as lipid peroxidation (which is influenced by the ratio of adipose tissue to fat-soluble antioxidant intake).

3. The power calculation estimates the sample size needed to obtain significance dependent upon the observed change between variables and standard deviation. For example, 100 animals will provide approximately 100 % statistical power to observe a 10 % change in average lifespan with an assumed 10 % standard deviation, but using 50 animals drops the statistical power to approximately 80 %. Additional mice should be set up in parallel to the lifespan study, typically 6–12 animals, in order to collect tissues at various time points throughout the experiment or perform testing that may interfere with the lifespan study, such as immune challenges or cancer resistance.

4. Longevity effects are proportional to the duration and proportion of CR regardless of when CR was initiated. In a compilation of 24 published survival studies, the increase in survival appears to be inversely proportional to the decrease in calories [27].

5. The mode of food delivery has an impact on the amount of food consumed. Delivery of food in the wire cage hopper reduces food consumption by more than 10 % (Dawn Boyer and Rafael de Cabo, unpublished). Our CR mice are typically floor fed.
6. The amount of food spilled into bedding does not differ between the CR and AL groups; however, the amount of food spilled does increase with age leading to an overestimation of food consumption [28]. Food spillage is also dependent upon the composition of the food and possibly the genetic or strain background of the animal.
7. Genetically modified animals may present with differences in food consumption, possibly due to changes in body weight. It is important to have an AL control for each experimental variable.
8. Gene expression and metabolic regulation are coordinately regulated by circadian rhythm. AL mice tend to eat during periods of darkness, while CR mice will consume their food at the time it is provided. Furthermore, in CR cohorts, the changes between post-prandial and post-absorptive states become exaggerated. Circadian rhythms usually differ between AL and CR groups, and accidental comparisons between fasted, active, and hungry cohorts and somnolent, inactive, just-fed cohorts should be avoided. It is important to be consistent with feeding time in your study and to report the time of sample collection relative to each cohort's diurnal cycle.

Acknowledgements

Dawn Boyer, Dawn Nines, and Justine Lucas for their continuous support and assistance with our animal studies. Dr. Anson for providing helpful feedback on this manuscript. This work was supported by the Intramural Research Program of the National Institute on Aging, NIH.

References

1. Reaven E, Wright D, Mondon CE, Solomon R, Ho H, Reaven GM (1983) Effect of age and diet on insulin secretion and insulin action in the rat. Diabetes 32(2):175–180
2. Taylor A, Zuliani AM, Hopkins RE et al (1989) Moderate caloric restriction delays cataract formation in the Emory mouse. FASEB J 3(6):1741–1746
3. Young JB, Mullen D, Landsberg L (1978) Caloric restriction lowers blood pressure in the spontaneously hypertensive rat. Metab Clin Exp 27(12):1711–1714
4. Yu ZF, Mattson MP (1999) Dietary restriction and 2-deoxyglucose administration reduce focal ischemic brain damage and improve behavioral outcome: evidence for a preconditioning mechanism. J Neurosci Res 57(6):830–839
5. Means LW, Higgins JL, Fernandez TJ (1993) Mid-life onset of dietary restriction extends life and prolongs cognitive functioning. Physiol Behav 54(3):503–508
6. Mattson MP (2010) The impact of dietary energy intake on cognitive aging. Front Aging Neurosci 2:5

7. McCay CM, Crowell MF, Maynard LA (1935) The effect of retarded growth upon the length of life span and upon the ultimate body size. J Nutr 10(1):63–79
8. Spindler SR (2010) Caloric restriction: from soup to nuts. Ageing Res Rev 9(3):324–353
9. Nogueiras R, Habegger KM, Chaudhary N et al (2012) Sirtuin 1 and sirtuin 3: physiological modulators of metabolism. Physiol Rev 92(3):1479–1514
10. Picard F, Kurtev M, Chung N et al (2004) Sirt1 promotes fat mobilization in white adipocytes by repressing PPAR-gamma. Nature 429(6993):771–776
11. Rodgers JT, Lerin C, Haas W, Gygi SP, Spiegelman BM, Puigserver P (2005) Nutrient control of glucose homeostasis through a complex of PGC-1alpha and SIRT1. Nature 434(7029):113–118
12. Moynihan KA, Grimm AA, Plueger MM et al (2005) Increased dosage of mammalian Sir2 in pancreatic beta cells enhances glucose-stimulated insulin secretion in mice. Cell Metab 2(2):105–117
13. Lin SJ, Defossez PA, Guarente L (2000) Requirement of NAD and SIR2 for life-span extension by calorie restriction in Saccharomyces cerevisiae. Science 289(5487): 2126–2128
14. Wood JG, Rogina B, Lavu S et al (2004) Sirtuin activators mimic caloric restriction and delay ageing in metazoans. Nature 430(7000): 686–689
15. Cohen HY, Miller C, Bitterman KJ et al (2004) Calorie restriction promotes mammalian cell survival by inducing the SIRT1 deacetylase. Science 305(5682):390–392
16. Boily G, Seifert EL, Bevilacqua L et al (2008) SirT1 regulates energy metabolism and response to caloric restriction in mice. PLoS One 3(3):e1759
17. Cheng H-L, Mostoslavsky R, Saito S et al (2003) Developmental defects and p53 hyperacetylation in Sir2 homolog (SIRT1)-deficient mice. Proc Natl Acad Sci U S A 100(19): 10794–10799
18. McBurney MW, Yang X, Jardine K et al (2003) The mammalian SIR2alpha protein has a role in embryogenesis and gametogenesis. Mol Cell Biol 23(1):38–54
19. Bordone L, Cohen D, Robinson A et al (2007) SIRT1 transgenic mice show phenotypes resembling calorie restriction. Aging Cell 6(6): 759–767
20. Herranz D, Muñoz-Martin M, Cañamero M et al (2010) Sirt1 improves healthy ageing and protects from metabolic syndrome-associated cancer. Nat Commun 1:3
21. Pfluger PT, Herranz D, Velasco-Miguel S, Serrano M, Tschöp MH (2008) Sirt1 protects against high-fat diet-induced metabolic damage. Proc Natl Acad Sci USA 105(28): 9793–9798
22. Minor RK, Baur JA, Gomes AP et al (2011) SRT1720 improves survival and healthspan of obese mice. Sci Rep 1:70
23. Mattison JA, Roth GS, Beasley TM et al (2012) Impact of caloric restriction on health and survival in rhesus monkeys from the NIA study. Nature 489(7415):318–321
24. Colman RJ, Anderson RM, Johnson SC et al (2009) Caloric restriction delays disease onset and mortality in rhesus monkeys. Science 325(5937):201–204
25. Ayala JE, Samuel VT, Morton GJ et al (2010) Standard operating procedures for describing and performing metabolic tests of glucose homeostasis in mice. Dis Model Mech 3(9–10):525–534
26. Tschöp MH, Speakman JR, Arch JRS et al (2012) A guide to analysis of mouse energy metabolism. Nat Methods 9(1):57–63
27. Merry BJ (2002) Molecular mechanisms linking calorie restriction and longevity. Int J Biochem Cell Biol 34(11):1340–1354
28. Starr ME, Saito H (2012) Age-related increase in food spilling by laboratory mice may lead to significant overestimation of actual food consumption: implications for studies on dietary restriction, metabolism, and dose calculations. J Gerontol A Biol Sci Med Sci 67(10): 1043–1048

ERRATUM

Sirtuins: Methods and Protocols

Matthew D. Hirschey

Matthew D. Hirschey (ed.), *Sirtuins: Methods and Protocols*, Methods in Molecular Biology, vol. 1077,
DOI 10.1007/978-1-62703-637-5, pp. iv, © Springer Science+Business Media, LLC 2013

DOI 10.1007/978-1-62703-637-5_21

The publisher regrets that in the print and online versions of this book, the following cover illustration line was inadvertently omitted from the copyright page. It should read as follows.

Front Cover Image created and supplied by Dr. Donald S. Backos.

The online version of the original book can be found at
http://dx.doi.org/10.1007/978-1-62703-637-5

INDEX

Matthew D. Hirschey (ed.), *Sirtuins: Methods and Protocols*, Methods in Molecular Biology, vol. 1077,
DOI 10.1007/978-1-62703-637-5, © Springer Science+Business Media, LLC 2013

MIX
Papier aus verantwortungsvollen Quellen
Paper from responsible sources
FSC® C105338

If you have any concerns about our products,
you can contact us on
ProductSafety@springernature.com

In case Publisher is established outside the EU,
the EU authorized representative is:
Springer Nature Customer Service Center GmbH
Europaplatz 3, 69115 Heidelberg, Germany

Printed by Libri Plureos GmbH
in Hamburg, Germany